Shadowing in Dynamical Systems

Mathematics and Its Applications

Volume 501

Shadowing in Dynamical Systems

Theory and Applications

by

Ken Palmer

School of Mathematical & Statistical Sciences,
La Trobe University,
Bundoora, Victoria, Australia

KLUWER ACADEMIC PUBLISHERS
DORDRECHT / BOSTON / LONDON

A C.I.P. Catalogue record for this book is available from the Library of Congress.

ISBN 978-1-4419-4827-4

Published by Kluwer Academic Publishers,
P.O. Box 17, 3300 AA Dordrecht, The Netherlands.

Sold and distributed in North, Central and South America
by Kluwer Academic Publishers,
101 Philip Drive, Norwell, MA 02061, U.S.A.

In all other countries, sold and distributed
by Kluwer Academic Publishers,
P.O. Box 322, 3300 AH Dordrecht, The Netherlands.

Printed on acid-free paper

CONTENTS

PREFACE ix

1 HYPERBOLIC FIXED POINTS OF DIFFEOMORPHISMS AND THEIR STABLE AND UNSTABLE MANIFOLDS 1

1.1 Examples of Diffeomorphisms 1

1.2 Definition and Examples of Hyperbolic Fixed Points 2

1.3 Definition of Stable and Unstable Manifolds 3

1.4 The Saddle-Point Property 5

1.5 Smoothness of the Local Stable Manifold 12

1.6 Diffeomorphisms depending on a Parameter 17

2 HYPERBOLIC SETS OF DIFFEOMORPHISMS 21

2.1 Definition of Hyperbolic Set 21

2.2 Boundedness of the Projection 24

2.3 Continuity of the Projection 25

2.4 Exponential Dichotomies for Difference Equations 29

2.5 Expansivity Property of Hyperbolic Sets 41

2.6 Roughness of Hyperbolic Sets 44

3 TRANSVERSAL HOMOCLINIC POINTS OF DIFFEOMORPHISMS AND HYPERBOLIC SETS 57

3.1 The Hyperbolic Set associated with a Transversal Homoclinic Point 57

3.2 The Construction of Diffeomorphisms with Transversal Homoclinic Points 64

3.3 An Example from Differential Equations 72

4 THE SHADOWING THEOREM FOR HYPERBOLIC SETS OF DIFFEOMORPHISMS 77

4.1 The Shadowing Theorem 77

4.2 More on the Roughness of Hyperbolic Sets 83

4.3 Asymptotic Phase for Hyperbolic Sets 88

5 SYMBOLIC DYNAMICS NEAR A TRANSVERSAL HOMOCLINIC POINT OF A DIFFEOMORPHISM 91

6 HYPERBOLIC PERIODIC ORBITS OF ORDINARY DIFFERENTIAL EQUATIONS, STABLE AND UNSTABLE MANIFOLDS AND ASYMPTOTIC PHASE 99

6.1 The Poincaré Map 100

6.2 Hyperbolic Periodic Orbits and Their Stable and Unstable Manifolds 104

6.3 Asymptotic Phase and the Stable and Unstable Foliations 109

7 HYPERBOLIC SETS OF ORDINARY DIFFERENTIAL EQUATIONS 115

7.1 Definition of Hyperbolic Set 115

7.2 Boundedness of the Projections 118

7.3 Continuity of the Projections 121

7.4 Exponential Dichotomies for Differential Equations 126

7.5 Expansivity Property of Hyperbolic Sets 149

7.6 Roughness of Hyperbolic Sets 160

8 TRANSVERSAL HOMOCLINIC POINTS AND HYPERBOLIC SETS IN DIFFERENTIAL EQUATIONS 171

8.1 The Hyperbolic Set associated with a Transversal Homoclinic Orbit 171

8.2 The Construction of a Differential Equation with a Transversal Homoclinic Orbit 181

9 SHADOWING THEOREMS FOR HYPERBOLIC SETS OF DIFFERENTIAL EQUATIONS 187

9.1 The Discrete Shadowing Theorem 187

9.2 The Continuous Shadowing Theorem 198

9.3 More on the Roughness of Hyperbolic Sets 206

9.4 Asymptotic Phase for Hyperbolic Sets 218

10 SYMBOLIC DYNAMICS NEAR A TRANSVERSAL HOMOCLINIC ORBIT OF A SYSTEM OF ORDINARY DIFFERENTIAL EQUATIONS 225

11 NUMERICAL SHADOWING 241

11.1 Finite Time Shadowing for Maps 241

11.2 Periodic Shadowing for Maps 248

11.3 Finite Time Shadowing for Differential Equations 255

11.4 Periodic Shadowing for Differential Equations 263

11.5 Rigorous Verification of Chaotic Behaviour 270

REFERENCES 285

PREFACE

This book is an introduction to the theory of shadowing, both for diffeomorphisms and flows, with applications to proving theorems about hyperbolic sets and also to the development of techniques for making deductions from numerical simulations of dynamical systems.

First let me review the history of shadowing. Sinai [1972] stated the shadowing theorem for Anosov diffeomorphisms, the proof being a variation of ideas from Anosov [1967]. The first formal statement of the shadowing theorem for more general diffeomorphisms is given in Bowen [1975a]. However, Bowen states that the proof is already contained in proofs of results in Bowen [1970, 1972] concerning the specification property (conf. also Sigmund [1974]). Bowen [1978] used his shadowing theorem in his study of Markov partitions. Conley [1980] (conf. also Robinson [1977]) also used it to show that if the chain recurrent set of a diffeomorphism is hyperbolic, then the periodic points are dense in the chain recurrent set. Walter [1978] and Lanford [1983] used it to prove topological conjugacy results for perturbations of diffeomorphisms with hyperbolic sets. McGehee [1982] (conf. also Conley [1975]) used it to prove the existence of orbits with given itineraries. Lanford [1985] and Palmer [1984, 1988] (conf. also Anosov and Solodov [1995]) used the same idea to prove Smale's theorem on the embedding of the shift in a diffeomorphism with a transversal homoclinic point. Lin [1989] used shadowing in his study of singularly perturbed systems. Recently (conf. Chapter 11) shadowing has also been used to give global error estimates for numerically computed orbits of dynamical systems and to rigorously prove the existence of periodic orbits and chaotic behaviour.

There is an old idea in the theory of differential equations closely related to shadowing. It appears in modern form in Coppel [1965]. If the linear system

$$\dot{\mathbf{x}} = A(t)\mathbf{x} \tag{1}$$

has an *exponential dichotomy*, $f(t,\mathbf{x})$ is small Lipschitzian with $f(t,\mathbf{0}) = \mathbf{0}$ and $\mathbf{p}(t)$ is small, then the nonlinear system

$$\dot{\mathbf{x}} = A(t)\mathbf{x} + f(t,\mathbf{x}) + \mathbf{p}(t)$$

has a unique small solution. The key hypothesis here is that the linear system (1) has an exponential dichotomy. When $A(t)$ is bounded, this hypothesis is equivalent to the invertibility of the linear operator which takes the bounded C^1 function $\mathbf{x}(t)$ to the bounded continuous function

$$\mathbf{x}(t) \to \dot{\mathbf{x}}(t) - A(t)\mathbf{x}(t).$$

This given, the theorem stated above is a simple consequence of the contraction mapping principle.

Essentially the same idea is used to prove the shadowing theorem for diffeomorphisms. The fundamental step is to show that the linear operator associated with the pseudo orbit is invertible. This can be shown directly by constructing an approximate inverse as in Coomes, Koçak and Palmer [1996] or by showing that the associated linear difference equation has an exponential dichotomy. It is the latter approach which is employed in this book.

Shadowing is a property of hyperbolic sets. The idea of hyperbolicity is closely related to that of exponential dichotomy. In fact, a compact invariant set of a diffeomorphism is hyperbolic if and only if the "variational equation" along each orbit in the set has an exponential dichotomy with projection of constant rank or, equivalently, the linearised flow on the tangent bundle has an exponential dichotomy (as defined by Sacker and Sell [1974] for linear skew product flows). My original intention was to develop the theory of exponential dichotomy for linear difference and differential equations and then use it to develop the theory of hyperbolic sets and shadowing. However, it occurred to me that this could alienate readers not familiar with the exponential dichotomy machinery. So next I tried to develop the theory of hyperbolicity without using exponential dichotomy. However I discovered that at certain points the use of exponential dichotomies was, as far as I was concerned, just the most effective way of proceeding. So now I develop the theory of exponential dichotomy as it is needed. No previous familiarity with the theory is assumed.

If there is a key idea in the theory of hyperbolicity, I would say it is the "roughness" (to use the translation of the Russian term) of exponential dichotomy, that is, its preservation under perturbation of the coefficient matrix. The classical result in this area (see Massera and Schäffer [1966], Coppel [1978]) is that exponential dichotomy is preserved under small uniform perturbations of the coefficient matrix. However, there are theorems (see Sacker and Sell [1978], Henry [1981], Palmer [1987], Sakamoto [1994], Johnson [1987]) which show that exponential dichotomy is considerably more robust than this. It is this additional robustness which ensures that shadowing works and also the fact that hyperbolicity is preserved under perturbation.

In Chapter 2, I develop the theory of hyperbolic sets for diffeomorphisms, proving their expansiveness and roughness or preservation under perturbation. Then I prove the shadowing theorem in Chapter 4 and use it to show that isolated hyperbolic sets have the asymptotic phase property; also that there is a topological conjugacy between the dynamics on the perturbed and unperturbed hyperbolic sets. A nontrivial example of a hyperbolic set is that associated with a transversal homoclinic point. To prove that this set is indeed hyperbolic (see Chapter 3), we need to study the stable and unstable manifolds of a hyperbolic fixed point (see Chapter 1). This is also needed in Chapter 3 in order to use Melnikov theory to construct examples of diffeomorphisms with transversal homoclinic points. In Chapter 5 we also use shadowing to show the existence of chaotic behaviour near a transversal homoclinic point. Here I have followed Steinlein and Walther [1990] and given a symbolic dynamical description of the maximal invariant set in a neighbourhood of a transversal homoclinic orbit. Smale's theorem on the embedding of the shift is easily seen to follow from this.

I develop the theory of hyperbolic sets and shadowing both for diffeomorphisms and flows. The shadowing theorem for diffeomorphisms has been proved many times as in Katok [1971], Bowen [1975a], Hirsch, Pugh and Shub [1977], Robinson [1977, 1995], Guckenheimer, Moser and Newhouse [1980], Ekeland [1983], Lanford [1983], Shub [1987], Palmer [1988], Chow, Lin and Palmer [1989a], Meyer and Hall [1992], Akin [1993], Pilyugin [1994a], Katok and Hasselblatt [1995], Coomes, Koçak and Palmer [1996] and Fenichel [1996]. For flows different versions of the shadowing theorem have been proved in Franke and Selgrade [1977], Nadzieja [1991], Katok and Hasselblatt [1995], Coomes, Koçak and Palmer [1995b] and Pilyugin [1997]. I consider here the case only of hyperbolic sets without equilibrium points. In Chapters 6 through 9, the theory for flows is developed along the same lines as that for diffeomorphisms but it is rather more complicated than that for diffeomorphisms. It is not simply a corollary of that for diffeomorphisms except perhaps the theory of stable and unstable manifolds of a hyperbolic periodic orbit (see Chapter 6), which can be deduced from that for diffeomorphisms via the Poincaré map, although not without further nontrivial reasoning in order to prove the property of asymptotic phase. For flows it is not even clear what the definition of a pseudo orbit should be. Should they be sequences or functions of time? I have considered both possibilities here and proved both "discrete" and "continuous" shadowing theorems. Again exponential dichotomies are a major tool (I decided against using exponential trichotomies as in Palmer [1996]) and they are used in two ways. First, if $\mathbf{x}(t)$ is a solution in a hyperbolic set for the system

$$\dot{\mathbf{x}} = F(\mathbf{x}),$$

then the linear system

$$\dot{\mathbf{x}} = DF(\mathbf{x}(t))\mathbf{x}, \tag{2}$$

does not have an exponential dichotomy since $\dot{\mathbf{x}}(t)$ is a nonzero bounded solution of this equation. However, if we project Eq.(2) onto the normal bundle to $\dot{\mathbf{x}}(t)$, the resulting linear system does have an exponential dichotomy. This is one way in which exponential dichotomy enters the picture. The other way is through the observation that if $\lambda \neq 0$ is sufficiently small and nonzero, then the shifted equation

$$\dot{\mathbf{x}} = [DF(\mathbf{x}(t)) - \lambda]\mathbf{x}$$

has an exponential dichotomy.

In the flow context, the analogous theorem to Smale's is that of Sil'nikov [1967], which describes the dynamics in the neighbourhood of a transversal homoclinic orbit associated with a hyperbolic periodic orbit of a flow. As stated in Hale and Lin [1986], this could be deduced from a version of the diffeomorphism theory for noninvertible maps applied to a certain return map (note, however, that this return map would not simply be the Poincaré map associated with the periodic orbit) but I have preferred to work directly with the flow and use the discrete shadowing theorem for flows (see Chapter 10).

A technical point here is that a convexity assumption is frequently made to simplify the proofs. How this convexity assumption can be avoided is demonstrated in Chapter 6. We have also made certain boundedness assumptions which could always be met by taking a suitable neighbourhood of the compact invariant set.

In dynamical systems the focus is on attractors. When the system is dissipative and the attractor is a fixed point or periodic orbit, then it is hyperbolic. However, if the attractor is strange, then usually it is not hyperbolic like, for example, the Hénon map and the Lorenz equations for certain parameter values. So the theory developed as above is not directly applicable. Nonetheless, even though these attractors are not hyperbolic, they are close to being so and even though the theory is not directly applicable it does give us a guide in our study of the systems. For instance, in these systems infinite pseudo orbits may not be shadowed by true orbits. However, as first observed by Hammel, Yorke and Grebogi [1987, 1988], pseudo orbits of chaotic maps can still be shadowed for long times by true orbits. Chow and Palmer [1991, 1992] and Sauer and Yorke [1990, 1991] realised this problem could be attacked by shadowing techniques. Here the key idea is the construction of a right inverse of small norm for a linear operator similar to the one used for infinite time shadowing. The choice of this right inverse is guided by the infinite time case — one takes the formula for the inverse in the infinite time case and truncates it appropriately (see Chapter 11, Sec.1). However, the flow case is somewhat more complicated. It is not simply a matter of looking at the time-one map and applying the theory for the map case. One must somehow "quotient-out" the direction of the vector field as in Chapter 11, Sec.3. That this leads to much better shadowing results is shown in Coomes, Koçak and Palmer [1995a].

Normally one expects periodic orbits to be plentiful in chaotic invariant sets. One would like to be able to prove that periodic orbits are dense in attractors like those for the Hénon map or the Lorenz equations. This has not been done. However, Coomes, Koçak and Palmer [1996, 1997] were able to use shadowing techniques to prove the existence of periodic orbits of long period in these systems. Again the basic idea is the invertibility of a certain linear operator associated with an approximate periodic orbit or "pseudo periodic orbit" (see Chapter 11, Secs. 2 and 4).

In his dissertation, Stoffer [1988a] observed that if a system contains two hyperbolic periodic orbits, the starting points of which are sufficiently close, then the system exhibits chaotic behaviour in the sense that the shift can be embedded. Coomes, Koçak and Palmer [1997, 1998] developed a Newton's method which very efficiently finds periodic orbits in chaotic dynamical systems such as the Hénon map and the Lorenz equations. Stoffer and I [1999] used this method to find two periodic orbits of the Hénon map with closeby initial points. However we did not verify the hyperbolicity of the periodic orbits directly. Instead we used the computer to verify directly that the linear operator associated with a pseudo orbit corresponding to a given sequence of zeros and ones was invertible and also obtained an upper bound on the norm of its inverse. Thus we were able to embed the shift and hence verified that the Hénon map is indeed chaotic (see Chapter 11, Sec.5).

In this book I have restricted consideration to finite dimensions and to invertible maps and flows (except for Chapter 11 where the maps need not be invertible). A theory of shadowing for not necessarily invertible maps in infinite dimensions has been developed in Chow, Lin and Palmer [1989b], Steinlein and Walther [1989,1990] and Lani-Wayda [1995]. Shadowing in elliptic partial differential equations has been studied by Angenent [1987]. Shadowing in nonautonomous systems has been studied by Palmer [1984], Blazquez [1986], Scheurle [1986], Stoffer [1988], and Palmer and Stoffer [1989] (see also Lerman and Shil'nikov [1992]). A shadowing theorem for skew product flows has been developed by Meyer and Sell [1989] (conf. also Meyer and Hall [1992], Meyer and Zhang [1996]). Al-Nayef et al. [1996], Diamond, Kloeden et al. [1995, 1996] have defined a notion of semi-hyperbolicity for non-differentiable mappings in Banach spaces. They give up the requirement that the stable and unstable bundles be invariant and emphasise bishadowing, that is, in addition to the usual shadowing property they require that for every orbit of a nearby mapping there is a pseudo orbit of the original mapping which it shadows. Chow and Van Vleck [1992], Bogenschütz and Gundlach [1992-93] and Liu, Qian and Jang [1996] have proved shadowing theorems for random diffeomorphisms.

For flows I have restricted consideration to hyperbolic sets without equilibria. Pliss and Sell [1991] have studied hyperbolic sets with equilibria and Pilyugin [1997] has proved a shadowing theorem for such sets. Actually, like Bowen [1975a], Pilyugin does not insist that his sets be hyperbolic but only that they have an Axiom A sort of structure.

Hyperbolic sets have the shadowing property. The extent to which this statement has a converse has received much study even for systems which are not differentiable. In the case of such non-differentiable systems, an alternative definition of hyperbolicity has to be developed. It was proved by Ombach [1986], building on work of Bowen and Reddy, that a homeomorphism is hyperbolic in a certain sense if and only if it has the shadowing property (sometimes called the pseudo orbit tracing property or stochastic stability) and is expansive. A similar theory for flows has been developed by Thomas [1991]. Other authors who have explored the relationship between shadowing and other dynamical properties include Aoki, Dateyama, Harada, Hiraide, Kato, Kloeden, Komuro, Kurata, Liu, Morimoto, Moriyasu, Odani, Oka, Sakai, Sasakai, Sawada, Shimomura and Yano (see the references).

I was trained by Andrew Coppel in the classical qualitative theory of differential equations. Smale [1967] published his seminal paper in the year I started my doctorate. Much of the dynamical systems theory of the Smale school remained a mystery to me until I received a preprint from Urs Kirchgraber [1982] (later published in modified form as Kirchgraber and Stoffer [1990]), which used ideas of McGehee [1982] to show how solutions of the sinusoidally forced pendulum equation could be constructed to follow certain itineraries. After reading this preprint I was able to see how to use the shadowing lemma to prove Smale's theorem on the embedding of the shift in a diffeomorphism with a transversal homoclinic point (I only recently became aware that Lanford [1985] was doing a similar thing at about the same time). Also the shadowing theorem itself, at least in the case of period maps for differential equations depending periodically on time, could be proved using the classical result in Coppel's book [1965] mentioned above, the main point being the verification that the variational equation along the pseudo solution had an exponential dichotomy. To prove this, I used the rotation method of Millionscikov [1969ab], not being aware at the time that I could have used a result given by Henry [1981]. Sakamoto informed me of Henry's result much later.

I wish to thank Andrew Coppel for training me in the qualitative theory of ordinary differential equations and for his encouragement in my later work. Warmest thanks are also due to Urs Kirchgraber, Daniel Stoffer, Kaspar Nipp and all their colleagues at ETH in Zürich for their collaboration, for the interest they have shown in my work and for their hospitality, to Peter Kloeden and Phil Diamond for their support and hospitality at Deakin University and the University of Queensland, to Shui-Nee Chow for his collaboration and hospitality at Georgia Tech, to Brian Coomes and Huseyin Koçak for the pleasurable collaboration on numerical shadowing and to John Stillwell and Monash University for hospitality during the last few months of the writing of this book. Next I wish to thank Marilyn Castano, Hugo Giovanazzo and Monique Rosenberg for assistance with the typing. Finally, let me thank Alan Lazer and my colleagues in the differential equations group at the University of Miami, Shair Ahmad at the University of Texas in San Antonio and the National Science Foundation for their support.

1. HYPERBOLIC FIXED POINTS OF DIFFEOMORPHISMS AND THEIR STABLE AND UNSTABLE MANIFOLDS

The most common kind of dynamical system is that coming from a differential equation since it is most commonly differential equations which are used to model physical processes. However, as we shall see later in this book, the study of the dynamical properties of a differential equation can often be reduced to the study of the properties of a diffeomorphism. Diffeomorphisms are also of interest in themselves since they can be used as discrete models of physical and biological processes. Also in some ways they are easier to study than differential equations. For these reasons, we begin this book with the study of diffeomorphisms.

1.1 EXAMPLES OF DIFFEOMORPHISMS

We begin with a couple of examples.

Example 1.1. Consider the mapping $f : \mathbb{R}^2 \to \mathbb{R}^2$ given by

$$f(x, y) = (1 - ax^2 + y, -bx)$$

where a and b are real parameters. Clearly f is a C^∞ map. When $b \neq 0$, f has a C^∞ inverse given by

$$f^{-1}(x, y) = (-b^{-1}y, -1 + ab^{-2}y^2 + x).$$

Thus when $b \neq 0$, f is a C^∞ diffeomorphism called the *Hénon map* (conf. Hénon [1976]).

Example 1.2. Let $U \subset \mathbb{R}^n$ be an open set and let $F : \mathbb{R} \times U \to \mathbb{R}^n$ be continuous and have continuous partial derivative $\partial F/\partial \mathbf{x}(t, \mathbf{x})$. Suppose that there exists $T > 0$ such that

$$F(t + T, \mathbf{x}) = F(t, \mathbf{x})$$

for all $t \in \mathbb{R}$ and $\mathbf{x} \in U$. Denote by $\phi(t, \xi)$ the solution of the initial value problem

$$\dot{\mathbf{x}} = F(t, \mathbf{x}), \quad x(0) = \xi.$$

The set

$$V = \{\xi \in U \ : \ \phi(t, \xi) \ \text{is defined on} \ [0, T]\}$$

is an open set in $\mathbb{R}^n$. If V is nonempty, we define the C^1 mapping $f : V \to \mathbb{R}^n$ by

$$f(\xi) = \phi(T, \xi).$$

It follows from the existence and uniqueness theorem for ordinary differential equations that f is a one to one mapping of V onto an open set of $\mathbb{R}^n$. Also note that $X(t) = \partial\phi/\partial\xi(t,\xi)$ is the solution of the matrix differential equation

$$\dot{X} = \frac{\partial F}{\partial \mathbf{x}}(t, \phi(t, \xi))X$$

such that $X(0) = I$. Hence $X(T)$ is nonsingular and so $Df(\xi)$ is invertible. Thus f is a C^1 diffeomorphism called the *period map* associated with $\dot{\mathbf{x}} = F(t, \mathbf{x})$.

1.2 DEFINITION AND EXAMPLES OF HYPERBOLIC FIXED POINTS

Shadowing is a property possessed by hyperbolic sets. The simplest case of a hyperbolic set is a hyperbolic fixed point. Associated with a hyperbolic fixed point are its stable and unstable manifolds. Later we will build up more complicated hyperbolic sets from a hyperbolic fixed point and its stable and unstable manifolds.

Definition 1.3. Let $U \in \mathbb{R}^n$ be an open set and $f : U \to \mathbb{R}^n$ a C^1 diffeomorphism. A point $\mathbf{x}_0 \in U$ is said to be a *hyperbolic fixed point* of f if $f(\mathbf{x}_0) = \mathbf{x}_0$ and the eigenvalues of $Df(\mathbf{x}_0)$ lie off the unit circle. The sum of the generalized eigenspaces corresponding to the eigenvalues inside (resp. outside) the unit circle is called the *stable* (resp. *unstable*) subspace and is denoted E^s (resp. E^u).

Example 1.4. To find the fixed points of the Hénon map (conf. Example 1.1), we have to solve the equations

$$1 - ax^2 + y = x, \quad -bx = y.$$

We find that

$$x = \frac{-(b+1) \pm \sqrt{(b+1)^2 - 4a}}{2a}, \quad y = -bx$$

if $a \neq 0$; if $a = 0$ we find that

$$x = \frac{1}{b+1}, \quad y = \frac{-b}{b+1}.$$

So there are two fixed points if $a \neq 0$ and $(b+1)^2 > 4a$, one fixed point if $a \neq 0$ and $(b+1)^2 = 4a$ or if $a = 0$ and $b \neq -1$, and no fixed point otherwise. It is left as an exercise for the reader to show that these fixed points are hyperbolic except when $b = 1$ and $-3 < a < 1$ or when $b \neq -1$ and $a = -3(b+1)^2/4$ or $(b+1)^2/4$.

Example 1.5. Let F, U, ϕ, f and V be as in Example 1.2. Then we see that ξ is a hyperbolic fixed point of f if and only if $\phi(T,\xi) = \xi$ and the eigenvalues of $\partial\phi/\partial\xi(T,\xi)$ lie off the unit circle. By the periodicity of F in t and the existence and uniqueness theorem for the solution of initial value problems, it follows that $\phi(T,\xi) = \xi$ if and only if $\phi(t,\xi)$ is defined for all t and has period T in t. Also note that $\partial\phi/\partial\xi(T,\xi)$ is just the monodromy matrix for the variational system

$$\dot{\mathbf{x}} = \frac{\partial F}{\partial \mathbf{x}}(t, \phi(t,\xi))\mathbf{x}.$$

So the conclusion is that ξ is a hyperbolic fixed point of f if and only if $\phi(t,\xi)$ is a T–periodic solution of the differential equation

$$\dot{\mathbf{x}} = F(t, \mathbf{x})$$

for which the corresponding Floquet multipliers lie off the unit circle.

1.3 DEFINITION OF STABLE AND UNSTABLE MANIFOLDS

Now let $\mathbf{x}_0$ be a hyperbolic fixed point of the C^1 diffeomorphism $f : U \to \mathbb{R}^n$ as in Definition 1.3. Note that the corresponding stable and unstable subspaces E^s and E^u are invariant under $Df(\mathbf{x}_0)$. Moreover, by standard results in linear algebra (conf. Hirsch and Smale [1974, p.151]), if we let λ_1 and λ_2 be positive constants such that

$$|\lambda| < \lambda_1 < 1$$

for all eigenvalues λ of $Df(\mathbf{x}_0)$ with $|\lambda| < 1$ and

$$1 < \lambda_2^{-1} < |\lambda|$$

for all eigenvalues λ with $|\lambda| > 1$, then there exist positive constants K_1 and K_2 such that for $k \geq 0$

$$\|[Df(\mathbf{x}_0)]^k\xi\| \leq K_1\lambda_1^k\|\xi\| \quad \text{for} \quad \xi \in E^s \tag{1}$$

and

$$\|[Df(\mathbf{x}_0)]^{-k}\xi\| \leq K_2\lambda_2^k\|\xi\| \quad \text{for} \quad \xi \in E^u. \tag{2}$$

Hence

$$[Df(\mathbf{x}_0)]^k\xi \to \mathbf{0} \quad \text{as} \quad k \to \infty \quad \text{if and only if} \quad \xi \in E^s$$

and

$$[Df(\mathbf{x}_0)]^k\xi \to \mathbf{0} \quad \text{as} \quad k \to -\infty \quad \text{if and only if} \quad \xi \in E^u.$$

These properties give the reason for the terminology stable and unstable subspaces and also motivate the following definition in the nonlinear case.

Definition 1.6. Let $\mathbf{x}_0$ be a hyperbolic fixed point of the C^1 diffeomorphism $f : U \to \mathbb{R}^n$. Then the set

$$W^s(\mathbf{x}_0) = \{\mathbf{x} \in U \ : \ f^k(\mathbf{x}) \to \mathbf{x}_0 \text{ as } k \to \infty\}$$

is called the *stable manifold* of $\mathbf{x}_0$ and the set

$$W^u(\mathbf{x}_0) = \{\mathbf{x} \in U \ : \ f^k(\mathbf{x}) \to \mathbf{x}_0 \text{ as } k \to -\infty\}$$

is called the *unstable manifold* of $\mathbf{x}_0$.

We shall see that the stable manifold, despite its name, may not be a submanifold of $\mathbb{R}^n$. However, we can describe it in terms of a local stable manifold which is a submanifold.

Definition 1.7. Let $U \subset \mathbb{R}^n$ be open and let $f : U \to \mathbb{R}^n$ be a C^1 diffeomorphism with a hyperbolic fixed point $\mathbf{x}_0$. For given $\varepsilon > 0$ we define the *local stable manifold*

$$W^{s,\varepsilon}(\mathbf{x}_0) = \{\mathbf{x} \in U \ : \ f^k(\mathbf{x}) \to \mathbf{x}_0 \ \text{ as } k \to \infty \text{ and } \|f^k(\mathbf{x}) - \mathbf{x}_0\| < \varepsilon \text{ for } k \geq 0\}.$$

It is easy to see that for any $\varepsilon > 0$,

$$W^s(\mathbf{x}_0) = \bigcup_{k \geq 0} f^{-k}\left(W^{s,\varepsilon}(\mathbf{x}_0)\right).$$

Also note the invariance properties

$$f(W^s(\mathbf{x}_0)) = W^s(\mathbf{x}_0) \quad \text{and} \quad f(W^{s,\varepsilon}(\mathbf{x}_0)) \subset W^{s,\varepsilon}(\mathbf{x}_0).$$

In the next two sections, our aim is to show that for $\varepsilon > 0$ sufficiently small the set $W^{s,\varepsilon}(\mathbf{x}_0)$ is indeed a smooth submanifold of $\mathbb{R}^n$ containing $\mathbf{x}_0$ such that the tangent space

$$T_{\mathbf{x}_0} W^{s,\varepsilon}(\mathbf{x}_0) = E^s,$$

the stable subspace.

1.4 THE SADDLE–POINT PROPERTY

First we show the important so called "saddle–point property". This states that if $\mathbf{x}_0$ is a hyperbolic fixed point of a diffeomorphism f, there is a positive number Δ such that if $\mathbf{x}$ is a point satisfying

$$\|f^k(\mathbf{x}) - \mathbf{x}_0\| \le \Delta \quad \text{for} \quad k \ge 0,$$

then

$$f^k(\mathbf{x}) \to \mathbf{x}_0 \quad \text{as} \quad k \to \infty.$$

More precisely, we have the following proposition.

Proposition 1.8. *Let $U \subset \mathbb{R}^n$ be open and let $f : U \to \mathbb{R}^n$ be a C^1 diffeomorphism with a hyperbolic fixed point $\mathbf{x}_0$ and associated stable and unstable subspaces E^s and E^u such that the inequalities in Eqs.(1) and (2) are satisfied. Let P be the projection of $\mathbb{R}^n$ onto E^s along E^u and write*

$$M^s = \|P\|\,,\; M^u = \|I - P\|.$$

Suppose the positive number Δ satisfies the inequality

$$\sigma = [K_1 M^s (1-\lambda_1)^{-1} + K_2 M^u \lambda_2 (1-\lambda_2)^{-1}]\omega(\Delta) < 1, \tag{3}$$

where

$$\omega(\Delta) = \sup\{\|Df(\mathbf{x}) - Df(\mathbf{x}_0)\| : \|\mathbf{x} - \mathbf{x}_0\| \le \Delta\}.$$

Then if $\mathbf{x} \in U$ and

$$\|f^k(\mathbf{x}) - \mathbf{x}_0\| \le \Delta \quad \textit{for} \quad k \ge 0,$$

the inequality

$$\|f^k(\mathbf{x}) - \mathbf{x}_0)\| \le K_1 M^s (1-\sigma)^{-1} [\lambda_1 + K_1 M^s (1-\sigma)^{-1} \omega(\Delta)]^k \|\mathbf{x} - \mathbf{x}_0\|$$

holds for $k \ge 0$ so that if, in addition,

$$K_1 M^s \omega(\Delta) < (1-\sigma)(1-\lambda_1),$$

it follows that $f^k(\mathbf{x}) \to \mathbf{x}_0$ as $k \to \infty$.

Proof. Set

$$\mathbf{y}_k = f^k(\mathbf{x}) - \mathbf{x}_0.$$

Then for $k \ge 0$

$$\mathbf{y}_{k+1} = f(\mathbf{x}_0 + \mathbf{y}_k) - f(\mathbf{x}_0).$$

So for $k \geq 0$

$$\mathbf{y}_{k+1} = A\mathbf{y}_k + g(\mathbf{y}_k), \tag{4}$$

where

$$A = Df(\mathbf{x}_0)$$

and

$$g(\mathbf{y}) = f(\mathbf{x}_0 + \mathbf{y}) - f(\mathbf{x}_0) - Df(\mathbf{x}_0)\mathbf{y}.$$

Note that if $\|\mathbf{y}\| \leq \Delta$,

$$\|g(\mathbf{y})\| \leq \omega(\Delta)\|\mathbf{y}\|.$$

Next set

$$\mathbf{u}_k = P\mathbf{y}_k, \ \mathbf{v}_k = (I - P)\mathbf{y}_k.$$

Since A commutes with P, when we multiply Eq.(4) by P, we find that for $k \geq 0$,

$$\mathbf{u}_{k+1} = A\mathbf{u}_k + Pg(\mathbf{y}_k).$$

Repeatedly using this equation, we find that for $k \geq 0$

$$\mathbf{u}_k = A^k\mathbf{u}_0 + \sum_{m=0}^{k-1} A^{k-m-1}Pg(\mathbf{y}_m). \tag{5}$$

Similarly, when we multiply Eq.(4) by $I - P$, we find that for $k \geq 0$

$$\mathbf{v}_{k+1} = A\mathbf{v}_k + (I - P)g(\mathbf{y}_k).$$

By applying the equation

$$\mathbf{v}_k = A^{-1}\mathbf{v}_{k+1} - A^{-1}(I - P)g(\mathbf{y}_k)$$

repeatedly, we find that for $0 \leq k \leq m$

$$\mathbf{v}_k = A^{-(m-k)}\mathbf{v}_m - \sum_{\ell=k}^{m-1} A^{-(\ell-k+1)}(I - P)g(\mathbf{y}_\ell). \tag{6}$$

Now note that for $0 \leq k \leq m$

$$\|A^{-(m-k)}\mathbf{v}_m\| = \|A^{-(m-k)}(I - P)\mathbf{y}_m\| \leq K_2\lambda_2{}^{m-k}M^u\Delta,$$

which tends to zero as m tends to infinity, and also that

$$\begin{aligned}\sum_{\ell=k}^{\infty}\|A^{-(\ell-k+1)}(I-P)g(\mathbf{y}_\ell)\| &\leq \sum_{\ell=k}^{\infty}K_2{\lambda_2}^{\ell-k+1}M^u\omega(\Delta)\Delta\\ &= K_2M^u\lambda_2(1-\lambda_2)^{-1}\omega(\Delta)\Delta.\end{aligned}$$

Hence we may let $m\to\infty$ in Eq. (6) to obtain

$$\mathbf{v}_k = -\sum_{m=k}^{\infty}A^{-(m-k+1)}(I-P)g(\mathbf{y}_m). \tag{7}$$

Adding Eqs.(5) and (7) together, we conclude that for $k\geq 0$

$$\mathbf{y}_k = A^kP\mathbf{y}_0+\sum_{m=0}^{k-1}A^{k-m-1}Pg(\mathbf{y}_m)-\sum_{m=k}^{\infty}A^{-(m-k+1)}(I-P)g(\mathbf{y}_m).$$

Then it follows that for $k\geq 0$

$$\begin{aligned}\|\mathbf{y}_k\| &\leq K_1M^s{\lambda_1}^k\|\mathbf{y}_0\|+\sum_{m=0}^{k-1}K_1M^s{\lambda_1}^{k-m-1}\omega(\Delta)\|\mathbf{y}_m\|\\ &\quad+\sum_{m=k}^{\infty}K_2M^u\lambda_2^{m-k+1}\omega(\Delta)\|\mathbf{y}_m\|.\end{aligned} \tag{8}$$

To deal with this inequality, we require the following two lemmas, the first of which can be regarded as a discrete Gronwall lemma (conf. Aulbach [1984, p.116]). Actually, for later use, we prove a little more than what is needed here.

Lemma 1.9. (i) *Let* $\{\mu_k\}_{k=a}^b$ *be a sequence of nonnegative numbers such that there exist positive numbers* c,d *for which*

$$\mu_k\leq c+d\sum_{m=a}^{k-1}\mu_m$$

for $a\leq k\leq b$. *Then*

$$\mu_k\leq c(1+d)^{k-a}$$

for $a\leq k\leq b$.

(ii) *Let* $\{\mu_k\}_{k=a}^b$ *be a sequence of nonnegative numbers such that there exist positive numbers* c,d *with* $d<1$ *for which*

$$\mu_k\leq c+d\sum_{m=k}^{b-1}\mu_m$$

for $a \le k \le b$. *Then*

$$\mu_k \le c(1-d)^{k-b}$$

for $a \le k \le b$.

Proof. (i) For $a \le k \le b$, set

$$v_k = \sum_{m=a}^{k-1} \mu_m.$$

Then for $a \le k \le b-1$,

$$v_{k+1} - v_k \le c + dv_k$$

and so

$$v_{k+1} \le c + (1+d)v_k.$$

Next, observing that $v_a = 0$, we obtain for $a \le k \le b$ that

$$v_k \le c[1 + (1+d) + \ldots + (1+d)^{k-a-1}] = cd^{-1}[(1+d)^{k-a} - 1]$$

and so

$$\mu_k \le c + dv_k \le c(1+d)^{k-a}.$$

(ii) For $a \le k \le b$, set

$$v_k = \sum_{m=k}^{b-1} \mu_m.$$

Then for $a \le k \le b-1$,

$$v_k - v_{k+1} \le c + dv_k,$$

and hence

$$v_k \le (1-d)^{-1}c + (1-d)^{-1}v_{k+1}.$$

So, observing that $v_b = 0$, we obtain for $a \le k \le b$ that

$$\begin{aligned} v_k &\le (1-d)^{-1}c[1 + (1-d)^{-1} + (1-d)^{-2} + \ldots + (1-d)^{-(b-k-1)}] \\ &= cd^{-1}[(1-d)^{k-b} - 1] \end{aligned}$$

and hence that

$$\mu_k \le c + dv_k \le c(1-d)^{k-b}.$$

Lemma 1.10. (i) *Let a, b be fixed integers with $-\infty < a \le b \le \infty$. Suppose $\{\mu_k\}_{k=a}^{b}$ is a sequence of nonnegative numbers, bounded when $b = \infty$, for which there exist positive numbers $K_1, K_2, \lambda_1, \lambda_2, \alpha_1, \alpha_2$ with $\lambda_1 < 1$, $\lambda_2 < 1$ such that for $a \le k < b+1$*

$$\mu_k \le K_1\lambda_1^{k-a}\mu_a + \sum_{m=a}^{k-1} \alpha_1\lambda_1^{k-m-1}\mu_m + \sum_{m=k}^{b-1} \alpha_2\lambda_2^{m-k+1}\mu_m. \tag{9}$$

Then if

$$\sigma = \alpha_1(1-\lambda_1)^{-1} + \alpha_2\lambda_2(1-\lambda_2)^{-1} < 1,$$

the inequality

$$\mu_k \le K_1(1-\sigma)^{-1}[\lambda_1 + \alpha_1(1-\sigma)^{-1}]^{k-a}\mu_a \tag{10}$$

holds for $a \le k < b+1$.

(ii) *Let a, b be fixed integers with $-\infty \le a \le b < \infty$. Suppose $\{\mu_k\}_{k=a}^{b}$ is a sequence of nonnegative numbers, bounded when $a = -\infty$, for which there exist positive numbers $K_1, K_2, \lambda_1, \lambda_2, \alpha_1, \alpha_2$ with $\lambda_1 < 1$, $\lambda_2 < 1$ such that for $a-1 < k \le b$*

$$\mu_k \le K_2\lambda_2^{b-k}\mu_b + \sum_{m=a}^{k-1} \alpha_1\lambda_1^{k-m-1}\mu_m + \sum_{m=k}^{b-1} \alpha_2\lambda_2^{m-k+1}\mu_m. \tag{11}$$

Then if

$$\sigma = \alpha_1(1-\lambda_1)^{-1} + \alpha_2\lambda_2(1-\lambda_2)^{-1} < 1 \quad \text{and} \quad \alpha_2\lambda_2 < 1-\sigma,$$

the inequality

$$\mu_k \le K_2(1-\sigma)^{-1}[(1-\sigma-\alpha_2\lambda_2)^{-1}(1-\sigma)\lambda_2]^{b-k}\mu_b$$

holds for $a-1 < k \le b$.

Proof. (i) We follow the discrete version of the proof of Theorem 13 in Coppel [1965, p.80] as given in Palmer [1988]. Put

$$\theta_k = \sup_{k \le m < b+1} \mu_m.$$

For given k in $a \le k < b+1$ and $\varepsilon > 0$ (of course, when $b < \infty$, we could have taken $\varepsilon = 0$), there is $k' \ge k$ such that

$$\mu_{k'} \geq \theta_k - \varepsilon.$$

By Eq.(9) with k' in place of k,

$$\begin{aligned}
\theta_k - \varepsilon &\leq K_1\lambda_1^{k'-a}\mu_a + \sum_{m=a}^{k'-1} \alpha_1\lambda_1^{k'-m-1}\mu_m + \sum_{m=k'}^{\infty} \alpha_2\lambda_2^{m-k'+1}\mu_m \\
&\leq K_1\lambda_1^{k-a}\mu_a + \alpha_1\sum_{m=a}^{k-1} \lambda_1^{k-m-1}\mu_m + \alpha_1\sum_{m=k}^{k'-1} \lambda_1^{k'-m-1}\mu_m \\
&\qquad + \alpha_2\sum_{m=k'}^{\infty} \lambda_2^{m-k'+1}\mu_m \\
&\leq K_1\lambda_1^{k-a}\mu_a + \alpha_1\sum_{m=a}^{k-1} \lambda_1^{k-m-1}\mu_m \\
&\qquad + [\alpha_1(1-\lambda_1)^{-1} + \alpha_2\lambda_2(1-\lambda_2)^{-1}]\theta_k.
\end{aligned}$$

This holds for all $\varepsilon > 0$ and so for $a \leq k < b+1$

$$\theta_k \leq K_1(1-\sigma)^{-1}\lambda_1^{k-a}\mu_a + \alpha_1(1-\sigma)^{-1}\sum_{m=a}^{k-1}\lambda_1^{k-m-1}\theta_m.$$

Then $\phi_k = \lambda_1^{-k}\theta_k$ satisfies

$$\phi_k \leq K_1(1-\sigma)^{-1}\lambda_1^{-a}\mu_a + \alpha_1(1-\sigma)^{-1}\lambda_1^{-1}\sum_{m=a}^{k-1}\phi_m$$

for $a \leq k < b+1$. By Lemma 1.9(i), it follows that for $a \leq k < b+1$

$$\phi_k \leq K_1(1-\sigma)^{-1}\lambda_1^{-a}\mu_a[1 + \alpha_1(1-\sigma)^{-1}\lambda_1^{-1}]^{k-a}.$$

Hence

$$\theta_k \leq K_1(1-\sigma)^{-1}[\lambda_1 + \alpha_1(1-\sigma)^{-1}]^{k-a}\mu_a$$

for $a \leq k < b+1$ and so Eq.(10) follows, completing the proof of part (i).

(ii) Set

$$\theta_k = \sup_{a-1<m\leq k} \mu_m.$$

For given k in $a-1 < k \leq b$ and $\varepsilon > 0$, there is $k' \leq k$ such that

$$\mu_{k'} \geq \theta_k - \varepsilon.$$

By Eq.(11) with k' in place of k,

$$\begin{aligned}
\theta_k - \varepsilon &\leq K_2\lambda_2^{b-k'}\mu_b + \sum_{m=a}^{k'-1} \alpha_1\lambda_1^{k'-m-1}\mu_m + \sum_{m=k'}^{b-1} \alpha_2\lambda_2^{m-k'+1}\mu_m \\
&\leq K_2\lambda_2^{b-k}\mu_b + \sum_{m=a}^{k'-1} \alpha_1\lambda_1^{k'-m-1}\mu_m + \sum_{m=k'}^{k-1} \alpha_2\lambda_2^{m-k'+1}\mu_m \\
&\qquad + \sum_{m=k}^{b-1} \alpha_2\lambda_2^{m-k'+1}\mu_m \\
&\leq K_2\lambda_2^{b-k}\mu_b + [\alpha_1(1-\lambda_1)^{-1} + \alpha_2\lambda_2(1-\lambda_2)^{-1}]\theta_k \\
&\qquad + \sum_{m=k}^{b-1} \alpha_2\lambda_2^{m-k+1}\theta_m.
\end{aligned}$$

Hence, letting $\varepsilon \to 0$, we have

$$\theta_k \leq K_2(1-\sigma)^{-1}\lambda_2^{b-k}\mu_b + \alpha_2(1-\sigma)^{-1}\sum_{m=k}^{b-1} \lambda_2^{m-k+1}\theta_m$$

for $a-1 < k \leq b$. Then $\phi_k = \lambda_2^k\theta_k$ satisfies

$$\phi_k \leq K_2(1-\sigma)^{-1}\lambda_2^b\mu_b + \alpha_2(1-\sigma)^{-1}\lambda_2\sum_{m=k}^{b-1}\phi_m$$

for $a-1 < k \leq b$. Applying Lemma 1.9(ii), it follows that

$$\phi_k \leq K_2(1-\sigma)^{-1}\lambda_2^b\mu_b[1-\alpha_2(1-\sigma)^{-1}\lambda_2]^{k-b}$$

for $a-1 < k \leq b$. Hence, for these same k,

$$\theta_k \leq K_2(1-\sigma)^{-1}\lambda_2^b\mu_b[1-\alpha_2(1-\sigma)^{-1}\lambda_2]^{k-b}\lambda_2^{-k}$$

and so part (ii) of the lemma follows.

To complete the proof of Proposition 1.8, we apply Lemma 1.10(i) to Eq.(8) to conclude that

$$\|\mathbf{y}_k\| \leq K_1 M^s (1-\sigma)^{-1} [\lambda_1 + K_1 M^s (1-\sigma)^{-1} \omega(\Delta)]^k \|\mathbf{y}_0\|$$

for $k \geq 0$. Thus the proposition follows.

1.5 SMOOTHNESS OF THE LOCAL STABLE MANIFOLD

In this section we show that for $\varepsilon > 0$ sufficiently small the local stable manifold $W^{s,\varepsilon}(\mathbf{x}_0)$ for a hyperbolic fixed point $\mathbf{x}_0$ of a diffeomorphism is a smooth submanifold of $\mathbb{R}^n$.

Theorem 1.11. *Let $U \subset \mathbb{R}^n$ be an open set and let $f : U \to \mathbb{R}^n$ be a C^r $(r \geq 1)$ diffeomorphism with hyperbolic fixed point $\mathbf{x}_0$ and associated stable subspace E^s. Then for $\varepsilon > 0$ sufficiently small, $W^{s,\varepsilon}(\mathbf{x}_0)$ is a C^r submanifold of $\mathbb{R}^n$ containing $\mathbf{x}_0$ with $T_{\mathbf{x}_0} W^{s,\varepsilon}(\mathbf{x}_0) = E^s$.*

Proof. Proposition 1.8 implies that if $\Delta > 0$ is sufficiently small, then if the inequality $\|f^k(\mathbf{x}) - \mathbf{x}_0\| \leq \Delta$ holds for $k \geq 0$ it follows that $f^k(\mathbf{x}) \to \mathbf{x}_0$ as $k \to \infty$. So to find points on the local stable manifold $W^{s,\varepsilon}(\mathbf{x}_0)$ with $\varepsilon = \Delta$, we look for points $\mathbf{x}$ for which $\|f^k(\mathbf{x}) - \mathbf{x}_0\| \leq \Delta$ for $k \geq 0$. Suppose for all $k \geq 0$

$$\|f^k(\mathbf{x}) - \mathbf{x}_0\| \leq \Delta.$$

We set

$$\mathbf{y}_k = f^k(\mathbf{x}) - \mathbf{x}_0.$$

Then it follows as in the proof of Proposition 1.8 that for $k \geq 0$

$$\mathbf{y}_k = A^k P \mathbf{y}_0 + \sum_{m=0}^{k-1} A^{k-m-1} P g(\mathbf{y}_m) - \sum_{m=k}^{\infty} A^{-(m-k+1)} (I - P) g(\mathbf{y}_m), \tag{12}$$

where P is the projection onto E^s along the unstable subspace E^u,

$$A = Df(\mathbf{x}_0)$$

and

$$g(\mathbf{y}) = f(\mathbf{x}_0 + \mathbf{y}) - f(\mathbf{x}_0) - Df(\mathbf{x}_0)\mathbf{y}.$$

So we look for solutions $\mathbf{y}_k$ of Eq.(12) such that $\|\mathbf{y}_k\| \leq \Delta$ for $k \geq 0$. It turns out that these solutions are not unique. However, provided Δ and $\|\xi\|$ are sufficiently small, they are unique if we impose a condition $P\mathbf{y}_0 = \xi \in E^s$.

We assume $\Delta > 0$ satisfies

$$\sigma = [K_1 M^s (1-\lambda_1)^{-1} + K_2 M^u \lambda_2 (1-\lambda_2)^{-1}]\omega(\Delta) < 1, \tag{13}$$

where the notations are as in Proposition 1.8, and that $\xi \in E^s$ satisfies

$$\|\xi\| < \Delta(1-\sigma)/K_1. \tag{14}$$

We shall show that there is a unique point $\mathbf{x}$ such that

$$P(\mathbf{x} - \mathbf{x}_0) = \xi$$

and

$$\|f^k(\mathbf{x}) - \mathbf{x}_0\| \leq \Delta \quad \text{for} \quad k \geq 0.$$

To this end, let us denote by $\mathcal{E}$ the Banach space of bounded $\mathbb{R}^n$-valued sequences $\mathbf{y} = \{\mathbf{y}_k\}_{k=0}^{\infty}$ with norm

$$\|\mathbf{y}\| = \sup_{k=0}^{\infty} \|\mathbf{y}_k\|.$$

We define an operator T on the closed ball of radius Δ in $\mathcal{E}$. If $\mathbf{y} \in \mathcal{E}$ and $\|\mathbf{y}\| \leq \Delta$, we define $T\mathbf{y}$ as the sequence $\{(T\mathbf{y})_k\}_{k=-\infty}^{\infty}$ given by

$$(T\mathbf{y})_k = A^k \xi + \sum_{m=0}^{k-1} A^{k-m-1} P g(\mathbf{y}_m) - \sum_{m=k}^{\infty} A^{-(m-k+1)}(I-P) g(\mathbf{y}_m)$$

for $k \geq 0$. Notice that

$$\sum_{m=0}^{k-1} \|A^{k-m-1} P g(\mathbf{y}_m)\| + \sum_{m=k}^{\infty} \|A^{-(m-k+1)}(I-P) g(\mathbf{y}_m)\|$$

$$\leq \sum_{m=0}^{k-1} K_1 \lambda_1^{k-m-1} M^s \omega(\Delta)\Delta + \sum_{m=k}^{\infty} K_2 \lambda_2^{m-k+1} M^u \omega(\Delta)\Delta$$

$$\leq [K_1 M^s (1-\lambda_1)^{-1} + K_2 M^u \lambda_2 (1-\lambda_2)^{-1}]\omega(\Delta)\Delta.$$

From this it follows that $T\mathbf{y}$ is well-defined, $T\mathbf{y} \in \mathcal{E}$ and

$$\|T\mathbf{y}\| \leq K_1 \|\xi\| + \sigma\Delta \leq K_1 \Delta(1-\sigma)/K_1 + \sigma\Delta = \Delta.$$

Moreover, if $\mathbf{y}$ and $\overline{\mathbf{y}} \in \mathcal{E}$ with $\|\mathbf{y}\| \leq \Delta$, $\|\overline{\mathbf{y}}\| \leq \Delta$, then

$$\begin{aligned}(T\mathbf{y})_k - (T\overline{\mathbf{y}})_k &= \sum_{m=0}^{k-1} A^{k-m-1} P[g(\mathbf{y}_m) - g(\overline{\mathbf{y}}_m)] \\ &\quad - \sum_{m=k}^{\infty} A^{-(m-k+1)}(I-P)[g(\mathbf{y}_m) - g(\overline{\mathbf{y}}_m)]\end{aligned}$$

and so

$$\|(T\mathbf{y})_k - (T\overline{\mathbf{y}})_k\| \le \left[\sum_{m=0}^{k-1} K_1 M^s \lambda_1^{k-m-1} + \sum_{m=k}^{\infty} K_2 M^u \lambda_2^{m-k+1}\right] \omega(\Delta)\|\mathbf{y}-\overline{\mathbf{y}}\|$$

$$\le [K_1 M^s (1-\lambda_1)^{-1} + K_2 M^u \lambda_2 (1-\lambda_2)^{-1}]\omega(\Delta)\|\mathbf{y}-\overline{\mathbf{y}}\|.$$

Hence if $\mathbf{y}$ and $\overline{\mathbf{y}} \in \mathcal{E}$ with $\|\mathbf{y}\| \le \Delta$, $\|\overline{\mathbf{y}}\| \le \Delta$

$$\|T\mathbf{y} - T\overline{\mathbf{y}}\| \le \sigma\|\mathbf{y}-\overline{\mathbf{y}}\|.$$

Thus T is a contraction on the complete metric space

$$\{\mathbf{y} \in \mathcal{E} : \|\mathbf{y}\| \le \Delta\}$$

and so has a unique fixed point $\mathbf{y} = \{\mathbf{y}_k\}_{k=0}^{\infty}$. Then $\mathbf{y}_k$ satisfies the equation

$$\mathbf{y}_k = A^k\xi + \sum_{m=0}^{k-1} A^{k-m-1}Pg(\mathbf{y}_m) - \sum_{m=k}^{\infty} A^{-(m-k+1)}(I-P)g(\mathbf{y}_m) \tag{15}$$

for $k \ge 0$, with

$$P\mathbf{y}_0 = \xi \quad \text{and} \quad \|\mathbf{y}_k\| \le \Delta \ \text{ for } k \ge 0.$$

Also we see that for $k \ge 0$

$$\mathbf{y}_{k+1} = A^{k+1}\xi + \sum_{m=0}^{k} A^{k-m}Pg(\mathbf{y}_m) - \sum_{m=k+1}^{\infty} A^{-(m-k)}(I-P)g(\mathbf{y}_m)$$

$$= A[A^k\xi + \sum_{m=0}^{k-1} A^{k-m-1}Pg(\mathbf{y}_m) - \sum_{m=k}^{\infty} A^{-(m-k+1)}(I-P)g(\mathbf{y}_m)]$$

$$+Pg(\mathbf{y}_k) + (I-P)g(\mathbf{y}_k)$$

and so

$$\mathbf{y}_{k+1} = A\mathbf{y}_k + g(\mathbf{y}_k) = f(\mathbf{x}_0 + \mathbf{y}_k) - f(\mathbf{x}_0) = f(\mathbf{x}_0 + \mathbf{y}_k) - \mathbf{x}_0.$$

It follows that $\mathbf{x} = \mathbf{x}_0 + \mathbf{y}_0$ satisfies $f^k(\mathbf{x}) = \mathbf{x}_0 + \mathbf{y}_k$ for $k \ge 0$. So

$$P(\mathbf{x} - \mathbf{x}_0) = \xi \tag{16}$$

and

$$\|f^k(\mathbf{x}) - \mathbf{x}_0\| \le \Delta \ \text{ for } k \ge 0. \tag{17}$$

Next let $\overline{\mathbf{x}}$ be another point satisfying Eqs.(16) and (17). Then, as at the beginning of this proof, it follows that $\overline{\mathbf{y}}_k = f^k(\overline{\mathbf{x}}) - \mathbf{x}_0$ satisfies Eq.(15). Thus

$\bar{\mathbf{y}} = \{\bar{\mathbf{y}}_k\}_{k=0}^{\infty}$ is a fixed point of the operator T and so coincides with $\mathbf{y}$. Thus $\bar{\mathbf{x}} = \mathbf{x}_0 + \bar{\mathbf{y}}_0 = \mathbf{x}_0 + \mathbf{y}_0 = \mathbf{x}$. So if $\Delta > 0$ satisfies Eq.(13) and ξ satisfies Eq.(14), the point $\mathbf{x} = \mathbf{x}_0 + \mathbf{y}_0$ is the unique point satisfying Eqs.(16) and (17).

To continue the proof of Theorem 1.11, we note that $\mathbf{y}_0$ is a function of ξ and so write

$$\mathbf{y}_0 = \phi^s(\xi)$$

for $\xi \in E^s$, $\|\xi\| < \Delta(1-\sigma)/K_1$. Next we show ϕ^s is a C^r function if f is C^r. In order to do this, we use the following lemma which shows that if a contraction mapping depends smoothly on parameters, then so also does the fixed point. Since the lemma is well-known, we omit its proof.

Lemma 1.12. *Let B be a closed ball in a Banach space $\mathcal{E}$ and let V be an open set in another Banach space Λ. Suppose $T : B \times V \to B$ is a mapping such that for all $\mathbf{y}_1$, $\mathbf{y}_2 \in B$ and $\mu \in V$*

$$\|T(\mathbf{y}_1, \mu) - T(\mathbf{y}_2, \mu)\| \leq \sigma \|\mathbf{y}_1 - \mathbf{y}_2\|$$

where $0 \leq \sigma < 1$. Then for each $\mu \in V$ the equation

$$T(\mathbf{y}, \mu) = \mathbf{y}$$

has a unique solution $\mathbf{y} = \mathbf{y}(\mu)$ in B. Moreover, if T is C^r $(0 \leq r \leq \infty)$, then so also is $\mathbf{y}(\mu)$, and the derivative $D\mathbf{y}(\mu)$ satisfies the equation

$$D\mathbf{y}(\mu) = \frac{\partial T}{\partial \mathbf{y}}(\mathbf{y}(\mu), \mu) D\mathbf{y}(\mu) + \frac{\partial T}{\partial \mu}(\mathbf{y}(\mu), \mu).$$

(Note: T is C^r means that T is defined and C^r in an open set containing $B \times V$.)

Now we use Lemma 1.12 to prove the smoothness of the function ϕ^s. Let $\mathcal{E}$ be the Banach space of bounded sequences $\mathbf{y} = \{\mathbf{y}_k\}_{k=0}^{\infty}$ as used above, B the closed ball $\{\mathbf{y} \in \mathcal{E} : \|\mathbf{y}\| \leq \Delta\}$ and V the ball $\{\xi \in E^s : \|\xi\| < \Delta(1-\sigma)/K_1\}$. Then we define $T : B \times V \to B$ by

$$[T(\mathbf{y}, \xi)]_k = A^k \xi + \sum_{m=0}^{k-1} A^{k-m-1} P g(\mathbf{y}_m) - \sum_{m=k}^{\infty} A^{-(m-k+1)} (I - P) g(\mathbf{y}_m).$$

Since g is C^r, T is C^r. Also above, we showed that for fixed ξ, T is a contraction on B with contraction constant σ and the fixed point is $\mathbf{y} = \mathbf{y}(\xi) = \{\mathbf{y}_k(\xi)\}_{k=0}^{\infty}$ with $\mathbf{y}_0(\xi) = \phi^s(\xi)$. It follows from Lemma 1.12 that $\mathbf{y}(\xi)$ is C^r in ξ. Hence $\phi^s(\xi)$ is C^r in ξ also. Next we see that $T(\mathbf{0}, 0) = \mathbf{0}$ and so by uniqueness, $\mathbf{y}(0) = \mathbf{0}$ and thus $\phi^s(0) = \mathbf{0}$. Also from Lemma 1.12, $Y = D\mathbf{y}(0)$ satisfies

$$Y = \frac{\partial T}{\partial \mathbf{y}}(\mathbf{0},0)Y + \frac{\partial T}{\partial \xi}(\mathbf{0},0).$$

However

$$\frac{\partial T}{\partial \mathbf{y}}(\mathbf{0},0) = \mathbf{0}$$

and if $\nu \in E^s$

$$\left[\frac{\partial T}{\partial \xi}(\mathbf{0},0)\nu\right]_k = A^k\nu.$$

Hence

$$[D\mathbf{y}(0)\nu]_k = A^k\nu$$

and it follows that $D\phi^s(0)$ is the inclusion of E^s in $\mathbb{R}^n$. Note also that $P\phi^s(\xi) = \xi$ and so $PD\phi^s(\xi)\nu = \nu$ for all $\nu \in E^s$. These properties imply that ϕ^s and $D\phi^s(\xi)$ are one to one.

It follows from the properties of ϕ^s just derived that the map $\xi \to \mathbf{x}_0 + \phi^s(\xi)$ is a C^r immersion and so the set

$$\mathcal{M} = \{\mathbf{x}_0 + \phi^s(\xi) : \xi \in E^s \, , \, \|\xi\| < \Delta(1-\sigma)/K_1\}$$

is a C^r submanifold of $\mathbb{R}^n$ containing $\mathbf{x}_0$. Moreover, since $D\phi^s(0)$ is the inclusion of E^s in $\mathbb{R}^n$,

$$T_{\mathbf{x}_0}\mathcal{M} = E^s.$$

As shown above, if $\mathbf{x} = \mathbf{x}_0 + \phi^s(\xi) \in \mathcal{M}$, then

$$P(\mathbf{x} - \mathbf{x}_0) = \xi \quad \text{and} \quad \|f^k(\mathbf{x}) - \mathbf{x}_0\| \leq \Delta \quad \text{for} \quad k \geq 0.$$

Moreover, for given $\xi \in E^s$ with $\|\xi\| < \Delta(1-\sigma)/K_1$, $\mathbf{x}$ is the unique such point. (Note it follows from Proposition 1.8 that if Δ is sufficiently small, $\mathcal{M}$ is a subset of $W^{s,\Delta}(\mathbf{x}_0)$ but may not coincide with $W^{s,\Delta}(\mathbf{x}_0)$.)

Now fix a positive number Δ satisfying the inequality in Eq.(13) and such that

$$\lambda = \lambda_1 + K_1 M^s(1-\sigma)^{-1}\omega(\Delta) < 1.$$

Let $\mathcal{M}$ be the manifold constructed above corresponding to Δ. Suppose the positive number ε satisfies

$$\varepsilon \leq \Delta, \; \varepsilon < \Delta(1-\sigma)/K_1 M^s.$$

We show that $W^{s,\varepsilon}(\mathbf{x}_0)$ is an open subset of $\mathcal{M}$ and hence a C^r submanifold of $\mathcal{M}$.

First note that if $\mathbf{x} \in W^{s,\varepsilon}(\mathbf{x}_0)$, then

$$\|P(\mathbf{x}-\mathbf{x}_0)\| \leq M^s\varepsilon < \Delta(1-\sigma)/K_1$$

and for $k \geq 0$

$$\|f^k(\mathbf{x}) - \mathbf{x}_0\| < \varepsilon \leq \Delta.$$

Hence $\mathbf{x} \in \mathcal{M}$ and so $W^{s,\varepsilon}(\mathbf{x}_0) \subset \mathcal{M}$. Next let $\mathbf{x} \in W^{s,\varepsilon}(\mathbf{x}_0)$ and $\mathbf{y} \in \mathcal{M}$ with $\|\mathbf{y}-\mathbf{x}\| < \delta$, with $\delta > 0$ to be determined. Then it follows from Proposition 1.8 that

$$\|f^k(\mathbf{y}) - \mathbf{x}_0\| \leq K_1 M^s(1-\sigma)^{-1}\lambda^k\|\mathbf{y}-\mathbf{x}_0\| \leq K_1M^s(1-\sigma)^{-1}\lambda^k\delta < \varepsilon$$

for $k \geq N$, provided

$$\delta \leq 1$$

and we choose the positive integer N (recall $\lambda < 1$) so that

$$K_1M^s(1-\sigma)^{-1}\lambda^N < \varepsilon.$$

Next if $k = 0, \ldots, N-1$

$$\|f^k(\mathbf{y}) - \mathbf{x}_0\| \leq \|f^k(\mathbf{y}) - f^k(\mathbf{x})\| + \|f^k(\mathbf{x}) - \mathbf{x}_0\| < \varepsilon$$

provided δ is so small that

$$\max_{0\leq k<N} \|f^k(\mathbf{y}) - f^k(\mathbf{x})\| < \varepsilon - \max_{0\leq k<N} \|f^k(\mathbf{x}) - \mathbf{x}_0\|$$

when $\|\mathbf{y}-\mathbf{x}\| < \delta$. That means that if $\mathbf{y} \in \mathcal{M}$ and $\|\mathbf{y}-\mathbf{x}\| < \delta$, then $\mathbf{y} \in W^{s,\varepsilon}(\mathbf{x}_0)$. So $W^{s,\varepsilon}(\mathbf{x}_0)$ is an open subset of $\mathcal{M}$ and therefore a C^r manifold also. Moreover, $\mathbf{x}_0 \in W^{s,\varepsilon}(\mathbf{x}_0)$ and

$$T_{\mathbf{x}_0}W^{s,\varepsilon}(\mathbf{x}_0) = T_{\mathbf{x}_0}\mathcal{M} = E^s.$$

This completes the proof of Theorem 1.11.

1.6 DIFFEOMORPHISMS DEPENDING ON A PARAMETER

Finally we consider a diffeomorphism depending on a parameter and examine how the local stable and unstable manifolds of a hyperbolic fixed point vary with the parameter.

So let $f : U \times V \to I\!R^n$ be a $C^r(r \geq 1)$ function, where $U \subset I\!R^n$ and $V \subset I\!R^m$ are open sets such that $f_\mu : U \to I\!R^n$ defined by $f_\mu(\mathbf{x}) = f(\mathbf{x}, \mu)$ is a diffeomorphism onto its image for each fixed μ in V. Suppose for $\mu_0 \in V$ the diffeomorphism f_{μ_0} has a hyperbolic fixed point $\mathbf{x}_0$. By applying the implicit function theorem to the equation

$$f(\mathbf{x}, \mu) - \mathbf{x} = 0,$$

we see that when μ is sufficiently near μ_0, the diffeomorphism f_μ has a unique fixed point $\mathbf{x}(\mu)$ near $\mathbf{x}_0$. Moreover, $\mathbf{x}(\mu_0) = \mathbf{x}_0$ and $\mathbf{x}(\mu)$ is C^r in μ. Also since the matrix $\partial f/\partial \mathbf{x}(\mathbf{x}(\mu), \mu)$ depends continuously on μ, $\mathbf{x}(\mu)$ is also hyperbolic if μ is sufficiently near μ_0 and the stable and unstable subspaces have the same dimension as those for $\mathbf{x}_0$ (this would follow from matrix theory or from Lemma 2.8 in the next chapter). Our aim is to show that the stable manifold of $\mathbf{x}(\mu)$ depends smoothly on μ in the sense that the manifold $\mathcal{M}$ constructed in the previous section depends smoothly on μ.

Suppose $\mathbf{x}(\mu)$ is a hyperbolic fixed point of f_μ for $\|\mu - \mu_0\| \leq \rho$. We assume $\Delta > 0$ and $\rho > 0$ satisfy

$$\sigma = [K_1 M^s (1 - \lambda_1)^{-1} + K_2 M^u \lambda_2 (1 - \lambda_2)^{-1}] \omega(\Delta, \rho) < 1 \qquad (18)$$

and

$$[K_1 M^s (1 - \lambda_1)^{-1} + K_2 M^u \lambda_2 (1 - \lambda_2)^{-1}] \left\| \frac{\partial f}{\partial \mu}(\mathbf{x}_0, \mu) \right\| \rho \leq \Delta(1 - \sigma)/2 \quad (19)$$

for $\|\mu - \mu_0\| \leq \rho$, where

$$\omega(\Delta, \rho) = \sup \left\{ \left\| \frac{\partial f}{\partial \mathbf{x}}(\mathbf{x}, \mu) - \frac{\partial f}{\partial \mathbf{x}}(\mathbf{x}_0, \mu_0) \right\| \ : \ \|\mathbf{x} - \mathbf{x}_0\| \leq \Delta, \ \|\mu - \mu_0\| \leq \rho \right\}$$

and $K_1, \lambda_1, K_2, \lambda_2$ and M^s, M^u are the constants associated with the hyperbolic fixed point $\mathbf{x}_0$ of f_{μ_0}. The Banach space $\mathcal{E}$ is defined as in the proof of Theorem 1.11. We define the function

$$g(\mathbf{y}, \mu) = f(\mathbf{x}_0 + \mathbf{y}, \mu) - f(\mathbf{x}_0, \mu_0) - \frac{\partial f}{\partial \mathbf{x}}(\mathbf{x}_0, \mu_0)\mathbf{y}.$$

If $\mathbf{y} \in \mathcal{E}$ with $\|\mathbf{y}\| \leq \Delta$ and $\|\mu - \mu_0\| < \rho$, we define $T(\mathbf{y}, \mu)$ in $\mathcal{E}$ by

$$(T(\mathbf{y}, \mu))_k = A^k \xi + \sum_{m=0}^{k-1} A^{k-m-1} P g(\mathbf{y}_m, \mu) - \sum_{m=k}^{\infty} A^{-(m-k+1)} (I - P) g(\mathbf{y}_m, \mu),$$

where $\xi \in E^s$ satisfies

$$\|\xi\| < \Delta(1 - \sigma)/2K_1.$$

Proceeding as in the proof of Theorem 1.11, we show that for each fixed μ in $\|\mu - \mu_0\| < \rho$, the mapping $\mathbf{y} \to T(\mathbf{y}, \mu)$ is a contraction on the complete metric space $\{\mathbf{y} \in \mathcal{E} : \|\mathbf{y}\| \leq \Delta\}$ with contraction constant σ. Let $\mathbf{y}(\xi, \mu) = \{\mathbf{y}_k(\xi, \mu)\}_{k=0}^{\infty}$ be the fixed point. Then $\mathbf{x} = \mathbf{x}_0 + \mathbf{y}_0(\xi, \mu)$ is the unique point such that

$$P(\mathbf{x} - \mathbf{x}_0) = \xi \quad \text{and} \quad \|f_\mu^k(\mathbf{x}) - \mathbf{x}_0\| \leq \Delta \quad \text{for} \quad k \geq 0.$$

We write

$$\phi^s(\xi, \mu) = \mathbf{y}_0(\xi, \mu).$$

It follows from Lemma 1.12 that $\mathbf{y}(\xi, \mu)$ and hence ϕ^s are C^r functions. We define

$$\mathcal{M}_\mu = \{\mathbf{x}_0 + \phi^s(\xi, \mu) \ : \ \|\xi\| < \Delta(1 - \sigma)/2K_1\}$$

where $\|\mu - \mu_0\| < \rho$. Then $\mathcal{M}_\mu$ is, for each fixed μ, a C^r submanifold of $\mathbb{R}^n$ which depends C^r in μ. If $\mathbf{x} = \mathbf{x}_0 + \phi^s(\xi, \mu) \in \mathcal{M}_\mu$, then

$$P(\mathbf{x} - \mathbf{x}_0) = \xi$$

and

$$\|f_\mu^k(\mathbf{x}) - \mathbf{x}_0\| \leq \Delta \quad \text{for} \quad k \geq 0.$$

Moreover, $\mathbf{x}$ is the unique point with these properties.

We already know that for ε sufficiently small, $W^{s,\varepsilon}(\mathbf{x}(\mu))$ is a subset of $\mathcal{M}_\mu$. What we want to show now is that the size of ε can be chosen independently of μ. Suppose the positive numbers Δ and ρ satisfy Eqs.(18) and (19), let ε_0 be a positive number such that

$$\varepsilon_0 < \min\{(1 - \sigma)/2K_1M^s, 1\}\Delta$$

and suppose, in addition, that ρ is so small that for $\|\mu - \mu_0\| < \rho$

$$\|\mathbf{x}(\mu) - \mathbf{x}_0\| < \min\{(1 - \sigma)/2K_1M^s, 1\}\Delta - \varepsilon_0.$$

Let $\mathcal{M}_\mu$, $\|\mu - \mu_0\| < \rho$ be the family of manifolds constructed above corresponding to Δ and ρ. Then if $\|\mu - \mu_0\| < \rho$, $0 < \varepsilon < \varepsilon_0$ and $\mathbf{x} \in W^{s,\varepsilon}(\mathbf{x}(\mu))$,

$$\begin{aligned}
\|P(\mathbf{x} - \mathbf{x}_0)\| &\leq M^s[\|\mathbf{x} - \mathbf{x}(\mu)\| + \|\mathbf{x}(\mu) - \mathbf{x}_0\|] \\
&\leq M^s[\varepsilon + \|\mathbf{x}(\mu) - \mathbf{x}_0\|] \\
&< \Delta(1 - \sigma)/2K_1
\end{aligned}$$

and for $k \geq 0$

$$\begin{aligned} \|f_\mu^k(\mathbf{x}) - \mathbf{x}_0\| &\leq \|f_\mu^k(\mathbf{x}) - \mathbf{x}(\mu)\| + \|\mathbf{x}(\mu) - \mathbf{x}_0\| \\ &\leq \varepsilon + \|\mathbf{x}(\mu) - \mathbf{x}_0\| \\ &\leq \Delta. \end{aligned}$$

Hence $\mathbf{x} \in \mathcal{M}_\mu$. That is, we have proved that

$$W^{s,\varepsilon}(\mathbf{x}(\mu)) \subset \mathcal{M}_\mu$$

when $\|\mu - \mu_0\| < \rho$ and $0 < \varepsilon < \varepsilon_0$.

2. HYPERBOLIC SETS OF DIFFEOMORPHISMS

In the previous chapter, we discussed hyperbolic fixed points. In this chapter we want to consider more general invariant sets consisting of more than one point and perhaps infinitely many points. We give the appropriate definition of hyperbolicity for such sets and show that the continuity of the splitting into stable and unstable bundles follows from the other items in the definition. Then we expound the theory of exponential dichotomies for difference equations and use it to show that hyperbolic sets are expansive and that they are robust under perturbation.

2.1 DEFINITION OF HYPERBOLIC SET

Let $U \subset \mathbb{R}^n$ be an open set and let $f : U \to \mathbb{R}^n$ be a C^1 diffeomorphism onto its image. We begin with an example.

Example 2.1. Consider the Hénon map $f : \mathbb{R}^2 \to \mathbb{R}^2$ given by

$$f(x, y) = (1 - ax^2 + y, -bx)$$

where a and b are real parameters. When $a = 1.4$, $b = .3$ one can verify that the point

$$\mathbf{x}_0 = \left(\frac{13 + \sqrt{53}}{28}, \frac{-3(13 - \sqrt{53})}{280}\right)$$

has the property that

$$f(\mathbf{x}_0) = \mathbf{x}_1 = \left(\frac{13 - \sqrt{53}}{28}, \frac{-3(13 + \sqrt{53})}{280}\right)$$

but

$$f(\mathbf{x}_1) = f^2(\mathbf{x}_0) = \mathbf{x}_0.$$

$\mathbf{x}_0$ is said to be a point of period 2. Note that $\mathbf{x}_0$ is a fixed point of f^2 but not of f. Also we find that

$$Df^2(\mathbf{x}_0) = \begin{bmatrix} \frac{43}{50} & \frac{-(13-\sqrt{53})}{10} \\ \frac{3(13+\sqrt{53})}{100} & -\frac{3}{10} \end{bmatrix}$$

which has complex eigenvalues, the square of whose modulus is .09. So $\mathbf{x}_0$ is a hyperbolic fixed point of f^2.

This example motivates the following definition.

Definition 2.2. A point $\mathbf{x}_0$ in U is said to be a *periodic point* of period $m \geq 1$ if the points $\mathbf{x}_0$, $f(\mathbf{x}_0), f^2(\mathbf{x}_0), ..., f^{m-1}(\mathbf{x}_0)$ are distinct but $f^m(\mathbf{x}_0) = \mathbf{x}_0$. The periodic point $\mathbf{x}_0$ is said to be *hyperbolic* if $\mathbf{x}_0$ is hyperbolic as a fixed point of f^m, that is, the eigenvalues of $Df^m(\mathbf{x}_0)$ lie off the unit circle.

Definition 2.3. If $\mathbf{x}_0$ is a periodic point of period m, the set

$$\{\mathbf{x}_0, f(\mathbf{x}_0), ..., f^{m-1}(\mathbf{x}_0)\}$$

is called the *orbit* of x_0.

In order to motivate the definition of hyperbolic set to be given shortly, we consider the orbit

$$S = \{\mathbf{x}_0, f(\mathbf{x}_0), ..., f^{m-1}(\mathbf{x}_0)\}$$

of a periodic point of period m. Let E^s, E^u be the stable and unstable subspaces of $\mathbb{R}^n$ for $\mathbf{x}_0$ considered as a hyperbolic fixed point of f^m. Then we know that E^s and E^u are invariant, that is,

$$Df^m(\mathbf{x}_0)(E^s) = E^s,\ Df^m(\mathbf{x}_0)(E^u) = E^u. \tag{1}$$

Also we have the splitting

$$\mathbb{R}^n = E^s \oplus E^u$$

and there exist positive constants K_1, K_2 and $\lambda_1 < 1, \lambda_2 < 1$ such that for $k \geq 0$

$$||Df^{mk}(\mathbf{x}_0)\xi|| \leq K_1\lambda_1^k||\xi|| \quad \text{for } \xi \in E^s \tag{2}$$

and

$$||Df^{-mk}(\mathbf{x}_0)\xi|| \leq K_2\lambda_2^k||\xi|| \quad \text{for } \xi \in E^u. \tag{3}$$

Now for $\mathbf{x} = f^k(\mathbf{x}_0) \in S$ we define the splitting

$$\mathbb{R}^n = E^s(\mathbf{x}) \oplus E^u(\mathbf{x}),$$

where

$$E^s(\mathbf{x}) = Df^k(\mathbf{x}_0)(E^s), \quad E^u(\mathbf{x}) = Df^k(\mathbf{x}_0)(E^u).$$

Using Eq.(1) and the chain rule it is clear that this splitting has the invariance property that for all $\mathbf{x}$ in S

$$Df(\mathbf{x})(E^s(\mathbf{x})) = E^s(f(\mathbf{x})), \quad Df(\mathbf{x})(E^u(\mathbf{x})) = E^u(f(\mathbf{x})).$$

Next from Eqs.(2) and (3) we know that

$$\|Df^k(\mathbf{x})\xi\| \leq K_1(\lambda_1^{\frac{1}{m}})^k\|\xi\| \quad \text{for } \xi \in E^s(\mathbf{x}),$$

$$\|Df^{-k}(\mathbf{x})\xi\| \leq K_2(\lambda_2^{\frac{1}{m}})^k\|\xi\| \quad \text{for } \xi \in E^u(\mathbf{x})$$

when $\mathbf{x} = \mathbf{x}_0$ and $k \geq 0$ is a multiple of m. An easy argument shows that the same inequalities hold for all $\mathbf{x} \in S$ and $k \geq 0$ but perhaps with bigger constants K_1 and K_2. So we are motivated to make the following definition.

Definition 2.4. A compact set $S \subset U$ is said to be *hyperbolic* if

(i) S is invariant, that is, $f(S) = S$;

(ii) there is a continuous splitting

$$\mathbb{R}^n = E^s(\mathbf{x}) \oplus E^u(\mathbf{x}), \ \mathbf{x} \in S$$

such that the subspaces $E^s(\mathbf{x})$ and $E^u(\mathbf{x})$ have constant dimensions; moreover, these subspaces have the invariance properties

$$Df(\mathbf{x})(E^s(\mathbf{x})) = E^s(f(\mathbf{x})), \ Df(\mathbf{x})(E^u(\mathbf{x})) = E^u(f(\mathbf{x}))$$

and there are positive constants K_1, K_2 and $\lambda_1 < 1, \lambda_2 < 1$ such that for $k \geq 0$ and $\mathbf{x} \in S$

$$\|Df^k(\mathbf{x})\xi\| \leq K_1\lambda_1^k\|\xi\| \quad \text{for } \xi \in E^s(\mathbf{x}) \tag{4}$$

and

$$\|Df^{-k}(\mathbf{x})\xi\| \leq K_2\lambda_2^k\|\xi\| \quad \text{for } \xi \in E^u(\mathbf{x}). \tag{5}$$

K_1, K_2 are called *constants* and λ_1, λ_2 *exponents* for the hyperbolic set S.

Let us explain here what is meant by the continuity of the splitting. Denote by $\mathcal{P}(\mathbf{x})$ the projection of $\mathbb{R}^n$ onto $E^s(\mathbf{x})$ along $E^u(\mathbf{x})$. By the continuity of the splitting we mean that the projection function $\mathbf{x} \to \mathcal{P}(\mathbf{x})$ is continuous. Since S is compact, this implies, in particular, that $\mathcal{P}(\mathbf{x})$ is bounded. What we want to show next is that the continuity of $\mathcal{P}(\mathbf{x})$ is implied by the other assumptions in the definition of hyperbolicity (conf. Lanford [1985, p.78], Palmer [1988, p.281]). We show the boundedness first.

2.2 BOUNDEDNESS OF THE PROJECTION

Suppose S is a compact hyperbolic set for the C^1 diffeomorphism $f : U \to \mathbb{R}^n$ as in Definition 2.4 without the assumption that $\mathcal{P}(\mathbf{x})$ is continuous. We show that $\mathcal{P}(\mathbf{x})$ is bounded.

For $\mathbf{x}$ in S, $\xi \neq 0$ in $E^s(\mathbf{x})$ and $\eta \neq 0$ in $E^u(\mathbf{x})$, we define

$$\mathbf{u}_k = Df^k(\mathbf{x})\xi, \quad \mathbf{v}_k = Df^k(\mathbf{x})\eta \quad \text{for } k \in \mathbf{Z}.$$

Choose a positive integer N so that

$$\sigma = K_1^{-1}K_2^{-1}\lambda_1^{-N}\lambda_2^{-N} - 1 > 0.$$

Then

$$\begin{aligned}\left\|\frac{\mathbf{u}_N}{\|\xi\|} + \frac{\mathbf{v}_N}{\|\eta\|}\right\| &= \frac{\|\mathbf{u}_N\|}{\|\xi\|}\left\|\frac{\mathbf{u}_N}{\|\mathbf{u}_N\|} + \frac{\|\xi\|}{\|\mathbf{u}_N\|}\cdot\frac{\mathbf{v}_N}{\|\eta\|}\right\| \\ &\geq \frac{\|\mathbf{u}_N\|}{\|\xi\|}\left[\frac{\|\xi\|}{\|\mathbf{u}_N\|}\cdot\frac{\|\mathbf{v}_N\|}{\|\eta\|} - 1\right] \\ &\geq M_1^{-N}[K_1^{-1}\lambda_1^{-N}\cdot K_2^{-1}\lambda_2^{-N} - 1],\end{aligned}$$

where

$$M_1 = \max\left\{\sup_{\mathbf{x}\in S}\|Df(\mathbf{x})\|, \ \sup_{\mathbf{x}\in S}\|Df^{-1}(\mathbf{x})\|\right\}.$$

Hence

$$\left\|\frac{\mathbf{u}_N}{\|\xi\|} + \frac{\mathbf{v}_N}{\|\eta\|}\right\| \geq \sigma M_1^{-N}. \tag{6}$$

On the other hand,

$$\left\|\frac{\mathbf{u}_N}{\|\xi\|} + \frac{\mathbf{v}_N}{\|\eta\|}\right\| = \left\|Df^N(\mathbf{x})\left[\frac{\xi}{\|\xi\|} + \frac{\eta}{\|\eta\|}\right]\right\| \leq M_1^N\left\|\frac{\xi}{\|\xi\|} + \frac{\eta}{\|\eta\|}\right\|. \tag{7}$$

Thus, combining inequalities (6) and (7), we find that

$$\left\|\frac{\xi}{\|\xi} + \frac{\eta}{\|\eta\|}\right\| \geq \sigma M_1^{-2N}.$$

Hence using the inequality,

$$\|\xi\|\left\|\frac{\xi}{\|\xi\|} + \frac{\eta}{\|\eta\|}\right\| \leq 2\|\xi + \eta\|, \tag{8}$$

the proof of which is given below, we see that

$$\|\xi\| \le 2\sigma^{-1} M_1^{2N} \|\xi + \eta\|.$$

Since $\xi = \mathcal{P}(\mathbf{x})(\xi + \eta)$ it follows that

$$\|\mathcal{P}(\mathbf{x})\| \le 2\sigma^{-1} M_1^{2N}$$

and hence the projection is indeed bounded.

Finally note that inequality (8) follows from

$$\begin{aligned} \|\xi + \eta\| &= \|\xi\| \left\| \frac{\xi}{\|\xi\|} + \frac{\eta}{\|\eta\|} + \left(\frac{1}{\|\xi\|} - \frac{1}{\|\eta\|} \right) \eta \right\| \\ &\ge \|\xi\| \left[\left\| \frac{\xi}{\|\xi\|} + \frac{\eta}{\|\eta\|} \right\| - \left| \frac{1}{\|\xi\|} - \frac{1}{\|\eta\|} \right| \|\eta\| \right] \\ &= \|\xi\| \left\| \frac{\xi}{\|\xi\|} + \frac{\eta}{\|\eta\|} \right\| - |\, \|\xi\| - \|\eta\| \,| \\ &\ge \|\xi\| \left\| \frac{\xi}{\|\xi\|} + \frac{\eta}{\|\eta\|} \right\| - \|\xi + \eta\|. \end{aligned}$$

Note that the proof in this section was suggested by the proof that the projections associated with a dichotomy are bounded when the coefficient matrix is bounded, conf. Coppel [1978, p.14].

2.3 CONTINUITY OF THE PROJECTION

Suppose S is a compact hyperbolic set for a C^1 diffeomorphism $f : U \to \mathbb{R}^n$ as in Definition 2.4 without assuming $\mathcal{P}(\mathbf{x})$ is continuous. In the previous section we showed that $\mathcal{P}(\mathbf{x})$ is bounded. Hence there are constants M^s, M^u such that for all $\mathbf{x}$ in S

$$\|\mathcal{P}(\mathbf{x})\| \le M^s, \quad \|I - \mathcal{P}(\mathbf{x})\| \le M^u. \tag{9}$$

From Eqs.(4) and (5) in Definition 2.4 of hyperbolicity, it follows that for $\mathbf{x}$ in S and $k \ge 0$,

$$\|Df^k(\mathbf{x})\mathcal{P}(\mathbf{x})\| \le K_1 M^s \lambda_1^k$$

and

$$\|Df^{-k}(\mathbf{x})(I-\mathcal{P}(\mathbf{x}))\| \leq K_2 M^u \lambda_2^k.$$

Note also that the invariance of the splitting $I\!R^n = E^s(\mathbf{x}) \oplus E^u(\mathbf{x})$ implies that $\mathcal{P}(\mathbf{x})$ satisfies the identity

$$Df(\mathbf{x})\mathcal{P}(\mathbf{x}) = \mathcal{P}(f(\mathbf{x}))Df(\mathbf{x}). \tag{10}$$

It follows for all integers k that

$$Df^k(\mathbf{x})\mathcal{P}(\mathbf{x}) = \mathcal{P}(f^k(\mathbf{x}))Df^k(\mathbf{x}). \tag{11}$$

First we show that if $\{\mathbf{x}_k\}_{k=0}^N$ and $\{\mathbf{y}_k\}_{k=0}^N$ are two orbits in S such that for some positive number δ

$$\|Df(\mathbf{y}_k) - Df(\mathbf{x}_k)\| \leq \delta \quad \text{for } 0 \leq k \leq N-1, \tag{12}$$

then

$$\begin{aligned} &\|\mathcal{P}(\mathbf{x}_N)(I-\mathcal{P}(\mathbf{y}_N))\| \\ &\qquad \leq K_1K_2M^sM^u(\lambda_1\lambda_2)^N + K_1K_2M^sM^u\lambda_2(1-\lambda_1\lambda_2)^{-1}\delta. \end{aligned} \tag{13}$$

To this end, consider the matrix

$$U_k = Df^{k-N}(\mathbf{y}_N)(I-\mathcal{P}(\mathbf{y}_N)).$$

Note that for $k = 0, \ldots, N-1$

$$U_{k+1} = Df(\mathbf{y}_k)U_k$$

and so U_k is a solution of the difference equation

$$U_{k+1} = A_kU_k + B_kU_k,$$

where

$$A_k = Df(\mathbf{x}_k), \quad B_k = Df(\mathbf{y}_k) - Df(\mathbf{x}_k).$$

It follows that

$$U_N = A_{N-1}\cdots A_0U_0 + \sum_{k=1}^{N} A_{N-1}\cdots A_kB_{k-1}U_{k-1}. \tag{14}$$

Hence by the chain rule,

$$
\begin{aligned}
I - \mathcal{P}(\mathbf{y}_N) &= Df^N(\mathbf{x}_0)Df^{-N}(\mathbf{y}_N)(I - \mathcal{P}(\mathbf{y}_N)) \\
&\quad + \sum_{k=1}^{N} Df^{N-k}(\mathbf{x}_k)B_{k-1}Df^{k-1-N}(\mathbf{y}_N)(I - \mathcal{P}(\mathbf{y}_N)).
\end{aligned}
$$

Multiplying by $\mathcal{P}(\mathbf{x}_N)$ and using Eq.(11), we find that

$$
\begin{aligned}
\mathcal{P}(\mathbf{x}_N)(I - \mathcal{P}(\mathbf{y}_N)) &= Df^N(\mathbf{x}_0)\mathcal{P}(\mathbf{x}_0)Df^{-N}(\mathbf{y}_N)(I - \mathcal{P}(\mathbf{y}_N)) \\
&\quad + \sum_{k=1}^{N} Df^{N-k}(\mathbf{x}_k)\mathcal{P}(\mathbf{x}_k)B_{k-1}Df^{k-1-N}(\mathbf{y}_N)(I - \mathcal{P}(\mathbf{y}_N)).
\end{aligned}
$$

Hence

$$
\begin{aligned}
&\|\mathcal{P}(\mathbf{x}_N)(I - \mathcal{P}(\mathbf{y}_N))\| \\
&\quad \leq K_1 M^s \lambda_1^N \cdot K_2 M^u \lambda_2^N + \sum_{k=1}^{N} K_1 M^s \lambda_1^{N-k} \cdot \delta \cdot K_2 M^u \lambda_2^{N-k+1} \\
&\quad \leq K_1 K_2 M^s M^u (\lambda_1\lambda_2)^N + K_1 K_2 M^s M^u \lambda_2 (1 - \lambda_1\lambda_2)^{-1}\delta.
\end{aligned}
$$

Thus inequality (13) is proved.

Next we assert that when Eq.(12) holds then

$$
\begin{aligned}
&\|(I - \mathcal{P}(\mathbf{x}_0))\mathcal{P}(\mathbf{y}_0)\| \\
&\qquad \leq K_1 K_2 M^s M^u (\lambda_1\lambda_2)^N + K_1 K_2 M^s M^u \lambda_2 (1 - \lambda_1\lambda_2)^{-1}\delta.
\end{aligned}
\tag{15}
$$

This time we set

$$U_k = Df^k(\mathbf{y}_0)\mathcal{P}(\mathbf{y}_0).$$

As before, we obtain the relation (14). Multiplying it by $A_0^{-1}A_1^{-1}\cdots A_{N-1}^{-1}$ we obtain

$$U_0 = A_0^{-1}\cdots A_{N-1}^{-1}U_N - \sum_{k=0}^{N-1} A_0^{-1}\cdots A_k^{-1}B_kU_k.$$

Hence, by the chain rule,

$$\mathcal{P}(\mathbf{y}_0) = Df^{-N}(\mathbf{x}_N)Df^N(\mathbf{y}_0)\mathcal{P}(\mathbf{y}_0) - \sum_{k=0}^{N-1} Df^{-(k+1)}(\mathbf{x}_{k+1})B_kDf^k(\mathbf{y}_0)\mathcal{P}(\mathbf{y}_0).$$

Multiplying by $I - \mathcal{P}(\mathbf{x}_0)$ and using Eq.(11), we get

$$(I - \mathcal{P}(\mathbf{x}_0))\mathcal{P}(\mathbf{y}_0) = Df^{-N}(\mathbf{x}_N)(I - \mathcal{P}(\mathbf{x}_N))Df^N(\mathbf{y}_0)\mathcal{P}(\mathbf{y}_0)$$
$$- \sum_{k=0}^{N-1} Df^{-(k+1)}(\mathbf{x}_{k+1})(I - \mathcal{P}(\mathbf{x}_{k+1}))B_k Df^k(\mathbf{y}_0)\mathcal{P}(\mathbf{y}_0).$$

Hence

$$\begin{aligned}\|(I - \mathcal{P}(\mathbf{x}_0))\mathcal{P}(\mathbf{y}_0)\| &\leq K_2 M^u \lambda_2^N \cdot K_1 M^s \lambda_1^N + \sum_{k=0}^{N-1} K_2 M^u \lambda_2^{k+1} \cdot \delta \cdot K_1 M^s \lambda_1^k \\ &\leq K_1 K_2 M^s M^u (\lambda_1\lambda_2)^N + K_1 K_2 M^s M^u \lambda_2 \sum_{k=0}^{N-1} (\lambda_1\lambda_2)^k \delta.\end{aligned}$$

Then assertion (15) follows.

Now suppose $\{\mathbf{x}_k\}_{k=-\infty}^{\infty}$ and $\{\mathbf{y}_k\}_{k=-\infty}^{\infty}$ are orbits such that

$$\|Df(\mathbf{y}_k) - Df(\mathbf{x}_k)\| \leq \delta \quad \text{for } -N \leq k \leq N-1. \tag{16}$$

Then we apply Eq.(13) to the orbits $\{\mathbf{x}_k\}_{k=-N}^{0}$, $\{\mathbf{y}_k\}_{k=-N}^{0}$ and deduce that

$$\begin{aligned}&\|\mathcal{P}(\mathbf{x}_0)(I - \mathcal{P}(\mathbf{y}_0))\| \\ &\qquad \leq K_1 K_2 M^s M^u (\lambda_1\lambda_2)^N + K_1 K_2 M^s M^u \lambda_2 (1 - \lambda_1\lambda_2)^{-1}\delta.\end{aligned}$$

Further if we apply Eq.(15) to the orbits $\{\mathbf{x}_k\}_{k=0}^{N}$ and $\{\mathbf{y}_k\}_{k=0}^{N}$ we deduce that

$$\begin{aligned}&\|(I - \mathcal{P}(\mathbf{x}_0))\mathcal{P}(\mathbf{y}_0)\| \\ &\qquad \leq K_1 K_2 M^s M^u (\lambda_1\lambda_2)^N + K_1 K_2 M^s M^u \lambda_2 (1 - \lambda_1\lambda_2)^{-1}\delta.\end{aligned}$$

Hence if Eq.(16) holds,

$$\begin{aligned}&\|\mathcal{P}(\mathbf{y}_0) - \mathcal{P}(\mathbf{x}_0)\| \\ &\qquad = \|\mathcal{P}(\mathbf{y}_0) - \mathcal{P}(\mathbf{x}_0)\mathcal{P}(\mathbf{y}_0) + \mathcal{P}(\mathbf{x}_0)\mathcal{P}(\mathbf{y}_0) - \mathcal{P}(\mathbf{x}_0)\| \\ &\qquad \leq \|(I - \mathcal{P}(\mathbf{x}_0))\mathcal{P}(\mathbf{y}_0)\| + \|\mathcal{P}(\mathbf{x}_0)(I - \mathcal{P}(\mathbf{y}_0))\| \\ &\qquad \leq 2K_1 K_2 M^s M^u (\lambda_1\lambda_2)^N + 2K_1 K_2 M^s M^u \lambda_2 (1 - \lambda_1\lambda_2)^{-1}\delta.\end{aligned} \tag{17}$$

Now we can show that $\mathcal{P}(\cdot)$ is continuous on S. Let $\varepsilon > 0$ be given. Choose N as the smallest positive integer such that

$$2K_1K_2M^sM^u(\lambda_1\lambda_2)^N \leq \varepsilon/2$$

and set

$$\delta = \varepsilon(1-\lambda_1\lambda_2)/4K_1K_2M^sM^u\lambda_2.$$

Then we choose $\delta_1 > 0$ so that

$$\mathbf{y} \in S, \ \ \mathbf{x} \in S, \ \ \|\mathbf{y}-\mathbf{x}\| \leq \delta_1 \Rightarrow \|Df(f^k(\mathbf{y})) - Df(f^k(\mathbf{x}))\| \leq \delta$$

for $-N \leq k \leq N-1$. It follows from Eq.(17) that when $\mathbf{y} \in S$, $\mathbf{x} \in S$ and $\|\mathbf{y}-\mathbf{x}\| \leq \delta_1$

$$\begin{aligned} \|\mathcal{P}(\mathbf{y}) - \mathcal{P}(\mathbf{x})\| &\leq 2K_1K_2M^sM^u(\lambda_1\lambda_2)^N + 2K_1K_2M^sM^u\lambda_2(1-\lambda_1\lambda_2)^{-1}\delta \\ &\leq \varepsilon/2 + \varepsilon/2 \\ &= \varepsilon. \end{aligned}$$

Thus $\mathcal{P}(\cdot)$ is continuous on S.

Remark 2.5. Note that the proof of continuity of $\mathcal{P}(\mathbf{x})$ was suggested by Sakamoto [1990, p.74]. Note also that the proofs of the boundedness and continuity of $\mathcal{P}(\mathbf{x})$ in the last two sections used the compactness of S only to ensure that $Df(\mathbf{x})$ and $Df^{-1}(\mathbf{x})$ are both bounded on S and also the uniform continuity of f and Df on S. This allows the possibility of defining noncompact hyperbolic sets provided appropriate boundedness and uniform continuity assumptions are added (conf. Steinlein and Walther [1989, 1990]). In the compact case considered here, a much quicker proof of the continuity of the projections can be given as in Lanford [1985, p.78] and Palmer [1988, p.281].

2.4 EXPONENTIAL DICHOTOMIES FOR DIFFERENCE EQUATIONS

In order to assist us in proving additional properties of hyperbolic sets, we need to develop the perturbation theory of exponential dichotomies for linear difference equations (conf. Henry [1981], Slyusarchuk [1983]), Palmer [1988], Papaschinopoulos [1988]).

Definition 2.6. For $k \in J$, an interval in $\mathbf{Z}$, let A_k be an invertible $n \times n$ matrix. The difference equation

$$\mathbf{u}_{k+1} = A_k \mathbf{u}_k \tag{18}$$

is said to have an *exponential dichotomy* on J if there are projections P_k and positive constants $K_1, K_2, \lambda_1, \lambda_2$ with $\lambda_1 < 1, \lambda_2 < 1$ such that for $k, m \in J$ the projections satisfy the invariance conditions

$$\Phi(k,m)P_m = P_k\Phi(k,m),$$

and the inequalities

$$\|\Phi(k,m)P_m\| \leq K_1\lambda_1^{k-m}, \quad k \geq m$$

and

$$\|\Phi(k,m)(I - P_m)\| \leq K_2\lambda_2^{m-k}, \quad k \leq m$$

hold. Here $\Phi(k,m)$ is the *transition matrix* for Eq.(18) defined by

$$\Phi(k,m) = \begin{cases} A_{k-1} \dots A_m & \text{for } k > m, \\ I & \text{for } k = m, \\ \Phi(m,k)^{-1} & \text{for } k < m. \end{cases}$$

Note the *cocycle* property

$$\Phi(k,p)\Phi(p,m) = \Phi(k,m).$$

K_1, K_2 are called *constants* associated with the dichotomy and λ_1, λ_2 *exponents.*

When A_k is a constant matrix A with eigenvalues off the unit circle,

$$\Phi(k,m) = A^{k-m}$$

and it follows as at the beginning of Section 1.3 (reasoning for A as for $Df(\mathbf{x}_0)$) that Eq.(18) has an exponential dichotomy on $(-\infty, \infty)$ with projections all equal to P, the projection onto E^s along E^u, with E^s and E^u as in Definition 1.3.

We now give a characterisation of hyperbolicity in terms of exponential dichotomies.

Proposition 2.7. *If S is a compact invariant set for the diffeomorphism $f : U \to \mathbb{R}^n$, then S is hyperbolic if and only if for all $\mathbf{x} \in S$ the difference equation*

$$\mathbf{u}_{k+1} = Df(f^k(\mathbf{x}))\mathbf{u}_k \tag{19}$$

has an exponential dichotomy on $(-\infty, \infty)$ with constants, exponents and rank of projection independent of $\mathbf{x}$.

Proof. Suppose first that S is hyperbolic as in Definition 2.4. It follows from the chain rule that the transition matrix for Eq.(19) is given by

$$\Phi(k,m) = Df^{k-m}(f^m(\mathbf{x})).$$

As usual, we let $\mathcal{P}(\mathbf{x})$ be the projection of $\mathbb{R}^n$ onto $E^s(\mathbf{x})$ along $E^u(\mathbf{x})$. As noted previously, the invariance of the splitting implies the identity

$$Df^k(\mathbf{x})\mathcal{P}(\mathbf{x}) = \mathcal{P}(f^k(\mathbf{x}))Df^k(\mathbf{x}) \tag{20}$$

for $k \in \mathbf{Z}$ and $\mathbf{x} \in S$. Then we take the projection

$$P_k = \mathcal{P}(f^k(\mathbf{x}))$$

and the invariance of these projections relative to Eq.(19) follows immediately from Eq.(20). Also it follows from the inequalities (4) that for all $\xi \in \mathbb{R}^n$ and $k \geq m$

$$\|\Phi(k,m)P_m\xi\| = \|Df^{k-m}(f^m(\mathbf{x}))P_m\xi\| \leq K_1\lambda_1^{k-m}\|P_m\xi\|.$$

Now from Section 2.2 we know that there exist constants M^s, M^u such that for all $\mathbf{x} \in S$

$$\|\mathcal{P}(\mathbf{x})\| \leq M^s, \ \|I - \mathcal{P}(\mathbf{x})\| \leq M^u.$$

So

$$\|\Phi(k,m)P_m\| \leq K_1 M^s \lambda_1^{k-m}$$

for $k \geq m$. Similarly,

$$\|\Phi(k,m)(I - P_m)\| \leq K_2 M^u \lambda_2^{m-k}$$

for $m \geq k$. Thus we have shown that the difference equation (19) has an exponential dichotomy on $(-\infty, \infty)$ with constants K_1M^s, K_2M^u, exponents λ_1, λ_2 and with $\dim E^s(\mathbf{x})$ as rank of the associated projections.

Conversely, suppose for all $\mathbf{x} \in S$ the difference equation (19) has an exponential dichotomy on $(-\infty, \infty)$ with constants K_1, K_2, exponents λ_1, λ_2 and with the rank of the projection $P_k(\mathbf{x})$ independent of $\mathbf{x}$. We claim that the range of $P_0(\mathbf{x})$ is

$$\mathcal{R}(P_0(\mathbf{x})) = \{\xi \in \mathbb{R}^n : \sup_{k\geq 0} \|Df^k(\mathbf{x})\xi\| < \infty\}$$

and the nullspace is

$$\mathcal{N}(P_0(\mathbf{x})) = \{\xi \in \mathbb{R}^n : \sup_{k\leq 0} \|Df^k(\mathbf{x})\xi\| < \infty\}.$$

To prove the first claim, let $\xi \in \mathcal{R}(P_0(\mathbf{x}))$. Then if $k \geq 0$

$$\|Df^k(\mathbf{x})\xi\| = \|\Phi(k,0)P_0(\mathbf{x})\xi\| \leq K_1\lambda_1^k\|\xi\|,$$

which is bounded. On the other hand, if $k \geq 0$,

$$\begin{aligned}\|(I - P_0(\mathbf{x}))\xi\| &= \|(I - P_0(\mathbf{x}))\Phi(0,k)\Phi(k,0)\xi\| \\ &= \|\Phi(0,k)(I - P_k(\mathbf{x}))Df^k(\mathbf{x})\xi\| \\ &\leq K_2\lambda_2^k\|Df^k(\mathbf{x})\xi\|.\end{aligned}$$

Hence if $Df^k(\mathbf{x})\xi$ is bounded for $k \geq 0$, $(I - P_0(\mathbf{x}))\xi = 0$ and so $\xi \in \mathcal{R}(P_0(\mathbf{x})$. This proves the claim for $\mathcal{R}(P_0(\mathbf{x}))$ and the claim for $\mathcal{N}(P_0(\mathbf{x}))$ is similarly proved.

The claims just proved imply that

$$Df(\mathbf{x})(\mathcal{R}(P_0(\mathbf{x}))) = \mathcal{R}(P_0(f(\mathbf{x}))),\ Df(\mathbf{x})(\mathcal{N}(P_0(\mathbf{x}))) = \mathcal{N}(P_0(f(\mathbf{x}))).$$

So if we define

$$E^s(\mathbf{x}) = \mathcal{R}(P_0(\mathbf{x})),\ E^u(\mathbf{x}) = \mathcal{N}(P_0(\mathbf{x})),$$

we see that these subspaces have the invariance property required in Definition 2.4. Also, if $k \geq 0$, then

$$\|Df^k(\mathbf{x})\xi\| = \|\Phi(k,0)P_0(\mathbf{x})\xi\| \leq K_1\lambda_1^k\|\xi\| \quad \text{for} \quad \xi \in E^s(\mathbf{x})$$

and

$$\|Df^{-k}(\mathbf{x})\xi\| = \|\Phi(0,k)(I - P_0(\mathbf{x}))\xi\| \leq K_2\lambda_2^k\|\xi\| \quad \text{for} \quad \xi \in E^u(\mathbf{x}).$$

So we have proved that all the conditions in Definition 2.4 are satisfied except the continuity of the splitting. However, as shown in the previous section, this follows from the other conditions. Hence S is hyperbolic and the proof of Proposition 2.7 is complete.

We now prove the *roughness* theorem for exponential dichotomies, that is, we show that the exponential dichotomy property is robust under perturbation of the coefficient matrix.

Lemma 2.8. *Let the difference equation* (18) *have an exponential dichotomy on an interval $J = [a, b]$ (interpreted as $[a, \infty)$ when $b = \infty$, etc.) with constants K_1, K_2, exponents λ_1, λ_2 and projections P_k. Suppose β_1, β_2 are numbers satisfying*

$$\lambda_1 < \beta_1 < 1,\ \lambda_2 < \beta_2 < 1.$$

Then there exists a positive number $\delta_0 = \delta_0(K_1, K_2, \lambda_1, \lambda_2, \beta_1, \beta_2)$ such that if

$$\|B_k\| \leq \delta \leq \delta_0$$

and $A_k + B_k$ is invertible for $k \in J$, the perturbed difference equation

$$\mathbf{u}_{k+1} = (A_k + B_k)\mathbf{u}_k \tag{21}$$

has an exponential dichotomy on J with constants L_1, L_2, exponents β_1, β_2 and projections Q_k satisfying

$$\|Q_k - P_k\| \leq N\delta,$$

where L_1, L_2 and N are constants depending only on $K_1, K_2, \lambda_1, \lambda_2$.

Proof. We choose δ_0 as the least positive number satisfying the inequalities

$$2K_1\delta_0 \leq \beta_1 - \lambda_1, \; 2K_2\delta_0 \leq \beta_2 - \lambda_2$$

and

$$4K_1K_2[(1 - \lambda_1)^{-1} + \lambda_2(1 - \lambda_2)^{-1}]\delta_0 \leq 1.$$

First consider the case where $J = [a, b]$ is a finite interval. If $\mathbf{u}_k$ is a solution of Eq.(21), then it is easy to show that

$$\mathbf{u}_k = \Phi(k, a)\mathbf{u}_a + \sum_{m=a}^{k-1} \Phi(k, m+1)B_m\mathbf{u}_m \tag{22}$$

or

$$\mathbf{u}_k = \Phi(k, b)\mathbf{u}_b - \sum_{m=k}^{b-1} \Phi(k, m+1)B_m\mathbf{u}_m, \tag{23}$$

where $\Phi(k, m)$ is the transition matrix for Eq.(18). Multiplying (22) by P_k, (23) by $I - P_k$, using the invariance and adding, we obtain

$$\begin{aligned} \mathbf{u}_k \; &= \Phi(k, a)P_a\mathbf{u}_a + \Phi(k, b)(I - P_b)\mathbf{u}_b \\ &\quad + \sum_{m=a}^{k-1} \Phi(k, m+1)P_{m+1}B_m\mathbf{u}_m \\ &\quad - \sum_{m=k}^{b-1} \Phi(k, m+1)(I - P_{m+1})B_m\mathbf{u}_m, \end{aligned} \tag{24}$$

a representation of the solution $\mathbf{u}_k$ in terms of its boundary values $P_a\mathbf{u}_a$ and $(I - P_b)\mathbf{u}_b$.

We consider those solutions $\mathbf{u}_k$ for which $(I - P_b)\mathbf{u}_b = \mathbf{0}$ and show they satisfy certain exponential estimates. By Eq.(24), with m instead of a, we have

$$\mathbf{u}_k = \Phi(k,m)P_m\mathbf{u}_m + \sum_{p=m}^{k-1}\Phi(k,p+1)P_{p+1}B_p\mathbf{u}_p$$
$$-\sum_{p=k}^{b-1}\Phi(k,p+1)(I-P_{p+1})B_p\mathbf{u}_p \tag{25}$$

for $a \le m \le k \le b$. Hence for these same k and m,

$$\|\mathbf{u}_k\| \le K_1\lambda_1^{k-m}\|\mathbf{u}_m\| + \sum_{p=m}^{k-1} K_1\lambda_1^{k-p-1}\delta\|\mathbf{u}_p\| + \sum_{p=k}^{b-1} K_2\lambda_2^{p+1-k}\delta\|\mathbf{u}_p\|.$$

Then, since $\sigma\delta_0 \le \frac{1}{4}$, where

$$\sigma = K_1(1-\lambda_1)^{-1} + K_2\lambda_2(1-\lambda_2)^{-1},$$

it follows from Lemma 1.10(i) that

$$\|\mathbf{u}_k\| \le K_1(1-\sigma\delta)^{-1}(\lambda_1 + K_1(1-\sigma\delta)^{-1}\delta)^{k-m}\|\mathbf{u}_m\| \le 2K_1\beta_1^{k-m}\|\mathbf{u}_m\| \tag{26}$$

for $a \le m \le k \le b$.

Next we consider those solutions $\mathbf{u}_k$ of Eq.(21) for which $P_a\mathbf{u}_a = 0$. According to Eq.(24), with m instead of b,

$$\mathbf{u}_k = \Phi(k,m)(I-P_m)\mathbf{u}_m + \sum_{p=a}^{k-1}\Phi(k,p+1)P_{p+1}B_p\mathbf{u}_p$$
$$-\sum_{p=k}^{m-1}\Phi(k,p+1)(I-P_{p+1})B_p\mathbf{u}_p \tag{27}$$

for $a \le k \le m \le b$. Hence for these same k and m

$$\|\mathbf{u}_k\| \le K_2\lambda_2^{m-k}\|\mathbf{u}_m\| + \sum_{p=a}^{k-1} K_1\lambda_1^{k-p-1}\delta\|\mathbf{u}_p\| + \sum_{p=k}^{m-1} K_2\lambda_2^{p+1-k}\delta\|\mathbf{u}_p\|.$$

Then, since $(K_2\lambda_2 + \sigma)\delta_0 \le \frac{1}{2}$, it follows from Lemma 1.10(ii) that

$$\|\mathbf{u}_k\| \le K_2(1-\sigma\delta)^{-1}\left(\frac{\lambda_2}{1 - K_2(1-\sigma\delta)^{-1}\lambda_2\delta}\right)^{m-k}\|\mathbf{u}_m\|$$

and hence that

$$\|\mathbf{u}_k\| \le 2K_2\beta_2^{m-k}\|\mathbf{u}_m\| \tag{28}$$

for $a \le k \le m \le b$.

Now we show given vectors $\xi \in \mathcal{R}(P_a)$ and $\eta \in \mathcal{N}(P_b)$ there is a unique solution $\mathbf{u}_k$ of Eq.(21) such that

$$P_a\mathbf{u}_a = \xi, \ (I - P_b)\mathbf{u}_b = \eta. \tag{29}$$

To show this, let X be the Banach space of finite sequences $\mathbf{u} = \{\mathbf{u}_k\}_{k=a}^{b}$ with norm

$$\|\mathbf{u}\| = \sup_{a\le k\le b} \|\mathbf{u}_k\|.$$

We define the operator $T : X \to X$ by

$$\begin{aligned}(T\mathbf{u})_k &= \Phi(k,a)\xi + \Phi(k,b)\eta + \sum_{m=a}^{k-1}\Phi(k,m+1)P_{m+1}B_m\mathbf{u}_m \\ &\quad - \sum_{m=k}^{b-1}\Phi(k,m+1)(I - P_{m+1})B_m\mathbf{u}_m\end{aligned}$$

for $a \le k \le b$. If $\mathbf{u} = \{\mathbf{u}_k\}_{k=a}^{b}$ and $\mathbf{v} = \{\mathbf{v}_k\}_{k=a}^{b}$ are in X, then for $a \le k \le b$

$$\begin{aligned}&\|(T\mathbf{u})_k - (T\mathbf{v})_k\| \\ &\quad = \Bigg\| \sum_{m=a}^{k-1}\Phi(k,m+1)P_{m+1}B_m(\mathbf{u}_m - \mathbf{v}_m) \\ &\qquad - \sum_{m=k}^{b-1}\Phi(k,m+1)(I - P_{m+1})B_m(\mathbf{u}_m - \mathbf{v}_m)\Bigg\| \\ &\quad \le \sum_{m=a}^{k-1}K_1\lambda_1^{k-m-1}\delta\|\mathbf{u}_m - \mathbf{v}_m\| + \sum_{m=k}^{b-1}K_2\lambda_2^{m+1-k}\delta\|\mathbf{u}_m - \mathbf{v}_m\|.\end{aligned}$$

Hence

$$\|T\mathbf{u} - T\mathbf{v}\| \le \sigma\delta\|\mathbf{u} - \mathbf{v}\|.$$

Since $\sigma\delta < 1$, it follows that T is a contraction and hence has a unique fixed point $\mathbf{u} = \{\mathbf{u}_k\}_{k=a}^{b}$ which satisfies

$$\begin{aligned} \mathbf{u}_k &= \Phi(k,a)\xi + \Phi(k,b)\eta + \sum_{m=a}^{k-1} \Phi(k,m+1)P_{m+1}B_m\mathbf{u}_m \\ &\quad - \sum_{m=k}^{b-1} \Phi(k,m+1)(I - P_{m+1})B_m\mathbf{u}_m \end{aligned} \tag{30}$$

for $a \le k \le b$. Clearly Eq.(29) holds and for $a \le k \le b-1$

$$\begin{aligned} \mathbf{u}_{k+1} &= \Phi(k+1,a)\xi + \Phi(k+1,b)\eta + \sum_{m=a}^{k} \Phi(k+1,m+1)P_{m+1}B_m\mathbf{u}_m \\ &\quad - \sum_{m=k+1}^{b-1} \Phi(k+1,m+1)(I - P_{m+1})B_m\mathbf{u}_m \\ &= A_k\mathbf{u}_k + P_{k+1}B_k\mathbf{u}_k + (I - P_{k+1})B_k\mathbf{u}_k \\ &= A_k\mathbf{u}_k + B_k\mathbf{u}_k. \end{aligned}$$

So $\mathbf{u}_k$ is indeed a solution of Eq.(21) satisfying (29). On the other hand, it follows from Eq.(24) that any such solution must be a fixed point of T. Thus the uniqueness is established.

Now denote by E_k^s the subspace of $\mathbb{R}^n$ consisting of the values $\mathbf{u}_k$ of the solutions of Eq.(21) satisfying (29) with $\eta = \mathbf{0}$ and by E_k^u those satisfying (29) with $\xi = \mathbf{0}$. Clearly $E_k^s \cap E_k^u = \{\mathbf{0}\}$ since a solution in the intersection would satisfy (29) with $\xi = \mathbf{0}$, $\eta = \mathbf{0}$ and hence, by uniqueness, must be zero. Also, since the $\mathbf{u}_k$ are solutions, $E_k^s = \Psi(k,a)(E_a^s)$ and $E_k^u = \Psi(k,b)(E_b^s)$, where $\Psi(k,m)$ is the transition matrix corresponding to Eq.(21). Moreover, since the mapping $\xi \to \mathbf{u}_a$ of $\mathcal{R}(P_a)$ onto E_a^s is linear and one to one, dim $E_k^s =$ rank P_a. Similarly, dim $E_k^u = n -$ rank P_a and so $\mathbb{R}^n = E_k^s \oplus E_k^u$. Hence if we let Q_k be the projection of $\mathbb{R}^n$ onto E_k^s along E_k^u, we see that rank $Q_k =$ rank P_k and that Q_k has the invariance property

$$\Psi(k,m)Q_m = Q_k\Psi(k,m).$$

Also it follows from Eqs.(26) and (28) that

$$\|\Psi(k,m)Q_m\| \le 2K_1\beta_1^{k-m}\|Q_m\| \tag{31}$$

for $a \le m \le k \le b$ and

$$\|\Psi(k,m)(I - Q_m)\| \le 2K_2\beta_2^{m-k}\|I - Q_m\| \tag{32}$$

for $a \le k \le m \le b$.

To complete the proof for the case $J = [a, b]$, we need to estimate $\|Q_k - P_k\|$. We follow an argument from Coppel [1978, pp. 32-33]. If $\mathbf{u}_k$ is a solution of Eq.(21) in E_k^s, then it satisfies Eq.(25). Multiplying by $I - P_k$ and using Eq.(26), we find that

$$\begin{aligned}\|(I - P_k)\mathbf{u}_k\| &= \left\|\sum_{m=k}^{b-1} \Phi(k, m+1)(I - P_{m+1})B_m \mathbf{u}_m\right\| \\ &\le \sum_{m=k}^{b-1} K_2 \lambda_2^{m+1-k} \delta \|\mathbf{u}_m\| \\ &\le \sum_{m=k}^{b-1} K_2 \lambda_2^{m+1-k} \delta \cdot 2K_1 \|\mathbf{u}_k\| \\ &\le a_1 \delta \|\mathbf{u}_k\|,\end{aligned}$$

where

$$a_1 = 2K_1 K_2 \lambda_2 (1 - \lambda_2)^{-1}.$$

Hence, for all ξ in $\mathbb{R}^n$,

$$\|(I - P_k)\Psi(k, a)Q_a \xi\| \le a_1 \delta \|\Psi(k, a) Q_a \xi\|.$$

Replacing ξ by $\Psi(a, k)\xi$ we deduce that

$$\|(I - P_k)Q_k\| \le a_1 \delta \|Q_k\|. \tag{33}$$

Next if $\mathbf{u}_k$ is a solution of Eq.(21) in E_k^u, then it satisfies Eq.(27). Multiplying by P_k and using Eq.(28), we get

$$\begin{aligned}\|P_k \mathbf{u}_k\| &= \left\|\sum_{m=a}^{k-1} \Phi(k, m+1) P_{m+1} B_m \mathbf{u}_m\right\| \\ &\le \sum_{m=a}^{k-1} K_1 \lambda_1^{k-m-1} \delta \|\mathbf{u}_m\| \\ &\le \sum_{m=a}^{k-1} K_1 \lambda_1^{k-m-1} \delta \cdot 2K_2 \|\mathbf{u}_k\| \\ &\le a_2 \delta \|\mathbf{u}_k\|,\end{aligned}$$

where

$$a_2 = 2K_1K_2(1-\lambda_1)^{-1}.$$

Hence for all η in $\mathbb{R}^n$

$$\|P_k\Psi(k,b)(I-Q_b)\eta\| \le a_2\delta\|\Psi(k,b)(I-Q_b)\eta\|.$$

Replacing η by $\Psi(b,k)\eta$, we conclude that

$$\|P_k(I-Q_k)\| \le a_2\delta\|I-Q_k\|. \tag{34}$$

Then it follows from Eqs.(33) and (34) that

$$\begin{aligned}\|Q_k-P_k\| &= \|(I-P_k)Q_k - P_k(I-Q_k)\| \\ &\le a_1\delta\|Q_k\| + a_2\delta\|I-Q_k\| \\ &\le a_1\delta[\|P_k\| + \|Q_k-P_k\|] \\ &\qquad + a_2\delta[\|I-P_k\| + \|Q_k-P_k\|] \\ &\le a_1\delta[K_1+\|Q_k-P_k\|] + a_2\delta[K_2+\|Q_k-P_k\|].\end{aligned} \tag{35}$$

Hence

$$\begin{aligned}\|Q_k-P_k\| &\le [1-(a_1+a_2)\delta]^{-1}(K_1a_1+K_2a_2)\delta \\ &\le 4K_1K_2[K_2(1-\lambda_1)^{-1}+K_1\lambda_2(1-\lambda_2)^{-1}]\delta \\ &\le K_1+K_2.\end{aligned} \tag{36}$$

This also means that

$$\|Q_k\| \le 2K_1+K_2 \quad \text{and} \quad \|I-Q_k\| \le 2K_2+K_1.$$

So referring to Eqs.(31), (32) and (36), we see that the lemma has been established for the case $J=[a,b]$ with

$$L_1 = 2K_1(2K_1+K_2), \quad L_2 = 2K_2(2K_2+K_1)$$

and

$$N = 4K_1K_2[K_2(1-\lambda_1)^{-1}+K_1\lambda_2(1-\lambda_2)^{-1}].$$

Next consider the case $J=[a,\infty)$. In this situation, we take E_k^s as the subspace generated by the solutions $\mathbf{u}_k$ for which $\mu_a = \sup_{k\ge a}\|\mathbf{u}_k\| < \infty$.

Then, if $\mathbf{u}_k$ is such a solution, we multiply Eq.(24) by $I - P_k$ to obtain for $a \leq k \leq b$

$$(I - P_k)\mathbf{u}_k = \Phi(k,b)(I - P_b)\mathbf{u}_b - \sum_{m=k}^{b-1} \Phi(k, m+1)(I - P_{m+1})B_m\mathbf{u}_m. \quad (37)$$

Note that

$$\|\Phi(k,b)(I - P_b)\mathbf{u}_b\| \leq K_2\lambda_2^{b-k}\mu_a$$

and

$$\begin{aligned}\sum_{m=k}^{\infty} \|\Phi(k, m+1)(I - P_{m+1})B_m\mathbf{u}_m\| &\leq \sum_{m=k}^{\infty} K_2\lambda_2^{m+1-k}\delta\mu_a \\ &= K_2\lambda_2(1-\lambda_2)^{-1}\delta\mu_a \\ &< \infty\end{aligned}$$

and so we may let $b \to \infty$ in Eq.(37) to obtain

$$(I - P_k)\mathbf{u}_k = -\sum_{m=k}^{\infty} \Phi(k, m+1)(I - P_{m+1})B_m\mathbf{u}_m. \quad (38)$$

When we add this to the equation obtained by multiplying Eq.(22) by P_k, we obtain

$$\begin{aligned}\mathbf{u}_k &= \Phi(k,a)P_a\mathbf{u}_a + \sum_{m=a}^{k-1} \Phi(k, m+1)P_{m+1}B_m\mathbf{u}_m \\ &\quad - \sum_{m=k}^{\infty} \Phi(k, m+1)(I - P_{m+1})B_m\mathbf{u}_m.\end{aligned}$$

The same reasoning shows that for $a \leq m \leq k$

$$\mathbf{u}_k = \Phi(k,m)P_m\mathbf{u}_m + \sum_{p=m}^{k-1} \Phi(k, p+1)P_{p+1}B_p\mathbf{u}_p - \sum_{p=k}^{\infty} \Phi(k, p+1)(I - P_{p+1})B_p\mathbf{u}_p.$$

It follows that for $a \leq m \leq k$

$$\|\mathbf{u}_k\| \leq K_1\lambda_1^{k-m}\|\mathbf{u}_m\| + \sum_{p=m}^{k-1} K_1\lambda_1^{k-p-1}\delta\|\mathbf{u}_p\| + \sum_{p=k}^{\infty} K_2\lambda_2^{p+1-k}\delta\|\mathbf{u}_p\|.$$

Then Lemma 1.10(i) allows us to conclude that Eq.(26) holds for $a \leq m \leq k$.

Next we show that for $\xi \in \mathcal{R}(P_a)$ there is a unique bounded solution $\mathbf{u}_k$ of Eq.(21) on $k \geq a$ such that

$$P_a\mathbf{u}_a = \xi.$$

Reasoning as above, such a solution must satisfy

$$\begin{aligned}\mathbf{u}_k = \Phi(k,a)P_a\xi + \sum_{m=a}^{k-1}\Phi(k,m+1)P_{m+1}B_m\mathbf{u}_m \\ -\sum_{m=k}^{\infty}\Phi(k,m+1)(I-P_{m+1})B_m\mathbf{u}_m.\end{aligned} \tag{39}$$

Now let X be the Banach space of bounded sequences $\mathbf{u} = \{\mathbf{u}_k\}_{k=a}^{\infty}$ with norm

$$\|\mathbf{u}\| = \sup_{k\geq a}\|\mathbf{u}_k\|.$$

Define an operator T on X by taking $(T\mathbf{u})_k$ as the right hand side of Eq.(39). Then

$$\begin{aligned}\|(T\mathbf{u})_k\| &\leq K_1\lambda_1^{k-a}\|\xi\| + \left(\sum_{m=a}^{k-1}K_1\lambda_1^{k-m-1}\delta + \sum_{m=k}^{\infty}K_2\lambda_2^{m+1-k}\delta\right)\|\mathbf{u}\| \\ &\leq K_1\|\xi\| + [K_1(1-\lambda_1)^{-1} + K_2\lambda_2(1-\lambda_2)^{-1}]\delta\|\mathbf{u}\|.\end{aligned}$$

So T maps X into itself. The fact that T is a contraction follows as in the previous case.

So, summarizing for the case $J = [a,\infty)$, what we have shown is that if $\mathbf{u}_k$ is a solution of Eq.(21) in E_k^s, that is, $\mathbf{u}_k$ is a solution of Eq.(21) bounded on $k \geq a$, then for $a \leq m \leq k$

$$\|\mathbf{u}_k\| \leq 2K_1\beta_1^{k-m}\|\mathbf{u}_m\|.$$

Moreover, for all $\xi \in \mathcal{R}(P_a)$, there is a unique such solution satisfying

$$P_a\mathbf{u}_a = \xi.$$

The solutions $\mathbf{u}_k$ in E_k^u consist of those for which $P_a\mathbf{u}_a = 0$, that is, $\mathbf{u}_a \in \mathcal{N}(P_a)$. Then we can apply the reasoning for the first case on any interval $[a,b]$ to deduce that Eq.(28) holds for $a \leq k \leq m$. Also since $\mathbf{u}_k = \mathbf{0}$ is the unique solution bounded in $k \geq a$ with $P_a\mathbf{u}_a = \mathbf{0}$, $E_k^s \cap E_k^u = \{\mathbf{0}\}$. Q_k is defined as in the case $J = [a,b]$ and inequalities (31), (32) hold for $a \leq m \leq k$ and $a \leq k \leq m$ respectively. The remaining reasoning for the case $J = [a,\infty)$ is the same as for the case $J = [a,b]$ with obvious modifications. The case $J = (-\infty, b]$ is treated similarly.

Finally we come to the case $J = (-\infty, \infty)$. Then the solutions in E_k^s are those for which $\sup_{k\geq 0} \|\mathbf{u}_k\| < \infty$ and we can apply the reasoning from the second case on any interval $[a, \infty)$ to deduce that inequality (26) holds for $m \leq k$. Also for any $\xi \in \mathcal{R}(P_0)$ there is a unique such solution $\mathbf{u}_k$ satisfying

$$P_0\mathbf{u}_0 = \xi.$$

Similarly, we take the solutions in E_k^u to be those for which $\sup_{k\leq 0} \|\mathbf{u}_k\| < \infty$. Then inequality (28) holds for $k \leq m$, and for any $\eta \in \mathcal{N}(P_0)$ there is a unique such solution $\mathbf{u}_k$ satisfying

$$(I - P_0)\mathbf{u}_0 = \eta.$$

Next if $\mathbf{u}_k$ is a solution in $E_k^s \cap E_k^u$, then it is bounded on $(-\infty, \infty)$ and we may let $a \to -\infty$ and $b \to \infty$ in Eq.(24) to obtain

$$\mathbf{u}_k = \sum_{m=-\infty}^{k-1} \Phi(k, m+1)P_{m+1}B_m\mathbf{u}_m - \sum_{m=k}^{\infty} \Phi(k, m+1)(I - P_{m+1})B_m\mathbf{u}_m.$$

If $\|\mathbf{u}\| = \sup_{-\infty<k<\infty} \|\mathbf{u}_k\|$, it follows that

$$\|\mathbf{u}\| \leq \left[K_1(1-\lambda_1)^{-1} + K_2\lambda_2(1-\lambda_2)^{-1}\right]\delta\|\mathbf{u}\| \leq \frac{1}{4}\|\mathbf{u}\|.$$

So $\|\mathbf{u}\| = \mathbf{0}$ and $E_k^s \cap E_k^u = \{\mathbf{0}\}$.

Next if $\mathbf{u}_k$ is in E_k^s it follows as in the case $J = [a, \infty)$ that Eq.(38) holds and if $\mathbf{u}_k$ is in E_k^u it follows similarly that

$$P_k\mathbf{u}_k = \sum_{m=-\infty}^{k-1} \Phi(k, m+1)P_{m+1}B_m\mathbf{u}_m.$$

Then if we take Q_k as the projection onto E_k^s along E_k^u, the proof in this case can be completed as in the previous cases.

2.5 EXPANSIVITY PROPERTY OF HYPERBOLIC SETS

Let $U \subset \mathbb{R}^n$ be a convex open set and $f : U \to \mathbb{R}^n$ a C^1 diffeomorphism onto its image. Let S be a (compact) hyperbolic set for f as in Definition 2.4. Denote by $\mathcal{P}(\mathbf{x})$ the projection of $\mathbb{R}^n$ onto $E^s(\mathbf{x})$ along $E^u(\mathbf{x})$. Then $\mathcal{P}(\mathbf{x})$ is continuous and there exist constants M^s and M^u such that

$$\|\mathcal{P}(\mathbf{x})\| \leq M^s, \quad \|I - \mathcal{P}(\mathbf{x})\| \leq M^u \quad \text{for all } \mathbf{x} \in S. \tag{40}$$

Our aim in this section is to show that f is *expansive* on S. First we present the definition of this concept.

Definition 2.9. Let $f : U \to \mathbb{R}^n$ be a C^1 diffeomorphism and let $S \subset U$ be *invariant* for f, that is, $f(S) = S$. Then f is said to be *expansive* on S with *expansivity constant* $d > 0$ if when $\mathbf{x}, \mathbf{y} \in S$ and

$$\|f^k(\mathbf{x}) - f^k(\mathbf{y})\| \leq d$$

for all integers k, then $\mathbf{x} = \mathbf{y}$.

In order to prove the expansiveness of a hyperbolic set, we use the following proposition. The property embodied in this proposition is sometimes called *exponential expansivity* (see, for example, Diamond, Kloeden, Kozyakin and Pokrovskii [1995d]).

Proposition 2.10. *Let S be a compact hyperbolic set for the C^1 diffeomorphism $f : U \to \mathbb{R}^n$ as in Definition 2.4 with M^s, M^u as given in Eq.(40). Choose numbers β_1 and β_2 satisfying*

$$\lambda_1 < \beta_1 < 1, \;\; \lambda_2 < \beta_2 < 1.$$

Then if d is sufficiently small depending only on $K_1, K_2, \lambda_1, \lambda_2, M^s, M^u, \beta_1, \beta_2$ and the modulus of continuity

$$\omega(\delta) = \sup\{\|Df(\mathbf{y}) - Df(\mathbf{x})\| : \mathbf{x} \in S, \; \|\mathbf{y} - \mathbf{x}\| \leq \delta\},$$

there are constants L_1, L_2 depending only on K_1, K_2, M^s, M^u, λ_1, λ_2 such that if $\{\mathbf{x}_k\}_{k=a}^{b}$ and $\{\mathbf{y}_k\}_{k=a}^{b}$ are orbits of f with $\mathbf{x}_k \in S$ for $a \leq k \leq b$ and satisfying

$$\|\mathbf{x}_k - \mathbf{y}_k\| \leq d \;\; \text{for } a \leq k \leq b, \tag{41}$$

the inequality

$$\|\mathbf{x}_k - \mathbf{y}_k\| \leq L_1 \beta_1^{k-a} \|\mathbf{x}_a - \mathbf{y}_a\| + L_2 \beta_2^{b-k} \|\mathbf{x}_b - \mathbf{y}_b\| \tag{42}$$

holds for $a \leq k \leq b$.

Proof. Set

$$\mathbf{u}_k = \mathbf{y}_k - \mathbf{x}_k.$$

Then for $k = a, \ldots, b-1$

$$\mathbf{u}_{k+1} = f(\mathbf{y}_k) - f(\mathbf{x}_k) = Df(\mathbf{x}_k)\mathbf{u}_k + f(\mathbf{y}_k) - f(\mathbf{x}_k) - Df(\mathbf{x}_k)\mathbf{u}_k.$$

So for $k = a, \ldots, b-1$,

$$\mathbf{u}_{k+1} = [A_k + B_k]\mathbf{u}_k, \tag{43}$$

where

$$A_k = Df(\mathbf{x}_k)$$

and

$$B_k = \int_0^1 [Df(\theta \mathbf{y}_k + (1-\theta)\mathbf{x}_k) - Df(\mathbf{x}_k)]d\theta.$$

Note that

$$\|B_k\| \le \omega(d).$$

Also from the proof of Proposition 2.7, we know that the difference equation

$$\mathbf{u}_{k+1} = A_k \mathbf{u}_k$$

has an exponential dichotomy on $[a, b]$ with projections $P_k = \mathcal{P}(\mathbf{x}_k)$, constants $K_1 M^s$, $K_2 M^u$ and exponents λ_1, λ_2. So we may apply Lemma 2.8 to deduce that if d is sufficiently small depending only on $K_1, K_2, \lambda_1, \lambda_2, M^s, M^u, \beta_1, \beta_2$ and $\omega(\cdot)$, the difference equation (43) has an exponential dichotomy on $[a, b]$ with exponents β_1, β_2 and with constants L_1, L_2 depending only on K_1, K_2, M^s, M^u, λ_1 and λ_2. Then for $a \le k \le b$,

$$\begin{aligned} \|\mathbf{u}_k\| &\le \|Q_k \mathbf{u}_k\| + \|(I - Q_k)\mathbf{u}_k\| \\ &= \|\Phi(k, a) Q_a \mathbf{u}_a\| + \|\Phi(k, b)(I - Q_b)\mathbf{u}_b\|, \end{aligned}$$

where $\Phi(k, m)$ is the transition matrix for Eq.(43) and the Q_k are the projections associated with the dichotomy. Hence, for the same k,

$$\|\mathbf{u}_k\| \le L_1 \beta_1^{k-a} \|\mathbf{u}_a\| + L_2 \beta_2^{b-k} \|\mathbf{u}_b\|$$

and so (42) follows.

Remark 2.11. Suppose inequality (41) holds for $k \ge a$. Then inequality (42) holds for all $b \ge a$. Hence for fixed $k \ge a$ we may let $b \to \infty$ to obtain

$$\|\mathbf{x}_k - \mathbf{y}_k\| \le L_1 \beta_1^{k-a} \|\mathbf{x}_a - \mathbf{y}_a\|$$

for $k \ge a$.

From this proposition we can easily deduce the expansivity.

Corollary 2.12. *Let S be a compact hyperbolic set for the C^1 diffeomorphism $f : U \to \mathbb{R}^n$. Then f is expansive on S.*

Proof. For a given choice of β_1 and β_2, let d be the positive constant in Proposition 2.10. Suppose $\mathbf{x}$ and $\mathbf{y}$ are in S and

$$\|f^k(\mathbf{x}) - f^k(\mathbf{y})\| \leq d$$

for all integers k. For a fixed positive integer N, we apply Proposition 2.10 to the orbit segments $\{f^k(\mathbf{x})\}_{k=-N}^{N}$ and $\{f^k(\mathbf{y})\}_{k=-N}^{N}$, deducing that

$$\|\mathbf{x}-\mathbf{y}\| \leq L_1\beta_1^N\|f^{-N}(\mathbf{x})-f^{-N}(\mathbf{y})\|+L_2\beta_2^N\|f^N(\mathbf{x})-f^N(\mathbf{y})\| \leq [L_1\beta_1^N+L_2\beta_2^N]d.$$

Letting $N \to \infty$, we conclude that $\mathbf{x} = \mathbf{y}$. Hence f is expansive on S.

The expansiveness of hyperbolic sets ensures that shadowing orbits (conf. Chapter 4) are unique. Also Ombach [1986], building on work of Bowen and Reddy, has shown that a compact invariant set is hyperbolic if and only if it is expansive and has the "shadowing" (or "pseudo orbit tracing ") property.

2.6 ROUGHNESS OF HYPERBOLIC SETS

Let $\mathbf{x}_0$ be a hyperbolic fixed point of a C^1 diffeomorphism $f : U \to I\!R^n$. We show that $\{\mathbf{x}_0\}$ *is the maximal invariant set for* f *inside some neighborhood of* $\mathbf{x}_0$. Suppose $\mathbf{x}$ has the property that

$$\|f^k(\mathbf{x}) - \mathbf{x}_0\| < \Delta$$

for all integers k, where the positive number Δ will be determined shortly. Then by reasoning similar to that used in the proof of Proposition 1.8 we can show that for all k

$$f^k(\mathbf{x}) = \mathbf{x}_0 \quad + \sum_{m=-\infty}^{k-1} A^{k-m-1}Pg(f^m(\mathbf{x}))$$

$$- \sum_{m=k}^{\infty} A^{-(m-k+1)}(I-P)g(f^m(\mathbf{x})),$$

where

$$A = Df(\mathbf{x}_0),$$

$$g(\mathbf{z}) = f(\mathbf{z}) - f(\mathbf{x}_0) - Df(\mathbf{x}_0)(\mathbf{z} - \mathbf{x}_0)$$

and P is the projection onto the stable subspace along the unstable subspace. Hence

$$\|f^k(\mathbf{x}) - \mathbf{x}_0\| \le \sum_{m=-\infty}^{k-1} K_1\lambda_1^{k-m-1} M^s \omega(\Delta)\|f^m(\mathbf{x}) - \mathbf{x}_0\|$$

$$+\sum_{m=k}^{\infty} K_2\lambda_2^{m-k+1} M^u \omega(\Delta)\|f^m(\mathbf{x}) - \mathbf{x}_0\|,$$

where the constants $K_1, K_2, \lambda_1, \lambda_2, M^s, M^u$ and modulus of continuity $\omega(\Delta)$ are as in the statement of Proposition 1.8. Thus

$$\|f^k(\mathbf{x}) - \mathbf{x}_0\| \le [K_1 M^s(1-\lambda_1)^{-1} + K_2 M^u \lambda_2(1-\lambda_2)^{-1}]\omega(\Delta) \cdot \sup_{l\in \mathbb{Z}} \|f^l(\mathbf{x}) - \mathbf{x}_0\|$$

for all integers k. Hence if

$$[K_1 M^s(1-\lambda_1)^{-1} + K_2 M^u \lambda_2(1-\lambda_2)^{-1}]\omega(\Delta) < 1, \tag{44}$$

we have

$$\sup_{k\in \mathbb{Z}} \|f^k(\mathbf{x}) - \mathbf{x}_0\| = 0$$

and so $f^k(\mathbf{x}) = \mathbf{x}_0$ for all k. The conclusion is that *if the positive number* Δ *satisfies the inequality in Eq.*(44), *then* $\{\mathbf{x}_0\}$ *is the maximal invariant set for* f *in the open set* $\{\mathbf{x} : \|\mathbf{x} - \mathbf{x}_0\| < \Delta\}$.

Next let $\mathbf{x}_0$ be a hyperbolic periodic point with period m of the C^1 diffeomorphism $f : U \to \mathbb{R}^n$. Then, as just proved, there exists a positive number Δ such that $\{\mathbf{x}_0\}$ is the maximal invariant set for f^m in the open ball

$$B(\mathbf{x}_0, \Delta) = \{\mathbf{x} : \|\mathbf{x} - \mathbf{x}_0\| < \Delta\}.$$

Now choose $\delta > 0$ so that the balls $B(f^k(\mathbf{x}_0), \delta)$ are mutually disjoint for $k = 0, ..., m-1$ and so that $f(B(f^k(\mathbf{x}_0), \delta))$ intersects only the ball $B(f^{k+1}(\mathbf{x}_0), \delta)$. Also suppose that $\delta \le \Delta$. Now let $\{f^k(\mathbf{y})\}_{k=-\infty}^{\infty}$ be an orbit of f in the open set

$$O = \bigcup_{k=0}^{m-1} B(f^k(\mathbf{x}_0), \delta).$$

Because of the choice of δ, we can assume that $f^k(\mathbf{y}) \in B(f^k(\mathbf{x}_0), \delta)$ for all k. But then the sequence $\{f^{km}(\mathbf{y})\}_{k=-\infty}^{\infty}$ is an orbit of f^m lying in $B(\mathbf{x}_0, \Delta)$. Hence $f^{km}(\mathbf{y}) = \mathbf{x}_0$ for all k. This implies that $\mathbf{y} = \mathbf{x}_0$ and that the orbit $f^k(\mathbf{y})$ coincides with the orbit of $\mathbf{x}_0$. Hence the orbit of $\mathbf{x}_0$ is the maximal invariant set inside O.

Definition 2.13. Let $f : U \to \mathbb{R}^n$ be a C^1 diffeomorphism. An invariant subset S of U is said to be *isolated* if it is the maximal invariant set in some open set containing it.

What we have shown above is that hyperbolic fixed points and periodic orbits are isolated invariant sets. However, it turns out that not all hyperbolic sets are isolated. For example, as we shall see later, the hyperbolic set consisting of a hyperbolic fixed point and associated transversal homoclinic orbit is not isolated.

Let us suppose that we have a compact hyperbolic set for a diffeomorphism. Then we consider an open neighborhood of this set and examine the maximal invariant set inside the closure of this neighborhood. We shall show that if the neighborhood is sufficiently "tight", then this maximal invariant set is also hyperbolic. Moreover, this conclusion applies not only to the original diffeomorphism but also to C^1 perturbations of it.

Let $U \subset \mathbb{R}^n$ be a convex open set and let $S \subset U$ be a compact hyperbolic set for the C^1 diffeomorphism $f : U \to \mathbb{R}^n$ as in Definition 2.4 and let O be a bounded open neighborhood of S with $\overline{O} \subset U$. Also let $g : U \to \mathbb{R}^n$ be another C^1 diffeomorphism such that

$$\sup_{\mathbf{x}\in U} \|g(\mathbf{x}) - f(\mathbf{x})\| + \sup_{\mathbf{x}\in U} \|Dg(\mathbf{x}) - Df(\mathbf{x})\| \leq \sigma, \tag{45}$$

where σ is a positive number the size of which is to be determined. We define

$$S_O = \{\mathbf{x} \in \overline{O} : g^k(\mathbf{x}) \in \overline{O} \text{ for all } k \in \mathbb{Z}\}. \tag{46}$$

Clearly, S_O is a compact invariant set for g and S_O is its maximal invariant set in $\overline{O}$. Set

$$d = \max_{\mathbf{x}\in\overline{O}} \operatorname{dist}(\mathbf{x}, S). \tag{47}$$

We show the following theorem (conf. Katok [1971], Hirsch, Pugh and Shub [1977], Guckenheimer, Moser and Newhouse [1980], Lanford [1985], Shub [1987], Akin [1993]).

Theorem 2.14. *Let S be a compact hyperbolic set for the C^1 diffeomorphism $f : U \to \mathbb{R}^n$ as in Definition 2.4 with U convex. Choose numbers β_1, β_2 such that*

$$\lambda_1 < \beta_1 < 1 \quad \text{and} \quad \lambda_2 < \beta_2 < 1.$$

Then there exist positive numbers σ_0 and d_0 depending only on f, S, β_1 and β_2 such that if O is an open neighborhood of S with d in (47) satisfying $d \leq d_0$

and $g : U \to I\!R^n$ is a C^1 diffeomorphism satisfying Eq.(45) with $\sigma \le \sigma_0$, the set S_O defined in Eq.(46) is a compact hyperbolic set for g with exponents β_1 and β_2. Also the dimension of the stable bundle is the same as for f and S, and the constants associated with the hyperbolicity and bounds on the norms of the projections can be chosen to depend only on f, S, β_1 and β_2.

Proof. In view of Proposition 2.7, in order to establish the hyperbolicity of S_O, all we need show is that for all $\mathbf{x} \in S_O$ the difference equation

$$\mathbf{u}_{k+1} = Dg(g^k(\mathbf{x}))\mathbf{u}_k \tag{48}$$

has an exponential dichotomy on $(-\infty, \infty)$ with the exponents, constants and rank of projection independent of $\mathbf{x}$.

So let $\mathbf{x} \in S_O$ and write

$$\mathbf{x}_k = g^k(\mathbf{x}) \quad \text{for } k \in \boldsymbol{Z}.$$

Since $\operatorname{dist}(\mathbf{x}_k, S) \le d$, there exists $\mathbf{y}_k$ in S such that

$$\|\mathbf{x}_k - \mathbf{y}_k\| \le d.$$

Then

$$\begin{aligned} \|\mathbf{y}_{k+1} - f(\mathbf{y}_k)\| &\le \|\mathbf{y}_{k+1} - \mathbf{x}_{k+1}\| + \|g(\mathbf{x}_k) - g(\mathbf{y}_k)\| + \|g(\mathbf{y}_k) - f(\mathbf{y}_k)\| \\ &\le d + (M_1 + \sigma)d + \sigma, \end{aligned}$$

where

$$M_1 = \sup_{\mathbf{x} \in U} \|Df(\mathbf{x})\|.$$

Hence

$$\|\mathbf{y}_{k+1} - f(\mathbf{y}_k)\| \le \delta \quad \text{for} \quad k \in \boldsymbol{Z},$$

where

$$\delta = (1 + M_1 + \sigma)d + \sigma.$$

We make the following definition.

Definition 2.15. Let $f : U \to I\!R^n$ be a C^1 diffeomorphism. If δ is a positive number, a sequence $\{\mathbf{y}_k\}_{k=-\infty}^{\infty}$ of points in U is said to be a δ *pseudo orbit* of f if

$$\|\mathbf{y}_{k+1} - f(\mathbf{y}_k)\| \le \delta \quad \text{for } k \in \boldsymbol{Z}.$$

We regard difference equation (48) as a perturbation of

$$\mathbf{u}_{k+1} = Df(\mathbf{y}_k)\mathbf{u}_k. \tag{49}$$

In the following lemma, we show that this difference equation has an exponential dichotomy on $(-\infty, \infty)$ provided δ is sufficiently small. Note that this lemma also plays an important role in the proof of the shadowing theorem in Chapter 4. The methods used in the proof are derived from those used in Henry [1981], Palmer [1987] and Sakamoto [1994]. The basic tool is Lemma 2.8 but it cannot be applied directly since we do not know apriori that $\mathbf{y}_k$ stays uniformly close to an orbit of f.

Lemma 2.16. *Let $\{\mathbf{y}_k\}_{k=-\infty}^{\infty}$ be a δ pseudo orbit in S, a compact hyperbolic set for the diffeomorphism $f : U \to \mathbb{R}^n$ as in Definition 2.4 with U convex. Suppose α_1 and α_2 are numbers satisfying*

$$\lambda_1 < \alpha_1 < 1, \quad \lambda_2 < \alpha_2 < 1.$$

Then if δ is sufficiently small, depending on f, S, α_1 and α_2, the difference equation (49) has an exponential dichotomy on $(-\infty, \infty)$ with exponents α_1, α_2, with the rank of the projection equal to $\dim E^s$ and with constants depending only on f, S, α_1 and α_2.

Proof. First we show Eq.(49) has an exponential dichotomy on finite intervals. To this end, fix integers $a < b$. With a view to applying Lemma 2.8, for $a \le k \le b$, we write Eq.(49) as

$$\mathbf{u}_{k+1} = (A_k + B_k)\mathbf{u}_k,$$

where

$$A_k = Df(f^{k-a}(\mathbf{y}_a))$$

and

$$B_k = Df(\mathbf{y}_k) - Df(f^{k-a}(\mathbf{y}_a)).$$

Note that

$$\|B_k\| \le \omega(\|\mathbf{y}_k - f^{k-a}(\mathbf{y}_a)\|),$$

where

$$\omega(\Delta) = \sup\{\|Df(\mathbf{x}) - Df(\mathbf{y})\| : \mathbf{x}, \mathbf{y} \in S, \ \|\mathbf{x} - \mathbf{y}\| \le \Delta\}.$$

Now we assert that when $k \ge m$

$$\|\mathbf{y}_k - f^{k-m}(\mathbf{y}_m)\| \le (1 + M_1 + \ldots + M_1^{k-m-1})\delta, \tag{50}$$

where both sides are interpreted as zero when $k = m$ and

$$M_1 = \sup_{\mathbf{x} \in U} \|Df(\mathbf{x})\|.$$

This inequality follows by induction on k using the estimate

$$\begin{aligned}\|\mathbf{y}_{k+1} - f^{k+1-m}(\mathbf{y}_m)\| &\leq \|\mathbf{y}_{k+1} - f(\mathbf{y}_k)\| + \|f(\mathbf{y}_k) - f(f^{k-m}(\mathbf{y}_m))\| \\ &\leq \delta + M_1\|\mathbf{y}_k - f^{k-m}(\mathbf{y}_m)\|.\end{aligned}$$

Then it follows that for $a \leq k \leq b$

$$\|B_k\| \leq \omega((M_1 - 1)^{-1}(M_1^{k-a} - 1)\delta) \leq \omega((M_1 - 1)^{-1}(M_1^{b-a} - 1)\delta).$$

For use in the sequel, we set

$$L_1 = 2K_1M^s(2K_1M^s + K_2M^u),\ L_2 = 2K_2M^u(2K_2M^u + K_1M^s),$$

where M^s and M^u are as in Eq.(40). Also we set

$$\tilde{\alpha}_1 = \frac{\lambda_1 + \alpha_1}{2},\quad \tilde{\alpha}_2 = \frac{\lambda_2 + \alpha_2}{2}$$

and let n_0 be the least positive integer such that

$$L_1\tilde{\alpha}_1^{n_0} < \alpha_1^{n_0},\ L_2\tilde{\alpha}_1^{n_0} < \alpha_2^{n_0}. \tag{51}$$

Next note that by Proposition 2.7, the difference equation

$$\mathbf{u}_{k+1} = A_k\mathbf{u}_k$$

has an exponential dichotomy on $\boldsymbol{Z}$ with constants K_1M^s, K_2M^u and exponents λ_1, λ_2. Then, by Lemma 2.8 and its proof, provided

$$\delta_1 = \omega((M_1 - 1)^{-1}(M_1^{n_0} - 1)\delta)$$

is sufficiently small depending only on K_1, K_2, M^s, M^u, λ_1, λ_2, α_1, α_2 the difference equation (49) has an exponential dichotomy on $[(i-1)n_0,\ in_0]$, where $i \in \boldsymbol{Z}$, with constants L_1, L_2, exponents $\tilde{\alpha}_1, \tilde{\alpha}_2$ and projections $\{Q_k^{(i)}\}_{k=(i-1)n_0}^{in_0}$ satisfying

$$\|Q_k^{(i)} - \mathcal{P}(f^{k-(i-1)n_0}(\mathbf{y}_{(i-1)n_0}))\| \leq N\delta_1,$$

N being a constant depending only on K_1, K_2, M^s, M^u, λ_1, λ_2. Also the rank of $Q_k^{(i)}$ is equal to the rank of $\mathcal{P}$.

In order to finish the proof of Lemma 2.16, we need another lemma. This shows that if a difference equation has an exponential dichotomy on $[(i-1)n_0, in_0]$ for $i \in \boldsymbol{Z}$ and the associated projections do not "jump" too much at the endpoints, then the difference equation has an exponential dichotomy on $(-\infty, \infty)$ (conf. Henry [1981, p.234]).

Lemma 2.17. *Let* $\{A_k\}_{k=-\infty}^{\infty}$ *be a sequence of invertible matrices such that*

$$\|A_k\| \le M, \;\; \|A_k^{-1}\| \le M$$

for all k. *Suppose also that there is a positive integer* n_0 *such that for all* $i \in \boldsymbol{Z}$ *the difference equation*

$$\mathbf{u}_{k+1} = A_k \mathbf{u}_k \tag{52}$$

has an exponential dichotomy on $[(i-1)n_0, in_0]$ *with constants* K_1, K_2, *exponents* λ_1, λ_2 *and projections* $\{Q_k^{(i)}\}_{k=(i-1)n_0}^{in_0}$ *satisfying*

$$\|Q_{in_0}^{(i+1)} - Q_{in_0}^{(i)}\| \le \delta.$$

Suppose, moreover, that β_1 *and* β_2 *are numbers satisfying*

$$\lambda_1 < \beta_1 < 1 \;\; , \;\; \lambda_2 < \beta_2 < 1$$

and that n_0 *has the property*

$$K_1 \lambda_1^{n_0} < \beta_1^{n_0} \;\; , \;\; K_2 \lambda_2^{n_0} < \beta_2^{n_0}.$$

Then if δ *is sufficiently small depending only on* $K_1, K_2, \lambda_1, \lambda_2, \beta_1, \beta_2, M$ *and* n_0, Eq.(52) *has an exponential dichotomy on* $(-\infty, \infty)$ *with constants depending only on the same quantities, with exponents* β_1, β_2 *and with projections* R_k *satisfying*

$$\|R_k - Q_k^{(i)}\| \le N\delta \quad \text{for} \quad k \in [(i-1)n_0, in_0],\; i \in \boldsymbol{Z},$$

where the constant N *depends only on* K_1, K_2, λ_1, λ_2, β_1, β_2, M *and* n_0.

Proof. We examine the difference equation

$$\mathbf{u}_{i+1} = C_i \mathbf{u}_i, \;\; i \in \boldsymbol{Z}, \tag{53}$$

where

$$C_i = A_{in_0-1} \cdots A_{(i-1)n_0}.$$

Note that the transition matrix for this difference equation is

$$\Psi(i,j) = \tilde{\Psi}((i-1)n_0, (j-1)n_0),$$

where $\tilde{\Psi}(n,m)$ is the transition matrix for Eq.(52). If we can show Eq.(53) has an exponential dichotomy, then it will follow very easily that Eq.(52) has one also.

We first define the projections

$$\overline{Q}_i = Q^{(i)}_{(i-1)n_0}.$$

By the dichotomy property for Eq.(52),

$$\|C_i\overline{Q}_i\| = \|\tilde{\Psi}(in_0,(i-1)n_0)Q^{(i)}_{(i-1)n_0}\| \le K_1\lambda_1^{n_0}$$

and

$$\begin{aligned}\|(I-\overline{Q}_i)C_i^{-1}\| &= \|(I-Q^{(i)}_{(i-1)n_0})\tilde{\Psi}((i-1)n_0,in_0)\| \\ &= \|\tilde{\Psi}((i-1)n_0,in_0)(I-Q^{(i)}_{in_0})\| \\ &\le K_2\lambda_2^{n_0}.\end{aligned}$$

Moreover,

$$\|\overline{Q}_{i+1}C_i - C_i\overline{Q}_i\| = \|Q^{(i+1)}_{in_0}C_i - Q^{(i)}_{in_0}C_i\| \le \|Q^{(i+1)}_{in_0} - Q^{(i)}_{in_0}\|\|C_i\| \le M^{n_0}\delta.$$

Now we replace C_i by matrices $\overline{C}_i$ which are invariant with respect to the $\overline{Q}_i$. So we define

$$\overline{C}_i = \overline{Q}_{i+1}C_i\overline{Q}_i + (I-\overline{Q}_{i+1})C_i(I-\overline{Q}_i).$$

Note that the invariance property

$$\overline{Q}_{i+1}\overline{C}_i = \overline{C}_i\overline{Q}_i \tag{54}$$

holds for all i. Next we estimate how close $\overline{C}_i$ is to C_i and $\overline{C}_i^{-1}$ to C_i^{-1}. In fact,

$$\begin{aligned}\|\overline{C}_i - C_i\| &= \|(I-\overline{Q}_{i+1})C_i\overline{Q}_i + \overline{Q}_{i+1}C_i(I-\overline{Q}_i)\| \\ &= \|(\overline{Q}_{i+1}C_i - C_i\overline{Q}_i)(I-\overline{Q}_i) - (\overline{Q}_{i+1}C_i - C_i\overline{Q}_i)\overline{Q}_i\| \\ &\le (K_1+K_2)M^{n_0}\delta.\end{aligned}$$

This also implies that

$$\|(\overline{C}_i - C_i)C_i^{-1}\| \le (K_1+K_2)M_1^{2n_0}\delta.$$

So if δ satisfies

$$(K_1+K_2)M^{2n_0}\delta \le \frac{1}{2},$$

$\overline{C}_i$ is invertible and

$$
\begin{aligned}
\|\overline{C}_i^{-1} - C_i^{-1}\| &= \|C_i^{-1}[(I + (\overline{C}_i - C_i)C_i^{-1})^{-1} - I]\| \\
&\leq \|C_i^{-1}\|(1 - \|(\overline{C}_i - C_i)C_i^{-1}\|)^{-1}\|(\overline{C}_i - C_i)C_i^{-1}\| \\
&\leq M^{n_0}(1 - (K_1 + K_2)M^{2n_0}\delta)^{-1}(K_1 + K_2)M_1^{2n_0}\delta \\
&\leq 2M^{3n_0}(K_1 + K_2)\delta.
\end{aligned}
$$

Next we show that the difference equation

$$\mathbf{u}_{i+1} = \overline{C}_i \mathbf{u}_i \tag{55}$$

has an exponential dichotomy on $(-\infty, \infty)$. We note that

$$\|\overline{C}_i\overline{Q}_i\| \leq \|\overline{C}_i - C_i\|\|\overline{Q}_i\| + \|C_i\overline{Q}_i\| \leq K_1(K_1 + K_2)M^{n_0}\delta + K_1\lambda_1^{n_0} \tag{56}$$

and

$$
\begin{aligned}
\|(I - \overline{Q}_i)\overline{C}_i^{-1}\| &\leq \|I - \overline{Q}_i\|\|\overline{C}_i^{-1} - C_i^{-1}\| + \|(I - \overline{Q}_i)C_i^{-1}\| \\
&\leq 2K_2M^{3n_0}(K_1 + K_2)\delta + K_2\lambda_2^{n_0}.
\end{aligned} \tag{57}
$$

So, provided

$$\alpha_1 = K_1(K_1 + K_2)M^{n_0}\delta + K_1\lambda_1^{n_0} < 1$$

and

$$\alpha_2 = 2K_2M^{3n_0}(K_1 + K_2)\delta + K_2\lambda_2^{n_0} < 1,$$

inequalities (56) and (57) together with the inequalities

$$\|\overline{Q}_i\| \leq K_1, \ \|I - \overline{Q}_i\| \leq K_2$$

and the invariance (54) imply that the difference equation (55) has an exponential dichotomy on $(-\infty, \infty)$ with projections $\overline{Q}_i$, constants K_1, K_2 and exponents α_1 and α_2.

Then, using the conditions $K_1\lambda_1^{n_0} < \beta_1^{n_0}$ and $K_2\lambda_2^{n_0} < \beta_2^{n_0}$, it follows from Lemma 2.8 that, provided δ is small enough, depending on K_1, K_2, λ_1, λ_2, β_1, β_2, M and n_0, difference equation (53) has an exponential dichotomy on $(-\infty, \infty)$ with constants L_1, L_2, exponents $\beta_1^{n_0}, \beta_2^{n_0}$ and projections Q_i satisfying

$$\|Q_i - \overline{Q}_i\| \leq \tilde{N}\delta,$$

where L_1, L_2 and $\tilde{N}$ depend only on K_1, K_2, λ_1, λ_2, β_1, β_2 and n_0. Also the rank of Q_i equals the rank of $\overline{Q}_i$.

Finally, to show that Eq.(52) has an exponential dichotomy, let k be an arbitrary integer. Then there is an integer i such that

$$(i-1)n_0 \leq k < in_0.$$

We define the projection

$$R_k = \tilde{\Psi}(k,(i-1)n_0)Q_i\tilde{\Psi}((i-1)n_0,k).$$

It follows easily that the R_k are invariant with respect to Eq.(52) and have the same rank as Q_i. Moreover, we have the inequality

$$\begin{aligned}\|R_k - Q_k^{(i)}\| &= \|\tilde{\Psi}(k,(i-1)n_0)[Q_i - \overline{Q}_i]\tilde{\Psi}((i-1)n_0,k)\| \\ &\leq M^{k-(i-1)n_0}\tilde{N}\delta M^{k-(i-1)n_0} \\ &\leq M^{2(n_0-1)}\tilde{N}\delta.\end{aligned}$$

Now if $k \geq m$ there exists integers $i \geq j$ such that

$$(j-1)n_0 \leq m < jn_0,\ (i-1)n_0 \leq k < in_0.$$

Then

$$\begin{aligned}\|\tilde{\Psi}(k,m)R_m\| &= \|\tilde{\Psi}(k,(i-1)n_0)\Psi(i,j)Q_j\tilde{\Psi}((j-1)n_0,m)\| \\ &\leq M^{k-(i-1)n_0}L_1\beta_1^{(i-j)n_0}M^{m-(j-1)n_0} \\ &\leq L_1M^{2(n_0-1)}\beta_1^{1-n_0}\beta_1^{k-m}.\end{aligned}$$

Also

$$\begin{aligned}\|\tilde{\Psi}(m,k)(I-R_k)\| &= \|\tilde{\Psi}(m,(j-1)n_0)\Psi(j,i)(I-Q_i)\tilde{\Psi}((i-1)n_0,k)\| \\ &\leq M^{m-(j-1)n_0}L_2\beta_2^{(i-j)n_0}M^{k-(i-1)n_0} \\ &\leq L_2M^{2(n_0-1)}\beta_2^{1-n_0}\beta_2^{k-m}.\end{aligned}$$

Thus we have shown, subject to the conditions imposed on δ in the proof, that difference equation (52) has an exponential dichotomy on $(-\infty,\infty)$ with exponents β_1, β_2 and constants depending only on K_1, K_2, λ_1, λ_2, β_1, β_2, M and n_0. Also the rank of the projection R_k equals that of $Q_k^{(i)}$ and

$$\|R_k - Q_k^{(i)}\| \leq N\delta,$$

where

$$N = M^{2(n_0-1)}\tilde{N}.$$

Thus the proof of Lemma 2.17 is complete.

To complete the proof of Lemma 2.16, we apply Lemma 2.17 to Eq.(49), noting that for the $Q_k^{(i)}$ defined just before Lemma 2.17

$$\begin{aligned} \|Q_{in_0}^{(i+1)} - Q_{in_0}^{(i)}\| &\leq \|Q_{in_0}^{(i+1)} - \mathcal{P}(\mathbf{y}_{in_0})\| + \|\mathcal{P}(\mathbf{y}_{in_0}) - \mathcal{P}(f^{n_0}(\mathbf{y}_{(i-1)n_0}))\| \\ &\quad + \|\mathcal{P}(f^{n_0}(\mathbf{y}_{(i-1)n_0})) - Q_{in_0}^{(i)}\| \\ &\leq 2N\delta_1 + \omega^s((M_1-1)^{-1}(M_1^{n_0}-1)\delta), \end{aligned}$$

where $\omega^s(\cdot)$ is the modulus of continuity of $\mathcal{P}$ on S. For the application of Lemma 2.17, we also take

$$M = \max\left\{\sup_{\mathbf{x}\in S}\|Df(\mathbf{x})\|, \sup_{\mathbf{x}\in S}\|Df^{-1}(\mathbf{x})\|\right\}$$

and for $i = 1,2$ we take $\lambda_i = \tilde{\alpha}_1$, $\beta_i = \alpha_i$ and $K_i = L_i$. (Note that these K_i and λ_i are not the same as for the original hyperbolic set in Lemma 2.16.)

So we deduce from Lemma 2.17 that if δ is sufficiently small depending only on f, S, α_1 and α_2, Eq.(49) has an exponential dichotomy on $(-\infty,\infty)$ with constants depending only on f, S, α_1 and α_2, exponents α_1, α_2 and rank of the projection equal to $\dim E^s$. Thus the proof of Lemma 2.16 is complete.

To complete the proof of Theorem 2.14, we choose

$$\alpha_1 = \frac{\lambda_1+\beta_1}{2}, \quad \alpha_2 = \frac{\lambda_2+\beta_2}{2}.$$

Then it follows from Lemma 2.16 with

$$\delta = (1+M_1+\sigma)d+\sigma,$$

provided d and σ are sufficiently small depending on f, S, β_1 and β_2, that difference equation (49) has an exponential dichotomy on $(-\infty,\infty)$ with exponents α_1, α_2 and constants depending only on f, S, β_1 and β_2. Also the rank of the projection equals the dimension of E^s. Next

$$\begin{aligned} \|Dg(\mathbf{x}_k) - Df(\mathbf{y}_k)\| &\leq \|Dg(\mathbf{x}_k) - Df(\mathbf{x}_k)\| + \|Df(\mathbf{x}_k) - Df(\mathbf{y}_k)\| \\ &\leq \sigma + M_1 d. \end{aligned}$$

Then it follows from Lemma 2.8, provided d and σ are sufficiently small depending on f, S, β_1 and β_2, that difference equation (48) has an exponential dichotomy on $(-\infty,\infty)$ with exponents β_1, β_2 and constants depending only on

f, S, β_1 and β_2. Also the rank of the projection equals the dimension of E^s and note that the constants in the dichotomy give bounds on the projections. Thus, in view of Proposition 2.7, the proof of Theorem 2.14 is complete.

3. TRANSVERSAL HOMOCLINIC POINTS OF DIFFEOMORPHISMS AND HYPERBOLIC SETS

Up till now our only examples of hyperbolic sets have been hyperbolic fixed points and periodic orbits. In this chapter we give an example of a hyperbolic set which is infinite. Moreover, even though the set consists of only two distinct orbits, we shall show in Chapter 5 that in its neighbourhood the diffeomorphism has chaotic dynamics.

3.1 THE HYPERBOLIC SET ASSOCIATED WITH A TRANSVERSAL HOMOCLINIC POINT

We proceed with some definitions. Again we are considering a C^1 diffeomorphism $f : U \to \mathbb{R}^n$, where U is an open subset of $\mathbb{R}^n$. We suppose $\mathbf{x}_0$ is a hyperbolic fixed point of f with stable and unstable manifolds $W^s(\mathbf{x}_0)$ and $W^u(\mathbf{x}_0)$.

Definition 3.1. A point $\mathbf{y}_0$ is said to be a *homoclinic point* with respect to the fixed point $\mathbf{x}_0$ if $\mathbf{y}_0 \neq \mathbf{x}_0$ and $\mathbf{y}_0 \in W^s(\mathbf{x}_0) \cap W^u(\mathbf{x}_0)$. Moreover, if in addition

$$\mathbb{R}^n = T_{\mathbf{y}_0} W^s(\mathbf{x}_0) \oplus T_{\mathbf{y}_0} W^u(\mathbf{x}_0),$$

the homoclinic point is said to be *transversal.*

What is meant here by $T_{\mathbf{y}_0} W^s(\mathbf{x}_0)$ and $T_{\mathbf{y}_0} W^u(\mathbf{x}_0)$? Recall that

$$W^s(\mathbf{x}_0) = \bigcup_{k=0}^{\infty} f^{-k}\left(W^{s,\varepsilon}(\mathbf{x}_0)\right),$$

where $W^{s,\varepsilon}(\mathbf{x}_0)$ is a manifold for $\varepsilon > 0$ sufficiently small. Hence if $\mathbf{x}$ is in $W^s(\mathbf{x}_0)$, there is $N \geq 0$ such that $f^N(\mathbf{x})$ is in $W^{s,\varepsilon}(\mathbf{x}_0)$, and so we define

$$T_{\mathbf{x}} W^{s,\varepsilon}(\mathbf{x}_0) = T_{\mathbf{x}}(f^{-N}(W^{s,\varepsilon}(\mathbf{x}_0)) = Df^{-N}(f^N(\mathbf{x}))(T_{f^N(\mathbf{x})} W^{s,\varepsilon}(\mathbf{x}_0)).$$

It is easy to see that this definition is independent of both N and ε and also that the invariance property

$$Df(\mathbf{x})(T_{\mathbf{x}} W^s(\mathbf{x}_0)) = T_{f(\mathbf{x})} W^s(\mathbf{x}_0).$$

holds. $T_{\mathbf{x}} W^u(\mathbf{x}_0)$ is defined in analogous fashion.

Fontich [1990], Marotto [1979ab], Misuriewicz and Szewc [1980], Morinaka [1983] and Rogers and Marotto [1983] have given theorems guaranteeing the existence and examples of diffeomorphisms with transversal homoclinic points;

McGehee and Meyer [1974] and Devaney and Nitecki [1984] have given theorems about and examples of diffeomorphisms with not necessarily transversal homoclinic points; Neumaier and Rage [1993] give a numerical procedure to prove the existence of a transversal homoclinic point. In the next section, we give a general method for the construction of diffeomorphisms with transversal homoclinic points.

Our purpose in this section is to prove the following theorem.

Theorem 3.2. *Let $f : U \to \mathbb{R}^n$ be a C^1 diffeomorphism with hyperbolic fixed point $\mathbf{x}_0$ and associated transversal homoclinic point $\mathbf{y}_0$. Then the set*

$$S = \{\mathbf{x}_0\} \cup \{f^k(\mathbf{y}_0) : k \in \mathbf{Z}\}$$

is hyperbolic with splitting

$$\mathbb{R}^n = T_{\mathbf{x}}W^s(\mathbf{x}_0) \oplus T_{\mathbf{x}}W^u(\mathbf{x}_0), \quad \mathbf{x} \in S$$

and with exponents as close to those for the hyperbolic fixed point $\mathbf{x}_0$ as we like.

Proof. Clearly S is a compact invariant set, the given splitting is invariant and the dimensions of the subspaces are constant. So, since the continuity of the splitting follows from the other properties in Definition 2.4, we need only verify the exponential estimates and, indeed, only along the orbit of $\mathbf{y}_0$. To this end we prove the following proposition.

Proposition 3.3. *Let $\mathbf{x}_0$ be a hyperbolic fixed point of the C^1 diffeomorphism $f : U \to \mathbb{R}^n$ such that the inequalities in Eqs.(1), (2) in Chapter 1 hold. Choose numbers β_1, β_2 such that*

$$\lambda_1 < \beta_1 < 1, \quad \lambda_2 < \beta_2 < 1.$$

Then if the positive number ε is sufficiently small, given any $\mathbf{x}$ in $W^{s,\varepsilon}(\mathbf{x}_0)$ and any subspace V such that

$$T_{\mathbf{x}}W^{s,\varepsilon}(\mathbf{x}_0) \oplus V = \mathbb{R}^n, \tag{1}$$

there exist positive constants K_3, K_4 such that for $k \geq m \geq 0$

$$\|Df^k(\mathbf{x})\xi\| \leq K_3\beta_1^{k-m}\|Df^m(\mathbf{x})\xi\| \text{ for } \xi \in T_{\mathbf{x}}W^{s,\varepsilon}(\mathbf{x}_0) \tag{2}$$

and

$$\|Df^m(\mathbf{x})\xi\| \leq K_4\beta_2^{k-m}\|Df^k(\mathbf{x})\xi\| \text{ for } \xi \in V. \tag{3}$$

Proof. In Chapter 1 we constructed an open neighbourhood U_0 of $\mathbf{0}$ in the stable subspace E^s and a C^1 function $\phi^s : U_0 \to \mathbb{R}^n$ such that for $\varepsilon > 0$

sufficiently small, there is an open neighbourhood U_ε of $\mathbf{0}$ contained in U_0 such that

$$W^{s,\varepsilon}(\mathbf{x}_0) = \{\mathbf{x}_0 + \phi^s(\mathbf{u}) : \mathbf{u} \in U_\varepsilon\}.$$

Moreover, ϕ^s has the property $P\phi^s(\mathbf{u}) = \mathbf{u}$, where P is the projection onto E^s along the unstable subspace E^u. Also $\phi^s(\mathbf{0}) = \mathbf{0}$ and $D\phi^s(\mathbf{0})$ is the inclusion of E^s in $\mathbb{R}^n$. In fact, the mapping $H : U_\varepsilon \to W^{s,\varepsilon}(\mathbf{x}_0)$ defined by

$$H(\mathbf{u}) = \mathbf{x}_0 + \phi^s(\mathbf{u})$$

is a C^1 diffeomorphism which conjugates $f : W^{s,\varepsilon}(\mathbf{x}_0) \to W^{s,\varepsilon}(\mathbf{x}_0)$ to $g : U_\varepsilon \to U_\varepsilon$, where g is defined by

$$g = H^{-1} \circ f \circ H. \tag{4}$$

This means that $H \circ g = f \circ H$, that is,

$$\mathbf{x}_0 + \phi^s\,(g(\mathbf{u})) = f\,(\mathbf{x}_0 + \phi^s(\mathbf{u}))$$

and so for $\mathbf{u} \in U_\varepsilon$

$$g(\mathbf{u}) = P[f(\mathbf{x}_0 + \phi^s(\mathbf{u})) - \mathbf{x}_0].$$

Note also that if $\mathbf{u} \in U_\varepsilon$, then $\mathbf{u} = P\phi^s(\mathbf{u})$ and so

$$\|\mathbf{u}\| \leq \|P\|\varepsilon.$$

Having finished with these preliminaries, we first establish the inequality in Eq.(2), assuming ε is small as just described. So let $\mathbf{x} \in W^{s,\varepsilon}(\mathbf{x}_0)$ and $\xi \in T_{\mathbf{x}}W^{s,\varepsilon}(\mathbf{x}_0)$. Then

$$\mathbf{x} = H(\mathbf{u}), \quad \xi = DH(\mathbf{u})\zeta \tag{5}$$

where $\mathbf{u} \in U_\varepsilon$ and $\zeta \in E^s$. From the relation $H \circ g^k = f^k \circ H$, derived by induction from (4), it follows that

$$Df^k(\mathbf{x})\xi = DH(g^k(\mathbf{u}))Dg^k(\mathbf{u})\zeta \text{ for } k \geq 0. \tag{6}$$

We consider the sequence

$$\zeta_k = Dg^k(\mathbf{u})\zeta \text{ for } k \geq 0.$$

By the chain rule,

$$\zeta_{k+1} = Dg(g^k(\mathbf{u}))\zeta_k \text{ for } k \geq 0. \tag{7}$$

Next, noting that

$$Dg(\mathbf{0}) = Df(\mathbf{x}_0) \mid E^s,$$

we rewrite Eq.(7) as

$$\zeta_{k+1} = Df(\mathbf{x}_0)\zeta_k + [Dg(g^k(\mathbf{u})) - Dg(\mathbf{0})]\zeta_k \text{ for } k \geq 0.$$

By repeated application, it follows that

$$\zeta_k = [Df(\mathbf{x}_0)]^k \zeta + \sum_{m=0}^{k-1} [Df(\mathbf{x}_0)]^{k-m-1} [Dg(g^m(\mathbf{u})) - Dg(0)] \zeta_m \text{ for } k \geq 0,$$

where we observe that ζ and $[Dg(g^m(\mathbf{u})) - Dg(\mathbf{0})]\zeta_m$ are in E^s and $g^m(\mathbf{u})$ is in U_ε. Then if we define

$$\omega_g(\varepsilon) = \sup \{\|Dg(\mathbf{u}) - Dg(\mathbf{0})\| : \mathbf{u} \in U_\varepsilon\},$$

it follows from Eq.(1) in Chapter 1 that for $k \geq 0$

$$\|\zeta_k\| \leq K_1 \lambda_1^k \|\zeta\| + \sum_{m=0}^{k-1} K_1 \lambda_1^{k-m-1} \omega_g(\|P\|\varepsilon) \|\zeta_m\|. \tag{8}$$

Putting

$$\mu_k = \lambda_1^{-k} \|\zeta_k\|$$

for $k \geq 0$, Eq.(8) becomes

$$\mu_k \leq K_1 \|\zeta\| + K_1 \lambda_1^{-1} \omega_g(\|P\|\varepsilon) \sum_{m=0}^{k-1} \mu_m,$$

and it follows from Lemma 1.9 (i) that for $k \geq 0$

$$\mu_k \leq K_1 \|\zeta\| [1 + K_1 \lambda_1^{-1} \omega_g(\|P\|\varepsilon)]^k.$$

This means that

$$\|\zeta_k\| \leq K_1 [\lambda_1 + K_1 \omega_g(\|P\|\varepsilon)]^k \|\zeta\| \tag{9}$$

for $k \geq 0$.

Now we assume that ε is so small that for $\mathbf{u} \in U_\varepsilon$ and $\nu \in E^s$

$$\|D\phi^s(\mathbf{u})\nu - \nu\| \leq \frac{1}{2}\|\nu\|. \tag{10}$$

Then, for $DH(\mathbf{u}) : E^s \to T_\mathbf{x} W^{s,\varepsilon}(\mathbf{x}_0)$ with $\mathbf{x} = H(\mathbf{u})$, it follows that for $\mathbf{u} \in U_\varepsilon$

$$\|DH(\mathbf{u})\| \leq \frac{3}{2}, \quad \|DH(\mathbf{u})^{-1}\| \leq 2. \tag{11}$$

So if $\mathbf{x} = H(\mathbf{u}) \in W^{s,\varepsilon}(\mathbf{x}_0)$ and $\xi = DH(\mathbf{u})\zeta \in T_\mathbf{x} W^{s,\varepsilon}(\mathbf{x}_0)$, it follows from Eqs.(5), (6), (9) and (11) that

$$\|Df^k(\mathbf{x})\xi\| \leq 3K_1 [\lambda_1 + K_1 \omega_g(\|P\|\varepsilon)]^k \|\xi\| \text{ for } k \geq 0. \tag{12}$$

Finally we assume that ε is so small that

$$\lambda_1 + K_1 \omega_g(\|P\|\varepsilon) \leq \beta_1. \tag{13}$$

Then if we apply inequality (12) to $f^m(\mathbf{x})$ and $\xi \in T_{f^m(\mathbf{x})}W^{s,\varepsilon}(\mathbf{x}_0)$, we obtain the desired inequality (2) with $K_3 = 3K_1$.

To prove inequality (3), for fixed $\mathbf{x} \in W^{s,\varepsilon}(\mathbf{x}_0)$, with ε as above, and fixed $\xi \in V$ set

$$\mathbf{u}_k = Df^k(\mathbf{x})\xi.$$

Then for $k \geq 0$

$$\mathbf{u}_{k+1} = Df(f^k(\mathbf{x}))\mathbf{u}_k. \tag{14}$$

So $\mathbf{u}_k$ is a solution of the difference equation

$$\mathbf{u}_{k+1} = [A + B_k]\mathbf{u}_k \quad \text{for} \quad k \geq 0, \tag{15}$$

where

$$A = Df(\mathbf{x}_0) \quad \text{and} \quad B_k = Df(f^k(\mathbf{x})) - Df(\mathbf{x}_0).$$

Note that for $k \geq 0$

$$\|B_k\| \leq \omega(\varepsilon) = \sup\left\{\|Df(\mathbf{x}) - Df(\mathbf{x}_0\| : \|\mathbf{x} - \mathbf{x}_0\| \leq \varepsilon\right\}.$$

Now the difference equation

$$\mathbf{u}_{k+1} = A\mathbf{u}_k$$

has an exponential dichotomy on $(-\infty, \infty)$ with constant projection P, exponents λ_1, λ_2 and constants K_1, K_2 (see the beginning of Section 1.3). So, by Lemma 2.8 and its proof, provided ε is sufficiently small depending on K_1, K_2, λ_1, λ_2, $\omega(\cdot)$, β_1 and β_2, there is a constant N depending only on K_1, K_2, λ_1 and λ_2 such that Eq.(15) has an exponential dichotomy on $[0, \infty)$ with exponents β_1, β_2, constants $L_1 = 2K_1(2K_1 + K_2)$, $L_2 = 2K_2(2K_2 + K_1)$ and projections Q_k with $\operatorname{rank} Q_k = \operatorname{rank} P$ satisfying

$$\|Q_k - P\| \leq N\omega(\varepsilon).$$

Now it follows as in the proof of Proposition 2.7 that

$$\mathcal{R}(Q_0) = \{\xi \in I\!R^n : \sup_{k \geq 0} \|Df^k(\mathbf{x})\xi\| < \infty\}.$$

So by inequality (2), $\mathcal{R}(Q_0)$ must contain $T_{\mathbf{x}}W^{s,\varepsilon}(\mathbf{x}_0)$. However, these subspaces have the same dimension and so

$$\mathcal{R}(Q_0) = T_{\mathbf{x}}W^{s,\varepsilon}(\mathbf{x}_0).$$

Also note that in Lemma 2.8 the projection Q_0 was chosen so that

$$\mathcal{N}(Q_0) = \mathcal{N}(P).$$

To complete the proof, we use an argument which is a special case of the general result that for exponential dichotomies on $[0, \infty)$, the nullspace of the projection can be any subspace complementary to the uniquely determined range (conf. Palmer [1988, p.269], Coppel [1978, p.16]). So let R_0 be the projection with range $T_{\mathbf{x}}W^{s,\varepsilon}(\mathbf{x}_0)$ and nullspace V. Since R_0 has the same range as Q_0, we have

$$R_0Q_0 = Q_0, \quad Q_0R_0 = R_0.$$

We define the projection

$$R_k = \Phi(k,0)R_0\Phi(0,k),$$

where $\Phi(k,m) = Df^{k-m}(\mathbf{x})$ is the transition matrix for Eq.(14) or (15). Clearly we have the invariance property

$$\Phi(k,m)R_m = R_k\Phi(k,m) \quad \text{for } k \geq 0, m \geq 0.$$

Moreover if $0 \leq m \leq k$

$$\begin{aligned}
||\Phi(m,k)(I - R_k)|| &= ||\Phi(m,0)\Phi(0,k)(I - R_k)|| \\
&= ||\Phi(m,0)(I - R_0)\Phi(0,k)|| \\
&\leq ||\Phi(m,0)(I - Q_0)\Phi(0,k)|| \\
&\qquad + ||\Phi(m,0)(Q_0 - R_0)(I - Q_0)\Phi(0,k)|| \\
&= ||\Phi(m,k)(I - Q_k)|| \\
&\qquad + ||\Phi(m,0)Q_0(Q_0 - R_0)\Phi(0,k)(I - Q_k)|| \\
&\leq L_2\beta_2^{k-m} + L_1\beta_1^m||Q_0 - R_0||L_2\beta_2^k \\
&\leq [L_2 + L_1||Q_0 - R_0||]\,\beta_2^{k-m}.
\end{aligned}$$

So if $\xi \in V$ and $0 \leq m \leq k$

$$\begin{aligned}
||Df^m(\mathbf{x})\xi|| &= ||\Phi(m,0)\xi|| \\
&= ||\Phi(m,k)\Phi(k,0)(I - R_0)\xi|| \\
&= ||\Phi(m,k)(I - R_k)\Phi(k,0)\xi|| \\
&= ||\Phi(m,k)(I - R_k)Df^k(x)\xi|| \\
&\leq [L_2 + L_1||Q_0 - R_0||]\,\beta_2^{k-m}||Df^k(\mathbf{x})\xi||.
\end{aligned}$$

Hence Eq.(3) holds with

$$K_4 = L_2 + L_1\|Q_0 - R_0\|. \tag{16}$$

The conclusion is that if ε is so small that $W^{s,\varepsilon}(\mathbf{x}_0)$ is a manifold as in Chapter 1, the conditions in Eqs.(13) and (10) are satisfied and the conditions imposed in applying Lemma 2.8 are satisfied, then inequalities (2) and (3) hold with $K_3 = 3K_1$ and K_4 as given in Eq.(16). Thus the proof of Proposition 3.3 is complete.

Note also it is a consequence of Proposition 3.3 that if ε is sufficiently small and $\mathbf{x} \in W^{s,\varepsilon}(\mathbf{x}_0)$, then

$$T_{\mathbf{x}}W^{s,\varepsilon}(\mathbf{x}_0) = \{\xi : Df^k(\mathbf{x})\xi \to \mathbf{0} \quad \text{as} \quad k \to \infty\}.$$

Similarly, if ε is sufficiently small and $\mathbf{x} \in W^{u,\varepsilon}(\mathbf{x}_0)$,

$$T_{\mathbf{x}}W^{u,\varepsilon}(\mathbf{x}_0) = \{\xi : Df^k(\mathbf{x})\xi \to \mathbf{0} \quad \text{as} \quad k \to -\infty\}.$$

Now we complete the proof of Theorem 3.2. As noted at the beginning, we need only verify the exponential estimates along the orbit of $\mathbf{y}_0$. So let $\xi \in T_{\mathbf{y}_0}W^s(\mathbf{x}_0)$ and $\eta \in T_{\mathbf{y}_0}W^u(\mathbf{x}_0)$ and set

$$\xi_k = Df^k(\mathbf{y}_0)\xi, \quad \eta_k = Df^k(\mathbf{y}_0)\eta \quad \text{for } k \in \mathbf{Z}.$$

Let the positive number ε satisfy the conditions imposed in Proposition 3.3 and choose an integer k_1 such that

$$\mathbf{x} = f^{k_1}(\mathbf{y}_0) \in W^{s,\varepsilon}(\mathbf{x}_0).$$

Then it follows from Proposition 3.3 with $V = T_{\mathbf{x}}W^u(\mathbf{x}_0)$ that there exist positive constants K_3, K_4 such that for $k_1 \le m \le k$,

$$\|\xi_k\| \le K_3\beta_1^{k-m}\|\xi_m\|$$

and

$$\|\eta_m\| \le K_4\beta_2^{k-m}\|\eta_k\|.$$

Similarly, by the analogue of Proposition 3.3 for $W^u(\mathbf{x}_0)$, there exist an integer k_2 and positive constants K_5, K_6 such that for $m \le k \le k_2$

$$\|\xi_k\| \le K_5\beta_1^{k-m}\|\xi_m\|$$

and

$$\|\eta_m\| \le K_6\beta_2^{k-m}\|\eta_k\|.$$

Now if $k_2 \geq k_1$ it follows that for $m \leq k$

$$\|\xi_k\| \leq K_7 \beta_1^{k-m} \|\xi_m\| \tag{17}$$

and

$$\|\eta_m\| \leq K_8 \beta_2^{k-m} \|\eta_k\|, \tag{18}$$

where

$$K_7 = K_3 K_5, \quad K_8 = K_4 K_6.$$

Suppose $k_2 < k_1$. We want to establish inequality (17) for suitable K_7. Well, if $m \leq k$, there are four possibilities in addition to the two already considered: $k_2 < m < k_1 \leq k$, $m \leq k_2 < k_1 \leq k$, $k_2 < m \leq k < k_1$ or $m \leq k_2 < k < k_1$. We consider only the second possibility and leave the others to the reader. If $m \leq k_2 < k_1 \leq k$, then

$$\|\xi_k\| \leq K_3 \beta_1^{k-k_1} \|\xi_{k_1}\| \leq K_3 \beta_1^{k-k_1} M_1^{k_1-k_2} \|\xi_{k_2}\|,$$

where

$$M_1 = \max \left\{ \sup_{\mathbf{x} \in S} \|Df(\mathbf{x})\|, \sup_{\mathbf{x} \in S} \|Df^{-1}(\mathbf{x})\| \right\}.$$

So

$$\|\xi_k\| \leq K_3 \beta_1^{k-k_1} M_1^{k_1-k_2} K_5 \beta_1^{k_2-m} \|\xi_m\| = K_3 K_5 (M_1/\beta_1)^{k_1-k_2} \beta_1^{k-m} \|\xi_m\|.$$

Then, assuming without loss of generality that $M_1 \geq \beta_1$, we are able to establish the inequality in Eq.(17) with

$$K_7 = K_3 K_5 (M_1/\beta_1)^{k_1-k_2}.$$

Similarly, we are able to establish the inequality in Eq.(18) with

$$K_8 = K_4 K_6 (M_1/\beta_2)^{k_1-k_2}.$$

Thus the proof of Theorem 3.2 is completed.

3.2 THE CONSTRUCTION OF DIFFEOMORPHISMS WITH TRANSVERSAL HOMOCLINIC POINTS

We consider a family $f_\mu : U \to \mathbb{R}^2$ (U open in $\mathbb{R}^2$) of diffeomorphisms depending on the real parameter μ, where $f(\mathbf{x}, \mu) = f_\mu(\mathbf{x})$ is a C^2 function. We suppose f_{μ_0} has a hyperbolic fixed point $\mathbf{x}_0$ which is a saddle, that is, $Df_{\mu_0}(\mathbf{x}_0)$ has one eigenvalue λ outside the unit circle and the other μ inside the unit circle so that the associated stable and unstable manifolds are one-dimensional. Moreover, we suppose that f_{μ_0} has a C^1 one-parameter family $\zeta(\alpha)$ of homoclinic points, that is,

$$\zeta(\alpha) \in W^s(\mathbf{x}_0) \cap W^u(\mathbf{x}_0)$$

for all α in some interval I and $\zeta'(\alpha) \neq \mathbf{0}$ for all α. More precisely, we assume for any $\varepsilon > 0$, there is a positive integer N such that

$$f_{\mu_0}^N(\zeta(\alpha)) \in W^{s,\varepsilon}(\mathbf{x}_0) \quad \text{and} \quad f_{\mu_0}^{-N}(\zeta(\alpha)) \in W^{u,\varepsilon}(\mathbf{x}_0)$$

for all α in I. This means that

$$T_{\zeta(\alpha)}W^s(\mathbf{x}_0) \cap T_{\zeta(\alpha)}W^u(\mathbf{x}_0) = \text{span } \{\zeta'(\alpha)\}$$

and so the $\zeta(\alpha)$ are definitely not transversal homoclinic points.

Note that by applying the implicit function theorem to the equation

$$f(\mathbf{x}, \mu) - \mathbf{x} = \mathbf{0},$$

we can deduce the existence of a saddle $\mathbf{x}(\mu)$ for f_μ near $\mathbf{x}_0$ when μ is near μ_0. Moreover, $\mathbf{x}(\mu)$ is a C^2 function of μ with $\mathbf{x}(\mu_0) = \mathbf{x}_0$. Then $\mathbf{x}(\mu)$ has one-dimensional stable and unstable manifolds $W^s(\mathbf{x}(\mu))$ and $W^u(\mathbf{x}(\mu))$ and we want to determine conditions under which $W^s(\mathbf{x}(\mu))$ and $W^u(\mathbf{x}(\mu))$ intersect transversally when μ is near μ_0 but not equal to μ_0. The method we describe here is due, in the case of period maps from periodic systems of differential equations, to Melnikov [1963].

When $\mu = \mu_0$, $W^s(\mathbf{x}(\mu))$ and $W^u(\mathbf{x}(\mu))$ coincide in the curve $\zeta(\alpha)$ parametrised by α. It is convenient to parametrise both $W^s(\mathbf{x}(\mu))$ and $W^u(\mathbf{x}(\mu))$ by α, even when $\mu \neq \mu_0$. To this end, recall that in Chapter 1 we found that when $\varepsilon > 0$ is sufficiently small and $|\mu - \mu_0| < \rho$, the local stable manifold $W^{s,\varepsilon}(\mathbf{x}(\mu))$ is contained in the manifold

$$\mathcal{M}^{s,\mu} = \{\mathbf{x}_0 + \phi(\xi, \mu) : \xi \in E^s, \|\xi\| < \sigma\} \subset W^s(\mathbf{x}(\mu)),$$

where E^s is the stable subspace of $Df_{\mu_0}(\mathbf{x}_0)$, σ and ρ are positive constants and $\phi(\xi, \mu)$ is a C^2 function such that

$$P(\phi(\xi, \mu)) = \xi,$$

P being the projection onto E^s along the unstable subspace E^u. Also, for each fixed μ, $\xi \to \mathbf{x}_0 + \phi(\xi, \mu)$ is a C^2 diffeomorphism.

Now there exists $N > 0$ such that if $\alpha \in I$, $f_{\mu_0}^N(\zeta(\alpha))$ is in $\mathcal{M}^{s,\mu_0}$. So

$$f_{\mu_0}^N(\zeta(\alpha)) = \mathbf{x}_0 + \phi(\xi_0(\alpha), \mu_0),$$

where $\xi_0(\alpha)$ is a C^2 function. Then for fixed μ near μ_0, the function

$$\psi(\xi,\mu) = f_\mu^{-N}(\mathbf{x}_0 + \phi(\xi,\mu))$$

gives a parametrization of a piece of $W^s(\mathbf{x}(\mu))$ which lies near the curve $\zeta(\alpha)$.

We reparametrize using (α,μ) instead of (ξ,μ). To this end, we solve the equation

$$g(\xi,\alpha,\mu) = \langle \psi(\xi,\mu) - \zeta(\alpha), \zeta'(\alpha)\rangle = 0 \tag{19}$$

for ξ as a function of (α,μ), where $\langle\cdot,\cdot\rangle$ is the inner product in $\mathbb{R}^2$. That is, given α and μ we choose ξ so that $\psi(\xi,\mu) - \zeta(\alpha)$ is orthogonal to $\zeta'(\alpha)$. We apply the implicit function theorem. Note that if α is near α_0

$$\begin{aligned} g(\xi_0(\alpha),\alpha,\mu_0) &= \langle \psi(\xi_0(\alpha),\mu_0) - \zeta(\alpha), \zeta'(\alpha)\rangle \\ &= \langle f_{\mu_0}^{-N}(f_{\mu_0}^{N}(\zeta(\alpha))) - \zeta(\alpha), \zeta'(\alpha)\rangle \\ &= \langle \zeta(\alpha) - \zeta(\alpha), \zeta'(\alpha)\rangle \\ &= 0 \end{aligned}$$

and

$$\begin{aligned} \frac{\partial g}{\partial \xi}(\xi_0(\alpha),\alpha,\mu_0) &= \langle \frac{\partial \psi}{\partial \xi}(\xi_0(\alpha),\mu_0), \zeta'(\alpha)\rangle \\ &= \langle Df_{\mu_0}^{-N}(f_{\mu_0}^{N}(\zeta(\alpha)))\frac{\partial \phi}{\partial \xi}(\xi_0(\alpha),\mu_0), \zeta'(\alpha)\rangle \\ &\neq 0, \end{aligned}$$

since the fact that

$$\frac{\partial \phi}{\partial \xi}(\xi_0(\alpha),\mu_0) \in T_{f_{\mu_0}^{N}(\zeta(\alpha))} W^s(\mathbf{x}_0)$$

implies that

$$Df_{\mu_0}^{-N}(f_{\mu_0}^{N}(\zeta(\alpha)))\frac{\partial \phi}{\partial \xi}(\xi_0(\alpha),\mu_0) \in T_{\zeta(\alpha)} W^s(\mathbf{x}_0)$$

and so is a nonzero multiple of $\zeta'(\alpha)$. So we can apply the implicit function theorem to solve Eq.(19) for $\xi = \xi(\alpha,\mu)$, where $\xi(\alpha,\mu)$ is a C^2 function with $\xi(\alpha,\mu_0) = \xi_0(\alpha)$. Then when μ is near μ_0 the C^2 function

$$\theta^s(\alpha,\mu) = \psi(\xi(\alpha,\mu),\mu)$$

gives a parametrization of a piece of $W^s(\mathbf{x}(\mu))$ near $\zeta(\alpha)$ such that

$$\theta^s(\alpha,\mu_0) = \zeta(\alpha)$$

and

$$\langle \theta^s(\alpha,\mu) - \zeta(\alpha), \zeta'(\alpha)\rangle = 0.$$

In fact, $\theta^s(\alpha,\mu)$ gives the intersection of $W^s(\mathbf{x}(\mu))$ with the line through $\zeta(\alpha)$ and orthogonal to $\zeta'(\alpha)$.

Now we adapt an argument of Arrowsmith and Place [1990, p.73], originally developed for differential equations, to diffeomorphisms. We define

$$d^s(\alpha,\mu) = \zeta'(\alpha) \wedge [\theta^s(\alpha,\mu) - \zeta(\alpha)],$$

where "$\wedge$" is the wedge product given by

$$\begin{bmatrix} a_1 \\ a_2 \end{bmatrix} \wedge \begin{bmatrix} b_1 \\ b_2 \end{bmatrix} = a_1 b_2 - a_2 b_1.$$

Note that $d^s(\alpha,\mu)$ is a signed measure of the distance between $\theta^s(\alpha,\mu)$ and $\zeta(\alpha)$. Also note that

$$d^s(\alpha,\mu_0) = 0$$

for all α.

Next we calculate $\partial d^s/\partial\mu(\alpha,\mu_0)$, which is given by

$$\frac{\partial d^s}{\partial \mu}(\alpha,\mu_0) = \zeta'(\alpha) \wedge \frac{\partial \theta^s}{\partial \mu}(\alpha,\mu_0).$$

To this end, for fixed α and for $k \geq 0$, set

$$A_k = Df_{\mu_0}(f^k_{\mu_0}(\zeta(\alpha))),\ \phi_k = \frac{d}{d\alpha} f^k_{\mu_0}(\zeta(\alpha)),\ \mathbf{h}_k = \frac{\partial f}{\partial \mu}(f^k_{\mu_0}(\zeta(\alpha)),\mu_0),$$

$$\theta^s_k(\alpha,\mu) = f^k_\mu(\theta^s(\alpha,\mu)),\ \xi_k = \frac{\partial \theta^s_k}{\partial \mu}(\alpha,\mu_0).$$

We claim that ξ_k is bounded for $k \geq 0$. To see this, note that using the notation of the last section of Chapter 1,

$$\theta^s_k(\alpha,\mu) = f^k_\mu(\psi(\xi(\alpha,\mu),\mu)) = f^{k-N}_\mu(\mathbf{x}_0 + \phi(\xi(\alpha,\mu),\mu)) = \mathbf{x}_0 + \mathbf{y}_{k-N}(\xi(\alpha,\mu),\mu)$$

for $k \geq N$. So

$$\xi_k = \frac{\partial \theta^s_k}{\partial \mu}(\alpha,\mu_0) = \frac{\partial \mathbf{y}_{k-N}}{\partial \xi}(\xi(\alpha,\mu_0),\mu_0)\frac{\partial \xi}{\partial \mu}(\alpha,\mu_0) + \frac{\partial \mathbf{y}_{k-N}}{\partial \mu}(\xi(\alpha,\mu_0),\mu_0).$$

This is bounded for $k \geq N$ (and hence for $k \geq 0$) by the smoothness proof in the last section of Chapter 1. Moreover, since

$$\theta^s_{k+1}(\alpha,\mu) = f(\theta^s_k(\alpha,\mu),\mu),$$

it follows that

$$\xi_{k+1} = \frac{\partial f}{\partial \mathbf{x}}(f_{\mu_0}^k(\zeta(\alpha)), \mu_0)\xi_k + \frac{\partial f}{\partial \mu}(f_{\mu_0}^k(\zeta(\alpha)), \mu_0) = A_k \, \xi_k + \mathbf{h}_k.$$

Next note that

$$\phi_{k+1} = \frac{d}{d\alpha} f_{\mu_0}(f_{\mu_0}^k(\zeta(\alpha))) = Df_{\mu_0}(f_{\mu_0}^k(\zeta(\alpha)))\frac{d}{d\alpha} f_{\mu_0}^k(\zeta(\alpha)) = A_k \phi_k.$$

Now consider

$$\Delta_k = \phi_k \wedge \xi_k.$$

We see that

$$\Delta_{k+1} = A_k \phi_k \wedge (A_k \xi_k + \mathbf{h}_k) = \det A_k \cdot \Delta_k + \phi_{k+1} \wedge \mathbf{h}_k.$$

So

$$\Delta_{k+1} = \lambda_k \Delta_k + g_k, \tag{20}$$

where

$$\lambda_k = \det A_k, \quad g_k = \phi_{k+1} \wedge \mathbf{h}_k.$$

By repeated application of Eq.(20), we find that for $k \geq 0$

$$\Delta_k = \lambda_{k-1} \cdots \lambda_0 \, \Delta_0 + \sum_{m=1}^{k} \lambda_{k-1} \cdots \lambda_m \, g_{m-1}.$$

Hence

$$\Delta_0 = -\lambda_0^{-1} \cdots \lambda_{k-1}^{-1} \, \Delta_k - \sum_{m=0}^{k-1} \lambda_0^{-1} \cdots \lambda_m^{-1} \, g_m. \tag{21}$$

Now since $\phi_0 = \zeta'(\alpha) \in T_{\zeta(\alpha)} W^s(\mathbf{x}_0)$, it follows from Proposition 3.3 that $\phi_k = Df_{\mu_0}^k(\zeta(\alpha))\zeta'(\alpha)$ is bounded for $k \geq 0$. In fact,

$$||\phi_k|| \leq \text{constant} \cdot \beta_1^k \quad \text{for } k \geq 0,$$

where β_1 can be chosen as close to the eigenvalue μ of $Df_{\mu_0}(\mathbf{x}_0)$ inside the unit circle as we like. Hence since $||\xi_k||$ is bounded, we have

$$|\Delta_k| = |\phi^k \wedge \xi_k| \leq \text{constant} \cdot \beta_1^k \quad \text{for } k \geq 0.$$

Next note that

$$\lambda_k = \det Df_{\mu_0}(f_{\mu_0}^k(\zeta(\alpha))) \to \det Df_{\mu_0}(\mathbf{x}_0) = \lambda\mu \;\text{ as } k \to \infty,$$

where λ is the eigenvalue of $Df_{\mu_0}(\mathbf{x}_0)$ outside the unit circle. Hence for $k \geq 0$

$$|\lambda_0^{-1} \cdots \lambda_{k-1}^{-1}| \leq \text{constant} \cdot \gamma^k,$$

where γ can be chosen as close to $(\lambda\mu)^{-1}$ as we like. Thus for $k \geq 0$

$$|\lambda_0^{-1} \cdots \lambda_{k-1}^{-1} \Delta_k| \leq \text{ constant} \cdot (\gamma\beta_1)^k,$$

where $\gamma\beta_1$ is as close to λ^{-1} as we like. Also

$$|\lambda_0^{-1} \cdots \lambda_{m-1}^{-1} g_{m-1}| \leq \text{ constant} \cdot (\gamma\beta_1)^m$$

for $m \geq 0$. Now, since $|\lambda| > 1$, we can arrange that $|\gamma\beta_1| < 1$. So we may let $k \to \infty$ in Eq.(21) to obtain

$$\Delta_0 = -\sum_{k=0}^{\infty} \lambda_0^{-1} \ldots \lambda_k^{-1} g_k.$$

That is,

$$\begin{aligned}\Delta_0 &= -\sum_{k=0}^{\infty} \left[\det Df_{\mu_0}^{k+1}(\zeta(\alpha))\right]^{-1} \frac{d}{d\alpha} f_{\mu_0}^{k+1}(\zeta(\alpha)) \wedge \frac{\partial f}{\partial \mu}(f_{\mu_0}^k(\zeta(\alpha)), \mu_0) \\ &= -\sum_{k=0}^{\infty} \det Df_{\mu_0}^{-k-1}(f_{\mu_0}^{k+1}(\zeta(\alpha))) \frac{d}{d\alpha} f_{\mu_0}^{k+1}(\zeta(\alpha)) \wedge \frac{\partial f}{\partial \mu}(f_{\mu_0}^k(\zeta(\alpha)), \mu_0)\end{aligned}$$

and so

$$\frac{\partial d^s}{\partial \mu}(\alpha, \mu_0) = -\sum_{k=1}^{\infty} \det Df_{\mu_0}^{-k}(f_{\mu_0}^k(\zeta(\alpha))) \frac{d}{d\alpha} f_{\mu_0}^k(\zeta(\alpha)) \wedge \frac{\partial f}{\partial \mu}(f_{\mu_0}^{k-1}(\zeta(\alpha)), \mu_0).$$

Similarly, there is a C^1 function $\theta^u(\alpha, \mu)$ such that $\theta^u(\alpha, \mu_0) = \zeta(\alpha)$, which gives the intersection of $W^u(\mathbf{x}(\mu))$ with the line through $\zeta(\alpha)$ perpendicular to $\zeta'(\alpha)$. We find that the signed distance

$$d^u(\alpha, \mu) = \zeta'(\alpha) \wedge [\theta^u(\alpha, \mu) - \zeta(\alpha)]$$

has the properties that

$$d^u(\alpha, \mu_0) = 0$$

and

$$\frac{\partial d^u}{\partial \mu}(\alpha, \mu_0) = \sum_{k=-\infty}^{0} \det Df_{\mu_0}^{-k}(f_{\mu_0}^k(\zeta(\alpha))) \frac{d}{d\alpha} f_{\mu_0}^k(\zeta(\alpha)) \wedge \frac{\partial f}{\partial \mu}(f_{\mu_0}^{k-1}(\zeta(\alpha)), \mu_0).$$

Then the signed distance between the intersections of $W^s(\mathbf{x}(\mu))$ and $W^u(\mathbf{x}(\mu))$ with the line through $\zeta(\alpha)$ perpendicular to $\zeta'(\alpha)$ is

$$d(\alpha, \mu) = d^u(\alpha, \mu) - d^s(\alpha, \mu).$$

Thus $d(\alpha,\mu)$ is a C^2 function with

$$d(\alpha,\mu_0)=0$$

for all α near α_0. If we define

$$h(\alpha,\mu)=\begin{cases}\dfrac{d(\alpha,\mu)}{\mu-\mu_0} & \text{if } \mu\neq\mu_0\\ \dfrac{\partial d}{\partial\mu}(\alpha,\mu_0) & \text{if } \mu=\mu_0,\end{cases}$$

$h(\alpha,\mu)$ is a C^1 function with

$$h(\alpha,\mu_0)=\frac{\partial d}{\partial\mu}(\alpha,\mu_0).$$

Now we define the so-called Melnikov function

$$\begin{aligned}\Delta(\alpha) &= \frac{\partial d}{\partial\mu}(\alpha,\mu_0)\\ &= \sum_{k=-\infty}^{\infty}\det Df_{\mu_0}^{-k}(f_{\mu_0}^{k}(\zeta(\alpha)))\frac{d}{d\alpha}f_{\mu_0}^{k}(\zeta(\alpha))\wedge\frac{\partial f}{\partial\mu}(f_{\mu_0}^{k-1}(\zeta(\alpha)),\mu_0)).\end{aligned}$$

It follows from the implicit function theorem that if

$$\Delta(\alpha_0)=0,\ \ \Delta'(\alpha_0)\neq 0$$

the equation

$$h(\alpha,\mu)=0$$

has a unique solution $\alpha=\alpha(\mu)$ for α near α_0 when μ is near μ_0. Moreover $\alpha(\mu)$ is a C^1 function and $\alpha(\mu_0)=\alpha_0$. Thus when $\mu\neq\mu_0$,

$$\theta^s(\alpha(\mu),\mu)=\theta^u(\alpha(\mu),\mu)\in W^s(\mathbf{x}(\mu))\cap W^u(\mathbf{x}(\mu))$$

and the *intersection is transversal* as the following considerations will show:

$$\begin{aligned}\frac{d}{d\mu}\frac{\partial\theta^s}{\partial\alpha}(\alpha(\mu),\mu)\bigg|_{\mu=\mu_0} &= \frac{\partial^2\theta^s}{\partial\alpha^2}(\alpha_0,\mu_0)\,\alpha'(\mu_0)+\frac{\partial^2\theta^s}{\partial\alpha\partial\mu}(\alpha_0,\mu_0)\\ &= \zeta''(\alpha_0)\alpha'(\mu_0)+\frac{\partial^2\theta^s}{\partial\alpha\partial\mu}(\alpha_0,\mu_0)\end{aligned}$$

and, similarly,

$$\frac{d}{d\mu}\frac{\partial\theta^u}{\partial\alpha}(\alpha(\mu),\mu)\bigg|_{\mu=\mu_0}=\zeta''(\alpha_0)\alpha'(\mu_0)+\frac{\partial^2\theta^u}{\partial\alpha\partial\mu}(\alpha_0,\mu_0).$$

So

$$\frac{\partial\theta^s}{\partial\alpha}(\alpha(\mu),\mu) = \zeta'(\alpha_0) + \left[\zeta''(\alpha_0)\alpha'(\mu_0) + \frac{\partial^2\theta^s}{\partial\alpha\partial\mu}(\alpha_0,\mu_0)\right](\mu-\mu_0) + o(\mu-\mu_0) \tag{22}$$

and

$$\frac{\partial\theta^u}{\partial\alpha}(\alpha(\mu),\mu) = \zeta'(\alpha_0) + \left[\zeta''(\alpha_0)\alpha'(\mu_0) + \frac{\partial^2\theta^u}{\partial\alpha\partial\mu}(\alpha_0,\mu_0)\right](\mu-\mu_0) + o(\mu-\mu_0). \tag{23}$$

Now note that $\Delta'(\alpha_0)$ is given by

$$\zeta'(\alpha_0)\wedge\left[\frac{\partial^2\theta^u}{\partial\alpha\partial\mu}(\alpha_0,\mu_0) - \frac{\partial^2\theta^s}{\partial\alpha\partial\mu}(\alpha_0,\mu_0)\right] + \zeta''(\alpha_0)\wedge\left[\frac{\partial\theta^u}{\partial\mu}(\alpha_0,\mu_0) - \frac{\partial\theta^s}{\partial\mu}(\alpha_0,\mu_0)\right]$$

where to see that

$$\frac{\partial\theta^u}{\partial\mu}(\alpha_0,\mu_0) = \frac{\partial\theta^s}{\partial\mu}(\alpha_0,\mu_0) \tag{24}$$

note first that

$$\zeta'(\alpha_0)\wedge\left[\frac{\partial\theta^u}{\partial\mu}(\alpha_0,\mu_0) - \frac{\partial\theta^s}{\partial\mu}(\alpha_0,\mu_0)\right] = \Delta(\alpha_0) = 0.$$

However, we also know that for all α and μ,

$$\langle\theta^s(\alpha,\mu) - \zeta(\alpha), \zeta'(\alpha)\rangle = \langle\theta^u(\alpha,\mu) - \zeta(\alpha), \zeta'(\alpha)\rangle = 0.$$

Hence

$$\langle\frac{\partial\theta^u}{\partial\mu}(\alpha_0,\mu_0) - \frac{\partial\theta^s}{\partial\mu}(\alpha_0,\mu_0), \zeta'(\alpha_0)\rangle = 0.$$

Thus (24) is proved and so

$$\zeta'(\alpha_0)\wedge\left[\frac{\partial^2\theta^u}{\partial\alpha\partial\mu}(\alpha_0,\mu_0) - \frac{\partial^2\theta^s}{\partial\alpha\partial\mu}(\alpha_0,\mu_0)\right] = \Delta'(\alpha_0) \neq 0.$$

Together with Eqs.(22) and (23), the last relation implies that

$$\frac{\langle\frac{\partial\theta^s}{\partial\alpha}(\alpha(\mu),\mu), \zeta'(\alpha_0)\rangle}{\langle\frac{\partial\theta^u}{\partial\alpha}(\alpha(\mu),\mu), \zeta'(\alpha_0)\rangle} \to 1 \text{ as } \mu \to \mu_0$$

and

$$\frac{\zeta'(\alpha_0)\wedge\frac{\partial\theta^s}{\partial\alpha}(\alpha(\mu),\mu)}{\zeta'(\alpha_0)\wedge\frac{\partial\theta^u}{\partial\alpha}(\alpha(\mu),\mu)} \to \frac{\zeta'(\alpha_0)\wedge\left[\zeta''(\alpha_0)\alpha'(\mu_0) + \frac{\partial^2\theta^s}{\partial\alpha\partial\mu}(\alpha_0,\mu_0)\right]}{\zeta'(\alpha_0)\wedge\left[\zeta''(\alpha_0)\alpha'(\mu_0) + \frac{\partial^2\theta^u}{\partial\alpha\partial\mu}(\alpha_0,\mu_0)\right]} \neq 1$$

as $\mu \to \mu_0$. Hence $\partial\theta^s/\partial\alpha(\alpha(\mu),\mu)$ and $\partial\theta^u/\partial\alpha(\alpha(\mu),\mu)$ cannot be linearly dependent for $\mu \neq \mu_0$, μ sufficiently near μ_0. Thus, $W^s(\mathbf{x}(\mu))$ and $W^u(\mathbf{x}(\mu))$ intersect transversally at $\theta^s(\alpha(\mu),\mu) = \theta^u(\alpha(\mu),\mu)$. So we have proved the following theorem.

Theorem 3.4. *Let U be an open set in $\mathbb{R}^2$ and $f : U \times \mathbb{R} \to \mathbb{R}^2$ a C^2 function such that for fixed μ, the function $f_\mu(\mathbf{x}) = f(\mathbf{x},\mu)$ is a diffeomorphism of U onto its image. Suppose that f_{μ_0} has a saddle point $\mathbf{x}_0$ such that*

$$\zeta(\alpha) \in W^s(\mathbf{x}_0) \cap W^u(\mathbf{x}_0),$$

where $\zeta : I \subset \mathbb{R} \to \mathbb{R}^2$ is a C^1 function with $\zeta'(\alpha) \neq 0$ for all α in the interval I. More precisely, suppose for all $\varepsilon > 0$ there is a positive integer N such that

$$f_{\mu_0}^N(\zeta(\alpha)) \in W^{s,\varepsilon}(\mathbf{x}_0) \quad \text{and} \quad f_{\mu_0}^{-N}(\zeta(\alpha)) \in W^{u,\varepsilon}(\mathbf{x}_0)$$

for all α in I. Define the Melnikov function

$$\Delta(\alpha) = \sum_{k=-\infty}^{\infty} \det Df_{\mu_0}^{-k}(\zeta_k(\alpha))\, \zeta_k'(\alpha) \wedge \frac{\partial f}{\partial \mu}(\zeta_{k-1}(\alpha), \mu_0),$$

where

$$\zeta_k(\alpha) = f_{\mu_0}^k(\zeta(\alpha))$$

and "$\wedge$" denotes the wedge product of two vectors in $\mathbb{R}^2$. Then if μ is near μ_0, the diffeomorphism f_μ has a unique saddle point $\mathbf{x}(\mu)$ near $\mathbf{x}_0$, and if $\Delta(\alpha)$ has a simple zero at α_0, the stable and unstable manifolds of $\mathbf{x}(\mu)$ intersect transversally at a point near $\zeta(\alpha_0)$ when $\mu \neq \mu_0$.

Note that Glasser, Papageorgiou and Bountis [1989] calculate the Melnikov function for a certain family of two dimensional mappings. We look at a example coming from differential equations in the plane.

3.3 AN EXAMPLE FROM DIFFERENTIAL EQUATIONS

Let $g : \mathbb{R}^2 \to \mathbb{R}^2$ be a C^2 function such that the autonomous system

$$\dot{\mathbf{x}} = g(\mathbf{x})$$

has a *saddle* point $\mathbf{x}_0$ with associated homoclinic orbit $\zeta(t) \neq \mathbf{x}_0$, that is, $g(\mathbf{x}_0) = \mathbf{0}$ so that $\mathbf{x}_0$ is a constant solution and the eigenvalues of $Dg(\mathbf{x}_0)$ are real, one being positive and the other negative, and $\zeta(t)$ is a solution distinct from $\mathbf{x}_0$ satisfying $||\zeta(t) - \mathbf{x}_0|| \to 0$ as $t \to \pm\infty$.

Next let $h : \mathbb{R} \times \mathbb{R}^2 \times \mathbb{R} \to \mathbb{R}^2$ be a C^2 function such that $h(t, \mathbf{x}, \mu)$ has period $T > 0$ in t. Let $\phi(t, \xi, \mu)$ denote the solution of the equation

$$\dot{\mathbf{x}} = g(\mathbf{x}) + \mu h(t, \mathbf{x}, \mu) \tag{25}$$

satisfying the initial condition $\mathbf{x}(0) = \xi$. Then

$$f(\mathbf{x}, \mu) = \phi(T, \mathbf{x}, \mu)$$

is a C^2 function such that $f_\mu(\mathbf{x}) = f(\mathbf{x}, \mu)$ is a C^2 diffeomorphism on some open subset of $\mathbb{R}^2$ for each fixed μ (conf. Example 1.2). Note that

$$f(\mathbf{x}_0, 0) = \phi(T, \mathbf{x}_0, 0) = \mathbf{x}_0$$

and

$$\frac{\partial f}{\partial \mathbf{x}}(\mathbf{x}_0, 0) = \frac{\partial \phi}{\partial \mathbf{x}}(T, \mathbf{x}_0, 0) = e^{TDg(\mathbf{x}_0)}.$$

Hence $\mathbf{x}_0$ is a hyperbolic fixed point of f_0, in fact a saddle. Notice also that

$$\zeta(kT + \alpha) = f_0^k(\zeta(\alpha)).$$

Hence for all real α, $\zeta(\alpha) \in W^s(\mathbf{x}_0) \cap W^u(\mathbf{x}_0)$. In fact, for α in any compact interval, $f_0^k(\zeta(\alpha)) = \zeta(kT + \alpha) \to \mathbf{x}_0$ as $|k| \to \infty$ uniformly in α. Moreover

$$\zeta'(\alpha) = g(\zeta(\alpha)) \neq \mathbf{0}$$

for all α (since by uniqueness if $\zeta'(t) = g(\zeta(t)) = \mathbf{0}$ for some $t = t_0$, then for all t we have $\zeta(t) = \zeta(t_0)$ and so $\zeta(t) = \mathbf{x}_0$).

Now we calculate the Melnikov function

$$\Delta(\alpha) = \sum_{k=-\infty}^{\infty} \det Df_0^{-k}(\zeta_k(\alpha))\, \zeta_k'(\alpha) \wedge \frac{\partial f}{\partial \mu}(\zeta_{k-1}(\alpha), 0). \tag{26}$$

First note that

$$\zeta_k(\alpha) = f_0^k(\zeta(\alpha)) = \zeta(kT + \alpha)$$

and so

$$\zeta_k'(\alpha) = \zeta'(kT + \alpha) = g(\zeta(kT + \alpha)). \tag{27}$$

Next $X(t) = \dfrac{\partial \phi}{\partial \mathbf{x}}(t, \xi, 0)$ is the solution of the initial value problem

$$\dot{X} = Dg(\phi(t, \xi, 0))X, \quad X(0) = I.$$

Note that for all integers k

$$f_0^k(\xi) = \phi(kT, \xi, 0)$$

and so

$$Df_0^k(\xi) = \frac{\partial \phi}{\partial \mathbf{x}}(kT, \xi, 0).$$

Hence

$$Df_0^{-k}(\zeta_k(\alpha)) = \frac{\partial \phi}{\partial \mathbf{x}}(-kT, \zeta(kT+\alpha), 0).$$

By the Jacobi formula (conf. Coppel [1965, p.44]),

$$\det \frac{\partial \phi}{\partial \mathbf{x}}(t, \xi, 0) = e^{\int_0^t \operatorname{Tr} Dg(\phi(s,\xi,0))ds}.$$

So

$$\begin{aligned}
\det Df_0^{-k}(\zeta_k(\alpha)) &= e^{-\int_{-kT}^{0} \operatorname{Tr} Dg(\phi(s,\zeta(kT+\alpha),0))ds} \\
&= e^{-\int_{-kT}^{0} \operatorname{Tr} Dg(\zeta(s+kT+\alpha))ds} \qquad (28) \\
&= e^{-\int_0^{kT} \operatorname{Tr} Dg(\zeta(t+\alpha))dt}.
\end{aligned}$$

Next $\mathbf{y}(t) = \frac{\partial \phi}{\partial \mu}(t, \xi, 0)$ is the solution of the initial value problem

$$\dot{\mathbf{y}} = Dg(\phi(t,\xi,0))\mathbf{y} + h(t, \phi(t,\xi,0), 0), \quad \mathbf{y}(0) = \mathbf{0}.$$

So, by variation of constants,

$$\begin{aligned}
\frac{\partial f}{\partial \mu}(\zeta_{k-1}(\alpha), 0) &= \frac{\partial \phi}{\partial \mu}(T, \zeta_{k-1}(\alpha), 0) \\
&= \int_0^T \tilde{Y}(T)\tilde{Y}^{-1}(t)h(t, \phi(t, \zeta((k-1)T+\alpha), 0), 0)dt,
\end{aligned}$$

where $\tilde{Y}(t)$ is a fundamental matrix for

$$\dot{\mathbf{y}} = Dg(\phi(t, \zeta((k-1)T+\alpha), 0))\mathbf{y} = Dg(\zeta(t+(k-1)T+\alpha))\mathbf{y}.$$

However if $Y(t)$ is a fundamental matrix for

$$\dot{\mathbf{y}} = Dg(\zeta(t))\mathbf{y},$$

then we may take

$$\tilde{Y}(t) = Y(t+(k-1)T+\alpha)$$

and so

$$\frac{\partial f}{\partial \mu}(\zeta_{k-1}(\alpha), 0) = \int_0^T Y(kT+\alpha)Y^{-1}(t+(k-1)T+\alpha)h(t, \zeta(t+(k-1)T+\alpha), 0)dt. \qquad (29)$$

Next, using Eqs.(27) and (29), we have

$$\zeta_k'(\alpha)\wedge\frac{\partial f}{\partial\mu}(\zeta_{k-1}(\alpha),0)$$

$$= g(\zeta(kT+\alpha))\wedge\int_{(k-1)T}^{kT} Y(kT+\alpha)Y^{-1}(t+\alpha)h(t,\zeta(t+\alpha),0)dt$$

$$= \int_{(k-1)T}^{kT} Y(kT+\alpha)Y^{-1}(t+\alpha)g(\zeta(t+\alpha))\wedge$$

$$Y(kT+\alpha)Y^{-1}(t+\alpha)h(t,\zeta(t+\alpha),0)dt$$

$$= \int_{(k-1)T}^{kT} \det Y(kT+\alpha)Y^{-1}(t+\alpha)[g(\zeta(t+\alpha))\wedge h(t,\zeta(t+\alpha),0)]dt$$

$$= \int_{(k-1)T}^{kT} e^{\int_t^{kT}\operatorname{Tr} Dg(\zeta(s+\alpha))ds}[g(\zeta(t+\alpha))\wedge h(t,\zeta(t+\alpha),0)]dt.$$

Then, putting this together with Eqs.(26) and (28), we get

$$\Delta(\alpha)=\int_{-\infty}^{\infty} e^{-\int_0^t \operatorname{Tr} Dg(\zeta(s+\alpha))ds}[g(\zeta(t+\alpha))\wedge h(t,\zeta(t+\alpha),0)]\,dt.$$

So, applying Theorem 3.5 to $\zeta(\alpha)$ with α in an appropriate interval, we get the following theorem.

Theorem 3.7. *Let $g:\mathbb{R}^2\to\mathbb{R}^2$ and $h:\mathbb{R}\times\mathbb{R}^2\times\mathbb{R}\to\mathbb{R}^2$ be C^2 functions such that for a positive number T*

$$h(t+T,\mathbf{x},\mu)\equiv h(t,\mathbf{x},\mu)$$

and the autonomous system

$$\dot{\mathbf{x}}=g(\mathbf{x})$$

has a saddle point $\mathbf{x}_0$ with associated homoclinic orbit $\zeta(t)$.

Then for μ sufficiently small the period map of the system

$$\dot{\mathbf{x}}=g(\mathbf{x})+\mu h(t,\mathbf{x},\mu) \tag{30}$$

has a unique hyperbolic fixed point $\mathbf{x}(\mu)$ near $\mathbf{x}_0$. Moreover, if we set

$$\Delta(\alpha)=\int_{-\infty}^{\infty} e^{-\int_0^t \operatorname{Tr} Dg(\zeta(s+\alpha))ds}[g(\zeta(t+\alpha))\wedge h(t,\zeta(t+\alpha),0)]\,dt$$

and there exists α_0 *such that*

$$\Delta(\alpha_0) = 0, \quad \Delta'(\alpha_0) \neq 0,$$

then for μ *sufficiently small but nonzero the period map for the periodically perturbed system* (30) *has a transversal homoclinic point* $\mathbf{y}(\mu)$ *near* $\zeta(\alpha_0)$ *associated with the hyperbolic fixed point* $\mathbf{x}(\mu)$.

To obtain an explicit example of Theorem 3.7, consider the second order equation

$$\ddot{x} + 2x^3 - x = \mu \cos t.$$

As a system this has the form

$$\begin{aligned} \dot{x}_1 &= x_2 \\ \dot{x}_2 &= x_1 - 2x_1^3 + \mu \cos t. \end{aligned}$$

(0,0) is a saddle point for the autonomous system

$$\begin{aligned} \dot{x}_1 &= x_2 \\ \dot{x}_2 &= x_1 - 2x_1^3 \end{aligned}$$

and $(\xi(t), \dot{\xi}(t))$, where $\xi(t) = \operatorname{sech} t$, is an associated homoclinic orbit. The Melnikov function is

$$\Delta(\alpha) = \int_{-\infty}^{\infty} \dot{\xi}(t+\alpha) \cos t \, dt = \int_{-\infty}^{\infty} \xi(t+\alpha) \sin t \, dt = \int_{-\infty}^{\infty} \operatorname{sech} t \, \sin(t-\alpha) dt.$$

We see that $\Delta(0) = 0$ and $\Delta'(0) = -\int_{-\infty}^{\infty} \operatorname{sech} t \cos t \, dt \neq 0$. Hence the conclusions of Theorem 3.7 apply with $\alpha_0 = 0$.

Theorem 3.7 can also be proved using a functional analytic approach, conf. Chow, Hale and Mallet-Paret [1980] and Palmer [1984]. Actually this theorem is a special case of the general bifurcation theorem in Crandall and Rabinowitz [1971]. In this general setting, it can be seen that Theorem 3.7 is closely related to similar theorems about periodic solutions like Theorem 5 in Loud [1957] (conf. also Coppel [1963], Hale and Táboas [1978] and Albizatti [1983]).

4. THE SHADOWING THEOREM FOR HYPERBOLIC SETS OF DIFFEOMORPHISMS

In this chapter we prove the shadowing theorem for diffeomorphisms. Also we give two applications of it to hyperbolic sets. A further application follows in the next chapter.

4.1 THE SHADOWING THEOREM

Let U be a convex open set in $\mathbb{R}^n$ and let $f : U \to \mathbb{R}^n$ be a C^1 diffeomorphism onto its image. First we recall a definition from Chapter 2.

Definition 4.1. A sequence $\{\mathbf{y}_k\}_{k=-\infty}^{\infty}$ of points in U is said to be a δ *pseudo orbit* of f if

$$\|\mathbf{y}_{k+1} - f(\mathbf{y}_k)\| \leq \delta \text{ for } k \in \mathbb{Z}.$$

A δ pseudo orbit can be regarded as an approximate orbit of f. The question we ask is: is there a true orbit of f near the δ pseudo orbit in the sense of the following definition?

Definition 4.2. An orbit $\{\mathbf{x}_k\}_{k=-\infty}^{\infty}$ of f, that is, $\mathbf{x}_{k+1} = f(\mathbf{x}_k)$ for all k, is said to ε *shadow* the δ pseudo orbit $\{\mathbf{y}_k\}_{k=-\infty}^{\infty}$ if

$$\|\mathbf{x}_k - \mathbf{y}_k\| \leq \varepsilon \text{ for } k \in \mathbb{Z}.$$

Our aim in this section is to prove a theorem which gives sufficient conditions under which δ pseudo orbits of a diffeomorphism are shadowed by true orbits of the diffeomorphism or a nearby diffeomorphism.

Theorem 4.3. *Let S be a compact hyperbolic set for a C^1 diffeomorphism $f : U \to \mathbb{R}^n$. Then there exist positive constants δ_0, σ_0 and M depending only on f and S such that if $g : U \to \mathbb{R}^n$ is a C^1 diffeomorphism satisfying*

$$\|f(\mathbf{x}) - g(\mathbf{x})\| + \|Df(\mathbf{x}) - Dg(\mathbf{x})\| \leq \sigma \text{ for } \mathbf{x} \in U \tag{1}$$

with $\sigma \leq \sigma_0$, any δ pseudo orbit of f in S with $\delta \leq \delta_0$ is ε shadowed by a unique true orbit of g with $\varepsilon = M(\delta + \sigma)$.

Proof. Let $\{\mathbf{y}_k\}_{k=-\infty}^{\infty}$ be a δ pseudo orbit of f in the compact hyperbolic set S. We seek a solution $\{\mathbf{x}_k\}_{k=-\infty}^{\infty}$ of the difference equation

$$\mathbf{x}_{k+1} = g(\mathbf{x}_k) \text{ for } k \in \mathbb{Z}$$

such that

$$\|\mathbf{x}_k - \mathbf{y}_k\| \leq \varepsilon \text{ for all } k.$$

Denote by X the Banach space $l^\infty(\mathbb{Z}, \mathbb{R}^n)$ of bounded $\mathbb{R}^n$-valued sequences $\mathbf{x} = \{\mathbf{x}_k\}_{k=-\infty}^{\infty}$ with norm

$$\|\mathbf{x}\| = \sup_{k \in \mathbb{Z}} \|\mathbf{x}_k\|.$$

Let O be the open subset of X consisting of all those sequences $\mathbf{x} = \{\mathbf{x}_k\}_{k=-\infty}^{\infty}$ such that

$$\|\mathbf{x} - \mathbf{y}\| < \operatorname{dist}(S, \partial U),$$

where $\mathbf{y} = \{\mathbf{y}_k\}_{k=-\infty}^{\infty}$ and $\operatorname{dist}(S, \partial U)$ is taken as ∞ if $U = \mathbb{R}^n$. We define the C^1 mapping $\mathcal{G} : O \to X$ as follows: if $\mathbf{x} = \{\mathbf{x}_k\}_{k=-\infty}^{\infty} \in O$ then

$$[\mathcal{G}(\mathbf{x})]_k = \mathbf{x}_{k+1} - g(\mathbf{x}_k) \text{ for } k \in \mathbb{Z}. \tag{2}$$

The theorem will be proved if we can show there exists a unique solution $\mathbf{x}$ of the equation

$$\mathcal{G}(\mathbf{x}) = 0$$

such that $\|\mathbf{x} - \mathbf{y}\| \leq \varepsilon$. In order to do this, we use the following lemma, which can be regarded as a form of Newton's method.

Lemma 4.4. *Let X, Y be Banach spaces, let O be an open subset of X and let $\mathcal{G} : O \to Y$ be a C^1 function. Suppose $\mathbf{y}$ is an element of O for which*

$$\|\mathcal{G}(\mathbf{y})\| \leq \Delta,$$

where Δ is a positive number, and for which the derivative $L = D\mathcal{G}(\mathbf{y})$ is invertible with

$$\|L^{-1}\| \leq M/2,$$

for some positive constant M. Then if the closed ball of radius $M\Delta$ and centre $\mathbf{y}$ lies in O and if the inequality

$$\|D\mathcal{G}(\mathbf{x}) - D\mathcal{G}(\mathbf{y})\| \leq 1/M$$

holds for $\|\mathbf{x} - \mathbf{y}\| \leq M\Delta$, there is a unique solution of the equation

$$\mathcal{G}(\mathbf{x}) = 0$$

satisfying

$$\|\mathbf{x} - \mathbf{y}\| \leq M\Delta.$$

Proof. Define the operator $F : O \to X$ by

$$F(\mathbf{x}) = \mathbf{y} - L^{-1}[\mathcal{G}(\mathbf{x}) - D\mathcal{G}(\mathbf{y})(\mathbf{x} - \mathbf{y})].$$

Clearly $\mathcal{G}(\mathbf{x}) = 0$ if and only if $F(\mathbf{x}) = \mathbf{x}$. Moreover, if $\|\mathbf{x} - \mathbf{y}\| \leq \varepsilon = M\Delta$,

$$\begin{aligned}\|F(\mathbf{x}) - \mathbf{y}\| &\leq \|L^{-1}\|\|\mathcal{G}(\mathbf{x}) - \mathcal{G}(\mathbf{y}) - D\mathcal{G}(\mathbf{y})(\mathbf{x} - \mathbf{y}) + \mathcal{G}(\mathbf{y})\| \\ &\leq \frac{M}{2}\left(M^{-1}\varepsilon + \Delta\right) \\ &= \frac{1}{2}\varepsilon + \frac{1}{2}\varepsilon \\ &= \varepsilon.\end{aligned}$$

Moreover, if $\|\mathbf{x} - \mathbf{y}\| \leq \varepsilon$ and $\|\mathbf{z} - \mathbf{y}\| \leq \varepsilon$, then

$$\begin{aligned}\|F(\mathbf{x}) - F(\mathbf{z})\| &= \|L^{-1}[\mathcal{G}(\mathbf{x}) - \mathcal{G}(\mathbf{z}) - D\mathcal{G}(\mathbf{y})(\mathbf{x} - \mathbf{z})]\| \\ &\leq \frac{M}{2} M^{-1}\|\mathbf{x} - \mathbf{z}\| \\ &= \frac{1}{2}\|\mathbf{x} - \mathbf{z}\|.\end{aligned}$$

So F is a contraction on the closed ball of radius ε, centre $\mathbf{y}$, and thus the lemma follows from the contraction mapping principle.

We apply Lemma 4.4 to the C^1 map $\mathcal{G} : O \to X$ defined in Eq.(2) and to $\mathbf{y} = \{\mathbf{y}_k\}_{k=-\infty}^{\infty}$, a δ pseudo orbit of f in S. It is easy to see that $D\mathcal{G}(\mathbf{y})$ is the linear operator $L : l^{\infty}(\mathbb{Z}, \mathbb{R}^n) \to l^{\infty}(\mathbb{Z}, \mathbb{R}^n)$ defined as follows: if $\mathbf{u} = \{\mathbf{u}_k\}_{k=-\infty}^{\infty}$, then

$$(L\mathbf{u})_k = \mathbf{u}_{k+1} - Dg(\mathbf{y}_k)\mathbf{u}_k \quad \text{for } k \in \mathbb{Z}.$$

To apply the lemma, we must show that when δ and σ are sufficiently small, the operator L is invertible and we must also find a bound for $\|L^{-1}\|$.

Invertibility of L and estimation of $\|L^{-1}\|$: Let S satisfy the conditions of Definition 2.4. Denote by $\mathcal{P}(\mathbf{x})$ the projection of $\mathbb{R}^n$ onto $E^s(\mathbf{x})$ along $E^u(\mathbf{x})$. It follows from the invariance of $E^s(\mathbf{x})$ and $E^u(\mathbf{x})$ that the identity

$$Df(\mathbf{x})\mathcal{P}(\mathbf{x}) = \mathcal{P}(f(\mathbf{x}))Df(\mathbf{x}) \quad \text{for } \mathbf{x} \in S \tag{3}$$

holds. Also, since $\mathcal{P}(\mathbf{x})$ is bounded, there are constants M^s, M^u such that

$$\|\mathcal{P}(\mathbf{x})\| \le M^s, \quad \|I - \mathcal{P}(\mathbf{x})\| \le M^u \quad \text{for } \mathbf{x} \in S.$$

Now we give a sufficient condition that an operator like L is invertible (which also turns out to be necessary, conf. Henry [1981, p.230], Slyusarchuk [1983]).

Lemma 4.5. *Let $\{A_k\}_{k=-\infty}^{\infty}$ be a bounded sequence of $n \times n$ invertible matrices and let $T : \ell^\infty(\mathbf{Z}, \mathbb{R}^n) \to \ell^\infty(\mathbf{Z}, \mathbb{R}^n)$ be the operator defined by*

$$(T\mathbf{u})_k = \mathbf{u}_{k+1} - A_k \mathbf{u}_k \quad \text{for} \quad k \in \mathbf{Z}.$$

Then if the difference equation

$$\mathbf{u}_{k+1} = A_k \mathbf{u}_k \tag{4}$$

has an exponential dichotomy on $(-\infty, \infty)$ with exponents λ_1, λ_2 and constants K_1, K_2 the operator T is invertible and

$$\|T^{-1}\| \le K_1(1-\lambda_1)^{-1} + K_2\lambda_2(1-\lambda_2)^{-1}.$$

Proof. Let $\theta \in X = \ell^\infty(\mathbf{Z}, \mathbb{R}^n)$. If T^{-1} exists, $\mathbf{u} = T^{-1}\theta$ is the unique solution of the equation

$$T\mathbf{u} = \theta.$$

That is, if $\mathbf{u} = \{\mathbf{u}_k\}_{k=-\infty}^{\infty}$ and $\theta = \{\theta_k\}_{k=-\infty}^{\infty}$, then $\mathbf{u}_k$ is the unique bounded solution of the difference equation

$$\mathbf{u}_{k+1} = A_k \mathbf{u}_k + \theta_k, \quad k \in \mathbf{Z}. \tag{5}$$

Suppose $\mathbf{u}_k$ is a bounded solution of Eq.(5). Then we derive an expression for it as follows. First observe that we may write for $k \ge a$

$$\mathbf{u}_k = \Phi(k,a)\mathbf{u}_a + \sum_{m=a}^{k-1} \Phi(k, m+1)\theta_m,$$

where $\Phi(k,m)$ is the transition matrix for Eq.(4). Multiplying by the projection P_k associated with the dichotomy and using the invariance of the P_k, we obtain the equation

$$P_k\mathbf{u}_k = \Phi(k,a)P_a\mathbf{u}_a + \sum_{m=a}^{k-1} \Phi(k, m+1)P_{m+1}\theta_m. \tag{6}$$

Now note that

$$\|\Phi(k,a)P_a\mathbf{u}_a\| \le K_1\lambda_1^{k-a}\|\mathbf{u}_a\| \quad \text{for} \quad k \ge a$$

and

$$\sum_{m=-\infty}^{k-1} \|\Phi(k,m+1)P_{m+1}\theta_m\| \leq \sum_{m=-\infty}^{k-1} K_1\lambda_1^{k-m-1}\|\theta\| = K_1(1-\lambda_1)^{-1}\|\theta\|.$$

So, since $\mathbf{u}_k$ is bounded, we may let $a \to -\infty$ in Eq.(6) to obtain

$$P_k\mathbf{u}_k = \sum_{m=-\infty}^{k-1} \Phi(k,m+1)P_{m+1}\theta_m. \tag{7}$$

Next for $k \leq b$,

$$\mathbf{u}_k = \Phi(k,b)\mathbf{u}_b - \sum_{m=k}^{b-1} \Phi(k,m+1)\theta_m.$$

Multiplying by $I - P_k$ and using the invariance of the P_k, we obtain

$$(I-P_k)\mathbf{u}_k = \Phi(k,b)(I-P_b)\mathbf{u}_b - \sum_{m=k}^{b-1} \Phi(k,m+1)(I-P_{m+1})\theta_m. \tag{8}$$

Noting that

$$\|\Phi(k,b)(I-P_b)\mathbf{u}_b\| \leq K_2\lambda_2^{b-k}\|\mathbf{u}_b\| \quad \text{for} \quad k \leq b$$

and

$$\sum_{m=k}^{\infty} \|\Phi(k,m+1)(I-P_{m+1})\theta_m\| \leq \sum_{m=k}^{\infty} K_2\lambda_2^{m+1-k}\|\theta\| \leq K_2\lambda_2(1-\lambda_2)^{-1}\|\theta\|,$$

we may let $b \to \infty$ in Eq.(8) to obtain

$$(I-P_k)\mathbf{u}_k = -\sum_{m=k}^{\infty} \Phi(k,m+1)(I-P_{m+1})\theta_m. \tag{9}$$

Adding Eqs.(7) and (9), we obtain the result that if $\mathbf{u}_k$ is a bounded solution of Eq.(5), then it is unique and given by

$$\mathbf{u}_k = \sum_{m=-\infty}^{k-1} \Phi(k,m+1)P_{m+1}\theta_m - \sum_{m=k}^{\infty} \Phi(k,m+1)(I-P_{m+1})\theta_m. \tag{10}$$

Conversely, if $\mathbf{u}_k$ is defined for $k \in \mathbf{Z}$ by Eq.(10), then it is easily checked by direct substitution that $\mathbf{u}_k$ is a solution of Eq.(5) and, estimating as above, we find that $\mathbf{u}_k$ is bounded with

$$\|\mathbf{u}_k\| \leq [K_1(1-\lambda_1)^{-1} + K_2\lambda_2(1-\lambda_2)^{-1}]\|\theta\|.$$

Hence T is invertible with

$$\|T^{-1}\| \leq K_1(1-\lambda_1)^{-1} + K_2\lambda_2(1-\lambda_2)^{-1},$$

and the proof of the lemma is complete.

So to prove the invertibility of L, all we need show is that the difference equation

$$\mathbf{u}_{k+1} = Dg(\mathbf{y}_k)\mathbf{u}_k, \tag{11}$$

has an exponential dichotomy on $(-\infty, \infty)$. Note that for all k

$$\|Dg(\mathbf{y}_k) - Df(\mathbf{y}_k)\| \leq \sigma. \tag{12}$$

Now it follows from Lemma 2.16, provided δ is sufficiently small depending on f and S, that the difference equation

$$\mathbf{u}_{k+1} = Df(\mathbf{y}_k)\mathbf{u}_k$$

has an exponential dichotomy on $(-\infty, \infty)$ with exponents $(1+\lambda_1)/2$, $(1+\lambda_2)/2$ and constants depending only on f and S. Then, using Eq.(12), it follows from Lemma 2.8 that, provided σ is sufficiently small depending on f and S, the difference equation (11) has an exponential dichotomy on $(-\infty, \infty)$ with exponents $\alpha_1 = (3+\lambda_1)/4$, $\alpha_2 = (3+\lambda_2)/4$ and constants L_1, L_2 depending only on f and S. So, by Lemma 4.5, L is invertible and

$$\|L^{-1}\| \leq L_1(1-\alpha_1)^{-1} + L_2\alpha_2(1-\alpha_2)^{-1}.$$

Completion of the Proof of the Theorem: Let $\{\mathbf{y}_k\}_{k=-\infty}^{\infty}$ be a δ pseudo orbit of f in S and let $g: U \to \mathbb{R}^n$ be a C^1 diffeomorphism satisfying the condition in Eq.(1). Suppose δ and σ satisfy the conditions ensuring the invertibility of L. Consider the C^1 map $\mathcal{G}: O \to X$ defined in Eq.(2). Note first from the definition of δ pseudo orbit that the inequality

$$\|\mathcal{G}(\mathbf{y})\| \leq \delta + \sigma$$

holds for $\mathbf{y} = \{\mathbf{y}_k\}_{k=-\infty}^{\infty}$. We have shown that $L = D\mathcal{G}(\mathbf{y})$ is invertible and that $\|L^{-1}\| \leq M/2$, where

$$M = 2L_1(1-\alpha_1)^{-1} + 2L_2\alpha_2(1-\alpha_2)^{-1}. \tag{13}$$

Next if $\|\mathbf{x} - \mathbf{y}\| \leq M(\delta + \sigma)$, with $\mathbf{x} = \{\mathbf{x}_k\}_{k=-\infty}^{\infty}$, we have the estimate

$$\|D\mathcal{G}(\mathbf{x}) - D\mathcal{G}(\mathbf{y})\| \leq \sup_{k\in\mathbb{Z}} \|Df(\mathbf{x}_k) - Df(\mathbf{y}_k)\| + 2\sigma \leq \omega(M(\delta+\sigma)) + 2\sigma,$$

where

$$\omega(\varepsilon) = \sup\{\|Df(\mathbf{x}) - Df(\mathbf{y})\| : \mathbf{y} \in S, \|\mathbf{x} - \mathbf{y}\| \leq \varepsilon\}. \tag{14}$$

Hence if

$$M(\delta + \sigma) < \text{dist}(S, \partial U), \quad M[\omega(M(\delta + \sigma)) + 2\sigma] \leq 1,$$

Lemma 4.4 can be applied and Theorem 4.3 follows, with M given in Eq.(13) and δ_0, σ_0 positive numbers satisfying

$$M(\delta_0 + \sigma_0) < \text{dist}(S, \partial U), \quad M[\omega(M(\delta_0 + \sigma_0)) + 2\sigma_0] \leq 1$$

and the conditions on δ and σ ensuring the invertibility of L.

4.2 MORE ON THE ROUGHNESS OF HYPERBOLIC SETS

Let S be a compact hyperbolic set for the C^1 diffeomorphism $f : U \to I\!R^n$. It was shown in Section 2.6 that if O is a sufficiently "tight" neighborhood of S and $g : U \to I\!R^n$ is a C^1 diffeomorphism sufficiently close in the C^1 topology to f, then the maximal invariant set S_O of g in $\overline{O}$ is hyperbolic. In this section we want to use shadowing to prove that if S is an *isolated* invariant set for f, then S_O is also isolated and the homeomorphisms $f : S \to S$ and $g : S_O \to S_O$ are *topologically conjugate* in the sense of the following definition.

Definition 4.6. Let X, Y be metric spaces and let $f : X \to X$, $g : Y \to Y$ be mappings. Then f and g are said to be *topologically conjugate* if there is a homeomorphism $h : X \to Y$ such that

$$h \circ f = g \circ h.$$

Our aim is to prove the following theorem.

Theorem 4.7. *Let U be a convex open set in $I\!R^n$ and let $f : U \to I\!R^n$ be a C^1 diffeomorphism with an isolated compact hyperbolic set S. Then if O is a sufficiently tight open neighborhood of S and $g : U \to I\!R^n$ is a C^1 diffeomorphism satisfying*

$$\|f(\mathbf{x}) - g(\mathbf{x})\| + \|Df(\mathbf{x}) - Dg(\mathbf{x})\| \leq \sigma, \quad \mathbf{x} \in U \tag{15}$$

for σ sufficiently small, the maximal invariant set S_O of g in $\overline{O}$ is isolated hyperbolic. Moreover, the homeomorphisms $f : S \to S$, $g : S_O \to S_O$ are topologically conjugate with conjugacy $h : S \to S_O$ satisfying

$$\|h(\mathbf{x}) - \mathbf{x}\| \leq M\sigma,$$

where M is the constant in Theorem 4.3.

Proof. By Theorem 2.14 there are numbers $d_0 > 0$ and $\overline{\sigma}_0 > 0$ such that if $\text{dist}(\mathbf{x}, S) \leq d_0$ for all $\mathbf{x}$ in $\overline{O}$ and σ in Eq.(15) does not exceed $\overline{\sigma}_0$, then the maximal invariant set S_O of g in $\overline{O}$ is hyperbolic and the constants K_1, K_2, exponents λ_1, λ_2 and bounds M^s, M^u on the projections can be chosen independently of O and g. We assume these conditions on O and σ hold in the rest of the proof.

Suppose δ_0, M, σ_0 satisfy the conditions in Theorem 4.3 for f and S. Let $\mathbf{x} \in S$. Then $\{\mathbf{x}_k\}_{k=-\infty}^{\infty} = \{f^k(\mathbf{x})\}_{k=-\infty}^{\infty}$ is a δ pseudo orbit of f in S with $\delta = 0$. So by Theorem 4.3, if $\sigma \leq \sigma_0$, there is a true orbit $\{\mathbf{z}_k\}_{k=-\infty}^{\infty}$ of g such that

$$\|\mathbf{z}_k - \mathbf{x}_k\| \leq M\sigma \quad \text{for } k \in \mathbf{Z}. \tag{16}$$

Moreover, this is the unique true orbit of g such that

$$\|\mathbf{z}_k - \mathbf{x}_k\| \leq M\sigma_0 \quad \text{for} \quad k \in \mathbf{Z}.$$

Then we define

$$h(\mathbf{x}) = \mathbf{z}_0.$$

Provided σ is so small that

$$\{\mathbf{x} : \text{dist}(\mathbf{x}, S) \leq M\sigma\} \subset O,$$

it follows that h is a mapping of S into the maximal invariant set S_O of g in $\overline{O}$. Also it follows from Eq.(16) that

$$\|h(\mathbf{x}) - \mathbf{x}\| \leq M\sigma$$

for $\mathbf{x} \in S$. Moreover, since $\{\mathbf{x}_{k+1}\}_{k=-\infty}^{\infty}$ is the orbit of f corresponding to $f(\mathbf{x}) = \mathbf{x}_1$ and

$$\|\mathbf{z}_{k+1} - \mathbf{x}_{k+1}\| \leq M\sigma \quad \text{for} \quad k \in \mathbf{Z},$$

it follows, by uniqueness, that

$$h(f(\mathbf{x})) = h(\mathbf{x}_1) = \mathbf{z}_1 = g(\mathbf{z}_0) = g(h(\mathbf{x})).$$

That is,

$$h \circ f = g \circ h.$$

So all that remains to be shown is that h is a homeomorphism of S onto S_O and that S_O is isolated.

It is clear that provided $2M\sigma$ does not exceed a constant of expansivity of f on S that h is one to one.

To prove that h is surjective, let $\mathbf{z} \in S_O$ and set $\mathbf{z}_k = g^k(\mathbf{z})$ for $k \in \mathbf{Z}$. Choose $\mathbf{y}_k$ in S such that $\|\mathbf{y}_k - \mathbf{z}_k\| \leq d_0$. Then

$$\begin{aligned}\|\mathbf{y}_{k+1} - f(\mathbf{y}_k)\| &\leq \|\mathbf{y}_{k+1} - \mathbf{z}_{k+1}\| + \|g(\mathbf{z}_k) - f(\mathbf{z}_k)\| + \|f(\mathbf{z}_k) - f(\mathbf{y}_k)\| \\ &\leq d_0 + \sigma + M_1 d_0,\end{aligned}$$

where

$$M_1 = \sup_{\mathbf{x}\in U} \|Df(\mathbf{x})\|. \tag{17}$$

So, by Theorem 4.3, if

$$(1 + M_1)d_0 + \sigma \leq \delta_0,$$

there is a unique true orbit $\{\mathbf{x}_k\}_{k=-\infty}^{\infty}$ of f such that

$$\|\mathbf{x}_k - \mathbf{y}_k\| \leq M[(1 + M_1)d_0 + \sigma].$$

Then if d_0 and σ are so small that S is the maximal invariant set for f in

$$\{\mathbf{x} \in \mathbb{R}^n : \mathrm{dist}(\mathbf{x}, S) \leq M[(1 + M_1)d_0 + \sigma]\},$$

it follows that $\mathbf{x}_k$ is in S for all k. Moreover, we observe that

$$\begin{aligned}\|\mathbf{z}_k - \mathbf{x}_k\| &\leq \|\mathbf{z}_k - \mathbf{y}_k\| + \|\mathbf{y}_k - \mathbf{x}_k\| \\ &\leq d_0 + M[(1 + M_1)d_0 + \sigma] \\ &= [1 + M(1 + M_1)]d_0 + M\sigma.\end{aligned}$$

So if d_0 and σ are so small that

$$[1 + M(1 + M_1)]d_0 + M\sigma \leq M\sigma_0,$$

it follows from the uniqueness in the first part of the proof that

$$\mathbf{z} = h(\mathbf{x}_0).$$

Thus h is indeed surjective. Also, since h maps S into O, it follows that $S_O \subset O$ and hence is isolated.

Now we prove h is continuous. Let $\mathbf{x}, \overline{\mathbf{x}}$ be in S and set $\mathbf{x}_k = f^k(\mathbf{x})$, $\overline{\mathbf{x}}_k = f^k(\overline{\mathbf{x}})$ for $k \in \mathbf{Z}$. The corresponding orbits $\{\mathbf{z}_k\}_{k=-\infty}^{\infty}$, $\{\overline{\mathbf{z}}_k\}_{k=-\infty}^{\infty}$ of g satisfy

$$\|\mathbf{z}_k - \mathbf{x}_k\| \le M\sigma, \quad \|\overline{\mathbf{z}}_k - \overline{\mathbf{x}}_k\| \le M\sigma$$

for $k \in \boldsymbol{Z}$. Here $\mathbf{z}_0 = h(\mathbf{x})$, $\overline{\mathbf{z}}_0 = h(\overline{\mathbf{x}})$ and $\{\mathbf{z}_k\}_{k=-\infty}^{\infty}$, $\{\overline{\mathbf{z}}_k\}_{k=-\infty}^{\infty}$ are orbits of g in the hyperbolic set S_O. Take

$$\beta_1 = \frac{1+\lambda_1}{2}, \quad \beta_2 = \frac{1+\lambda_2}{2}. \tag{18}$$

By Proposition 2.10 applied to g and S_O, if σ and Δ are sufficiently small depending only on K_1, K_2, λ_1, λ_2, M^s, M^u and $\omega(\cdot)$ (as defined in Eq.(14)), there are constants L_1, L_2 depending only on K_1, K_2, M^s, M^u, λ_1, λ_2 such that if

$$\|\mathbf{z}_k - \overline{\mathbf{z}}_k\| \le \Delta \quad \text{for } |k| \le N,$$

then

$$\|\mathbf{z}_k - \overline{\mathbf{z}}_k\| \le [L_1\beta_1^{k+N} + L_2\beta_2^{N-k}]\Delta \quad \text{for } |k| \le N.$$

(Note that in the proof of Proposition 2.10 applied to g, $\omega(d)$ is replaced by $\omega(\Delta) + 2\sigma$, since

$$\|Dg(\mathbf{x}) - Dg(\mathbf{y})\| \le \|Dg(\mathbf{x}) - Df(\mathbf{x})\| + \|Df(\mathbf{x}) - Df(\mathbf{y})\| + \|Df(\mathbf{y}) - Dg(\mathbf{y})\|.)$$

Given $\varepsilon > 0$, we choose $N > 0$ so that

$$[L_1\beta_1^N + L_2\beta_2^N]\Delta < \varepsilon. \tag{19}$$

Now note that for $|k| \le N$

$$\begin{aligned}\|\mathbf{z}_k - \overline{\mathbf{z}}_k\| &\le \|\mathbf{z}_k - \mathbf{x}_k\| + \|\mathbf{x}_k - \overline{\mathbf{x}}_k\| + \|\overline{\mathbf{x}}_k - \overline{\mathbf{z}}_k\| \\ &\le 2M\sigma + \|\mathbf{x}_k - \overline{\mathbf{x}}_k\| \\ &\le \Delta,\end{aligned}$$

provided

$$4M\sigma \le \Delta$$

and $\|\mathbf{x} - \overline{\mathbf{x}}\|$ is so small that

$$\|\mathbf{x}_k - \overline{\mathbf{x}}_k\| \le \Delta/2 \quad \text{for } |k| \le N.$$

Then it follows from Proposition 2.10 as described above that

$$\|h(\mathbf{x}) - h(\overline{\mathbf{x}})\| = \|\mathbf{z}_0 - \overline{\mathbf{z}}_0\| \le [L_1\beta_1^N + L_2\beta_2^N]\Delta < \varepsilon.$$

So h is continuous.

Finally we prove h^{-1} is continuous. (Actually, this would follow from the continuity of h and the compactness of S. The proof here has the advantage that it could possibly be used for extensions to the non-compact case.) Let $\mathbf{z}, \overline{\mathbf{z}}$ be in S_O and set $\mathbf{z}_k = g^k(\mathbf{z})$, $\overline{\mathbf{z}}_k = g^k(\overline{\mathbf{z}})$ for $k \in \boldsymbol{Z}$. The corresponding orbits $\{\mathbf{x}_k\}_{k=-\infty}^{\infty}$ and $\{\overline{\mathbf{x}}_k\}_{k=-\infty}^{\infty}$ of f in S satisfy

$$\|\mathbf{z}_k - \mathbf{x}_k\| \leq M\sigma, \quad \|\overline{\mathbf{z}}_k - \overline{\mathbf{x}}_k\| \leq M\sigma$$

for $k \in \boldsymbol{Z}$. Here $\mathbf{z} = h(\mathbf{x}_0)$, $\overline{\mathbf{z}} = h(\overline{\mathbf{x}}_0)$. Choose β_1, β_2 as in Eq.(18). Then by Proposition 2.10 applied to f and S, there is a positive number Δ depending only on K_1, K_2, λ_1, λ_2, M^s, M^u and $\omega(\cdot)$ and constants L_1, L_2 depending only on K_1, K_2, λ_1, λ_2, M^s, M^u such that if

$$\|\mathbf{x}_k - \overline{\mathbf{x}}_k\| \leq \Delta \quad \text{for } |k| \leq N,$$

then

$$\|\mathbf{x}_k - \overline{\mathbf{x}}_k\| \leq [L_1\beta_1^{k+N} + L_2\beta_2^{N-k}]\Delta \quad \text{for } |k| \leq N.$$

Given $\varepsilon > 0$, we choose N as in Eq.(19). Note that for $|k| \leq N$

$$\begin{aligned}\|\mathbf{x}_k - \overline{\mathbf{x}}_k\| &\leq \|\mathbf{x}_k - \mathbf{z}_k\| + \|\mathbf{z}_k - \overline{\mathbf{z}}_k\| + \|\overline{\mathbf{z}}_k - \overline{\mathbf{x}}_k\| \\ &\leq 2M\sigma + \|\mathbf{z}_k - \overline{\mathbf{z}}_k\| \\ &\leq \Delta,\end{aligned}$$

provided

$$4M\sigma \leq \Delta$$

and $\|\mathbf{z} - \overline{\mathbf{z}}\|$ is so small that

$$\|\mathbf{z}_k - \overline{\mathbf{z}}_k\| \leq \Delta/2 \quad \text{for } |k| \leq N.$$

Then it follows that

$$\|h^{-1}(\mathbf{z}) - h^{-1}(\overline{\mathbf{z}})\| = \|\mathbf{x}_0 - \overline{\mathbf{x}}_0\| \leq [L_1\beta_1^N + L_2\beta_2^N]\Delta < \varepsilon.$$

So we have proved h and h^{-1} are continuous and the proof of Theorem 4.7 is complete.

Let us remark here that Walters [1978] was the first to use shadowing to prove topological conjugacy results. A similar result to Theorem 4.7 is in Guckenheimer, Moser and Newhouse [1980] and Akin [1993]. Katok [1971], Lanford

[1985] and Pilyugin [1992] prove a similar result for hyperbolic sets which are not isolated. We could also prove such a result by following through the above proof omitting the proof of surjectivity.

4.3 ASYMPTOTIC PHASE FOR HYPERBOLIC SETS

Let $f : U \to \mathbb{R}^n$ be a C^1 diffeomorphism and let S be a compact hyperbolic set for f. For the set S we can define its *stable manifold*

$$W^s(S) = \{\mathbf{x} \in U : \text{dist}(f^k(\mathbf{x}), S) \to 0 \text{ as } k \to \infty\}$$

and its *unstable manifold*

$$W^u(S) = \{\mathbf{x} \in U : \text{dist}(f^k(\mathbf{x}), S) \to 0 \text{ as } k \to -\infty\}.$$

We can also define the stable and unstable manifolds for points in S as follows. If $\mathbf{x} \in S$ its stable manifold is

$$W^s(\mathbf{x}) = \{\mathbf{y} \in U : \|f^k(\mathbf{y}) - f^k(\mathbf{x})\| \to 0 \text{ as } k \to \infty\}$$

and its unstable manifold is

$$W^u(\mathbf{x}) = \{\mathbf{y} \in U : \|f^k(\mathbf{y}) - f^k(\mathbf{x})\| \to 0 \text{ as } k \to -\infty\}.$$

It is obvious that

$$\bigcup_{\mathbf{x}\in S} W^s(\mathbf{x}) \subset W^s(S), \quad \bigcup_{\mathbf{x}\in S} W^u(\mathbf{x}) \subset W^u(S).$$

When S is an isolated invariant set, we will use shadowing to prove that the reverse inclusions hold. This means that an orbit which is asymptotic to the set S must be asymptotic to some orbit in S (conf. Hirsch, Palis, Pugh and Shub [1970], Katok [1971], Fenichel [1996]).

Theorem 4.8. *Let S be an isolated compact hyperbolic set for the C^1 diffeomorphism $f : U \to \mathbb{R}^n$. Then*

$$W^s(S) = \bigcup_{\mathbf{x}\in S} W^s(\mathbf{x}), \quad W^u(S) = \bigcup_{\mathbf{x}\in S} W^u(\mathbf{x}).$$

Proof. We just prove the first equation since the proof of the second is analogous. All we need show is that $W^s(S) \subset \bigcup_{\mathbf{x}\in S} W^s(\mathbf{x})$ since the other inclusion is obvious. So let $\mathbf{z} \in W^s(S)$ and set

$$\mathbf{z}_k = f^k(\mathbf{z}).$$

Then $\text{dist}(\mathbf{z}_k, S) \to 0$ as $k \to \infty$.

Now let

$$\beta_1 = (1+\lambda_1)/2, \ \beta_2 = (1+\lambda_2)/2,$$

where λ_1 and λ_2 are the exponents for S, and let Δ be a constant of expansivity for S (more precisely, Δ has the properties that d has in Proposition 2.10 with β_1 and β_2 as just chosen) and let Δ also have the property that S is the maximal invariant set for f in $\{\mathbf{x} \in U : \text{dist}(\mathbf{x}, S) \leq \Delta/2\}$. Choose a positive number δ such that

$$\delta \leq \delta_0 \ \text{ and } \ 2M\delta \leq \Delta,$$

where δ_0, M are the constants in Theorem 4.3. Next choose a positive number d so that

$$(1+M_1)d \leq \delta, \ \ 2d \leq \Delta,$$

where M_1 is defined in Eq.(17) (as usual, we assume for simplicity that U is convex).

Now there exists k_0 such that for $k \geq k_0$

$$\text{dist}(\mathbf{z}_k, S) \leq d$$

and so for $k \geq k_0$ there exists $\mathbf{y}_k \in S$ such that

$$\|\mathbf{z}_k - \mathbf{y}_k\| \leq d.$$

Note that if $k \geq k_0$

$$\begin{aligned} \|\mathbf{y}_{k+1} - f(\mathbf{y}_k)\| &\leq \|\mathbf{y}_{k+1} - \mathbf{z}_{k+1}\| + \|f(\mathbf{z}_k) - f(\mathbf{y}_k)\| \\ &\leq d + M_1 d \\ &\leq \delta. \end{aligned}$$

So, if we define $\mathbf{y}_k = f^{k-k_0}(\mathbf{y}_{k_0})$ for $k < k_0$, we see that $\{\mathbf{y}_k\}_{k=-\infty}^{\infty}$ is a δ pseudo orbit of f in S. Hence, by Theorem 4.3, there is a unique true orbit $\{\mathbf{x}_k\}_{k=-\infty}^{\infty}$ of f such that for all k

$$\|\mathbf{x}_k - \mathbf{y}_k\| \leq M\delta.$$

Since $M\delta \leq \Delta/2$, it follows that $\mathbf{x}_k \in S$ for all k. Also if $k \geq k_0$,

$$\|f^k(\mathbf{z}) - f^k(\mathbf{x}_0)\| \leq \|\mathbf{z}_k - \mathbf{y}_k\| + \|\mathbf{y}_k - \mathbf{x}_k\| \leq d + M\delta \leq \Delta.$$

Then it follows from Proposition 2.10 (see Remark 2.11) that

$$\|f^k(\mathbf{z}) - f^k(\mathbf{x}_0)\| \to 0 \quad \text{as} \quad k \to \infty.$$

So $\mathbf{z} \in W^s(\mathbf{x}_0)$ and the proof of the theorem is complete.

5. SYMBOLIC DYNAMICS NEAR A TRANSVERSAL HOMOCLINIC POINT OF A DIFFEOMORPHISM

In Chapter 3 we showed that the set consisting of a hyperbolic fixed point of a diffeomorphism and an associated transversal homoclinic orbit is hyperbolic. In this chapter we use *symbolic dynamics* to describe all the orbits in a neighbourhood of this set. In particular, we shall show that the dynamics in such a neighbourhood is chaotic. Our main tool in establishing the symbolic dynamics is the shadowing theorem (Theorem 4.3).

So let U be an open subset of $\mathbb{R}^n$ and let $f : U \to \mathbb{R}^n$ be a C^1 diffeomorphism onto its image. Suppose that $\mathbf{x}_0$ is a hyperbolic fixed point of f and that $\{\mathbf{y}_k\}_{k=-\infty}^{\infty} = \{f^k(\mathbf{y}_0)\}_{-\infty}^{\infty}$ is an associated transversal homoclinic orbit. Then we proved in Chapter 3 that the set

$$S = \{\mathbf{x}_0\} \cup \{\mathbf{y}_k : k \in \mathbf{Z}\}$$

is a compact hyperbolic set for f. In the following theorem, we characterise the maximal invariant sets in certain open neighbourhoods of S and also give a symbolic dynamical description of the dynamics of f on these invariant sets.

Theorem 5.1. *Let $\mathbf{x}_0$ be a hyperbolic fixed point of the C^1 diffeomorphism $f : U \to \mathbb{R}^n$ with associated transversal homoclinic orbit $\{\mathbf{y}_k = f^k(\mathbf{y}_0)\}_{k=-\infty}^{\infty}$. Then there is a positive integer J such that for any sufficiently large positive integer L, there is an open neighbourhood O of the set*

$$S = \{\mathbf{x}_0\} \cup \{\mathbf{y}_k : k \in \mathbf{Z}\}$$

such that the orbits of f which remain in O for all (discrete) time are in one to one correspondence with the set Y of bi-infinite sequences $\{a_k\}_{k=-\infty}^{\infty}$ with $a_k \in \{0, 1, \dots, J\}$ described by the following properties:

(i) if $a_k = 0 \neq a_{k+1}$ then $a_{k+1} = 1$;

(ii) if $a_k = j \in \{1, \dots, J-1\}$, then $a_{k+1} = j+1$;

(iii) if $a_k = J$, then $a_{k+\ell} = 0$ for $1 \leq \ell \leq L$.

Furthermore, if Y is endowed with the product discrete topology, there is a mapping $\phi : Y \to \mathbb{R}^n$ which is a homeomorphism onto its image such that

$$\phi \circ \sigma = f \circ \phi,$$

where $\sigma : Y \to Y$ is the shift defined by

$$\sigma(\{a_k\}_{k=--\infty}^{\infty}) = \{a_{k+1}\}_{k=-\infty}^{\infty}.$$

Proof. The proof given here follows that given by Steinlein and Walther [1990] (conf. also Silnikov [1967]). Let M and δ_0 be the constants associated with the hyperbolic set S as in Theorem 4.3 and set

$$\varepsilon_0 = M\delta_0.$$

We adjust δ_0 if necessary so that the maximal invariant set of f in the closed ε_0 neighbourhood of S is hyperbolic (conf. Theorem 2.14), and we let d be the constant of expansivity of this maximal invariant set associated with a fixed choice for β_1 and β_2 (see Proposition 2.10). Then set

$$\delta_1 = \min\{\delta_0, \varepsilon_0/4, d/4\}$$

and choose a positive integer k^+ and a negative integer k^- such that

$$\|\mathbf{y}_k - \mathbf{x}_0\| < \delta_1$$

if $k \geq k^+$ or $k \leq k^-$. Next set

$$J = k^+ - k^- - 1$$

and define

$$I_0 = \{\mathbf{x}_0\} \cup \{\mathbf{y}_k : k \leq k^- \text{ or } k \geq k^+\}.$$

Choose $\varepsilon > 0$ with

$$\varepsilon \leq \varepsilon_0/2 \quad \text{and} \quad \varepsilon \leq d/6$$

such that the closures of the open sets

$$V_0 = B(I_0, \varepsilon) = \{\mathbf{x} : \|\mathbf{x} - \mathbf{z}\| < \varepsilon \quad \text{for} \quad \text{some} \quad \mathbf{z} \in I_0\},$$

$$V_j = B(\mathbf{y}_{k^-+j}, \varepsilon) \ (j = 1, 2, \ldots, J)$$

are mutually disjoint and such that $f(V_0) \cap V_i$ is empty for $2 \leq i \leq J$ and $f(V_j) \cap V_i$ is empty for $1 \leq j \leq J-1$, $i \neq j+1$ and $j = J$, $i \neq 0$. Now set

$$\delta = \varepsilon/2M$$

so that $\varepsilon/2 = M\delta$ and choose a positive integer k^* such that for all $k \geq k^*$

$$\|\mathbf{y}_{k^- - k} - \mathbf{x}_0\| \leq \delta, \ \|\mathbf{y}_{k^+ + k} - \mathbf{x}_0\| \leq \delta.$$

Then take

$$L \geq 2k^* + 1$$

and replace V_J by its intersection with $\bigcap_{k=1}^{L} f^{-k}(V_0)$. Finally we define the open neighbourhood O of S by

$$O = \bigcup_{j=0}^{J} V_j.$$

Consider an orbit $\{\mathbf{w}_k\}_{k=-\infty}^{\infty}$ of f which lies entirely in O. From the properties of the sets V_j we see that

$$\mathbf{w}_k \in V_0,\ \mathbf{w}_{k+1} \notin V_0 \Rightarrow \mathbf{w}_{k+1} \in V_1;$$

$$\mathbf{w}_k \in V_j \quad \text{and} \quad 1 \le j \le J-1 \Rightarrow \mathbf{w}_{k+1} \in V_{j+1};$$

$$\mathbf{w}_k \in V_J \Rightarrow \mathbf{w}_{k+\ell} \in V_0 \quad \text{for} \quad 1 \le \ell \le L.$$

We define

$$a_k = j$$

if $\mathbf{w}_k \in V_j$. It is clear we obtain a bi-infinite sequence $\{a_k\}_{k=-\infty}^{\infty}$ of the type described in the statement of the theorem and we write

$$\{a_k\}_{k=-\infty}^{\infty} = \alpha(\{\mathbf{w}_k\}_{k=-\infty}^{\infty}).$$

We show that the map α just defined of the orbits in O into Y is one to one. So let $\alpha(\{\mathbf{w}_k\}_{k=-\infty}^{\infty}) = \{a_k\}_{k=-\infty}^{\infty}$. We define a δ_0 pseudo orbit $\{\mathbf{z}_k\}_{k=-\infty}^{\infty}$ of f in S by taking

$$\mathbf{z}_k = \begin{cases} \mathbf{y}_{k^-+j} & \text{if } a_k = j,\ 1 \le j \le J, \\ \mathbf{x}_0 & \text{if } a_k = a_{k+1} = 0, \\ \mathbf{y}_{k^-} & \text{if } a_k = 0,\ a_{k+1} = 1. \end{cases}$$

We see that $f(\mathbf{z}_k) = \mathbf{z}_{k+1}$ if $a_k = j$, $1 \le j \le J-1$ or if $a_k = 0$, $a_{k+1} = 1$ or if $a_k = a_{k+1} = a_{k+2} = 0$. If $a_k = J$, then

$$\|\mathbf{z}_{k+1} - f(\mathbf{z}_k)\| = \|\mathbf{x}_0 - \mathbf{y}_{k^+}\| < \delta_1 \le \delta_0$$

and if $a_k = a_{k+1} = 0$ but $a_{k+2} = 1$ then

$$\|\mathbf{z}_{k+1} - f(\mathbf{z}_k)\| = \|\mathbf{y}_{k^-} - \mathbf{x}_0\| < \delta_1 \le \delta_0.$$

So $\{\mathbf{z}_k\}_{k=-\infty}^{\infty}$ is a δ_0 pseudo orbit. Next we note that $\mathbf{w}_k \in V_j \subset B(\mathbf{y}_{k^-+j}, \varepsilon)$ and $\mathbf{z}_k = \mathbf{y}_{k^-+j}$ if $a_k = j$ and $1 \le j \le J$ so that $\|\mathbf{w}_k - \mathbf{z}_k\| \le \varepsilon < \varepsilon_0$. If $a_k = 0$, $\mathbf{w}_k \in V_0$ so that there exists ℓ with $\ell \le k^-$ or $\ell \ge k^+$ such that $\|\mathbf{w}_k - \mathbf{y}_\ell\| < \varepsilon$. Then, since $\mathbf{z}_k = \mathbf{x}_0$ or $\mathbf{y}_{k^-}$,

$$\|\mathbf{w}_k - \mathbf{z}_k\| \le \|\mathbf{w}_k - \mathbf{y}_\ell\| + \|\mathbf{y}_\ell - \mathbf{x}_0\| + \|\mathbf{z}_k - \mathbf{x}_0\| < \varepsilon + 2\delta_1 \le \varepsilon_0.$$

So $\{\mathbf{w}_k\}_{k=-\infty}^{\infty}$ ε_0-shadows this δ_0 pseudo orbit and hence, by Theorem 4.3 and the choice of ε_0 and δ_0, is unique. Thus α is, indeed, one to one.

Next we show that α is onto. Let a sequence $\{a_k\}_{k=-\infty}^{\infty}$ in the set Y be given. We define a δ pseudo orbit $\{\mathbf{z}_k\}_{k=-\infty}^{\infty}$ of f in S with $\mathbf{z}_k \in V_{a_k}$ as follows. If $a_k = j$, $1 \le j \le J$, we take

$$\mathbf{z}_k = \mathbf{y}_{k^-+j};$$

corresponding to a segment of (at least L) zeros between a J and a 1, we take the same number of points

$$\mathbf{y}_{k^+}, \ldots, \mathbf{y}_{k^+ + k^* - 1}, \mathbf{x}_0, \ldots, \mathbf{x}_0, \mathbf{y}_{k^- - k^*}, \mathbf{y}_{k^- - k^* + 1}, \ldots, \mathbf{y}_{k^-};$$

corresponding to infinitely many zeros before a 1, we take the infinitely many points

$$\ldots, \mathbf{x}_0, \ldots, \mathbf{x}_0, \mathbf{y}_{k^- - k^*}, \mathbf{y}_{k^- - k^* + 1}, \ldots, \mathbf{y}_{k^-};$$

corresponding to infinitely many zeros after a J, we take the infinitely many points

$$\mathbf{y}_{k^+}, \ldots, \mathbf{y}_{k^+ + k^* - 1}, \mathbf{x}_0, \ldots, \mathbf{x}_0, \ldots;$$

if $a_k = 0$ for all k, we take $\mathbf{z}_k = \mathbf{x}_0$ for all k. The only k for which $\mathbf{z}_{k+1} \neq f(\mathbf{z}_k)$ are the ones for which $\mathbf{z}_k = \mathbf{y}_{k^+ + k^* - 1}$, $\mathbf{z}_{k+1} = \mathbf{x}_0$ when

$$\|\mathbf{z}_{k+1} - f(\mathbf{z}_k)\| = \|\mathbf{x}_0 - \mathbf{y}_{k^+ + k^*}\| \leq \delta$$

and those for which $\mathbf{z}_k = \mathbf{x}_0$, $\mathbf{z}_{k+1} = \mathbf{y}_{k^- - k^*}$ when

$$\|\mathbf{z}_{k+1} - f(\mathbf{z}_k)\| = \|\mathbf{y}_{k^- - k^*} - \mathbf{x}_0\| \leq \delta.$$

So $\{\mathbf{z}_k\}_{k=-\infty}^{\infty}$ is indeed a δ pseudo orbit with $\mathbf{z}_k \in V_{a_k}$. By Theorem 4.3 and the choice of ε and δ, there is a unique orbit $\{\mathbf{w}_k\}_{k=-\infty}^{\infty}$ of f which $\varepsilon/2$-shadows this δ pseudo orbit. If $a_k = j$ with $j = 1, \ldots, J-1$, then $\mathbf{z}_k = \mathbf{y}_{k^- + j}$ and so $\|\mathbf{w}_k - \mathbf{y}_{k^- + j}\| \leq \varepsilon/2$, implying that $\mathbf{w}_k \in V_j$. If $a_k = 0$, $\mathbf{z}_k \in I_0$ and since $\|\mathbf{w}_k - \mathbf{z}_k\| \leq \varepsilon/2 < \varepsilon$, it follows that $\mathbf{w}_k \in V_0$. If $a_k = J$, $\mathbf{z}_k = \mathbf{y}_{k^- + J}$ and so $\mathbf{w}_k \in B(\mathbf{y}_{k^- + J}, \varepsilon)$. Also, since $a_{k+\ell} = 0$ for $\ell = 1, \ldots, L$, it follows that $\mathbf{w}_{k+\ell} = f^\ell(\mathbf{w}_k) \in V_0$ for $\ell = 1, \ldots, L$. Thus, if $a_k = J$, $\mathbf{w}_k \in V_J$. Hence the true orbit $\{\mathbf{w}_k\}_{k=-\infty}^{\infty}$ lies in O and generates $\{a_k\}_{k=-\infty}^{\infty}$. Thus α is onto, as asserted, and the proof of the first part of the theorem is complete.

If $\alpha(\{\mathbf{w}_k\}_{k=-\infty}^{\infty}) = \{a_k\}_{k=-\infty}^{\infty}$, we define

$$\phi(\{a_k\}_{k=-\infty}^{\infty}) = \mathbf{w}_0.$$

Clearly $\phi : Y \to I\!R^n$ so defined is one to one. Now we show ϕ is continuous. First note that it is well-known that the set of all bi-infinite sequences $\{a_k\}_{k=-\infty}^{\infty}$ with $a_k \in \{0, 1, \ldots, J\}$ is a compact metric space, when endowed with the product discrete topology. Clearly, Y is a closed subset and hence a compact metric space also.

Next let $\varepsilon_1 > 0$ be given and choose a positive integer N such that

$$[L_1 \beta_1^N + L_2 \beta_2^N] d < \varepsilon_1,$$

where L_1, L_2, β_1 and β_2 are the constants from Proposition 2.10 as determined at the beginning of this proof. Suppose $\{a_k^{(m)}\}_{k=-\infty}^{\infty} \to \{a_k\}_{k=-\infty}^{\infty}$ as $m \to$

∞, let $\{\mathbf{z}_k^{(m)}\}_{k=-\infty}^{\infty}$ and $\{\mathbf{z}_k\}_{k=-\infty}^{\infty}$ be the corresponding δ pseudo orbits as constructed in the proof that α is onto and let $\{\mathbf{w}_k^{(m)}\}_{k=-\infty}^{\infty}$ and $\{\mathbf{w}_k\}_{k=-\infty}^{\infty}$ be the corresponding $\varepsilon/2$-shadowing orbits so that

$$\phi(\{a_k^{(m)}\}_{k=-\infty}^{\infty}) = \mathbf{w}_0^{(m)}, \ \phi(\{a_k\}_{k=-\infty}^{\infty}) = \mathbf{w}_0.$$

Note that

$$\|\mathbf{w}_k^{(m)} - \mathbf{w}_k\| \le \varepsilon + \|\mathbf{z}_k^{(m)} - \mathbf{z}_k\| = 2M\delta + \|\mathbf{z}_k^{(m)} - \mathbf{z}_k\|$$

for all k. Now there is a positive integer M_0 such that if $m \ge M_0$, $a_k^{(m)} = a_k$ for $-N \le k \le N$ and hence $\mathbf{z}_k^{(m)}$ and $\mathbf{z}_k$ are in the same V_j for $-N \le k \le N$. So if $m \ge M_0$,

$$\|\mathbf{w}_k^{(m)} - \mathbf{w}_k\| \le 2M\delta + 4M\delta + 2\delta_1 \le d$$

for $-N \le k \le N$. Then, since both $\{\mathbf{w}_k^{(m)}\}_{k=-\infty}^{\infty}$ and $\{\mathbf{w}_k\}_{k=-\infty}^{\infty}$ are orbits of f in the closed ε_0 neighbourhood of S, it follows from Proposition 2.10 and our choice of d that

$$\|\mathbf{w}_0^{(m)} - \mathbf{w}_0\| \le [L_1\beta_1^N + L_2\beta_2^N]d.$$

Hence, if $m \ge M_0$,

$$\|\phi(\{a_k^{(m)}\}_{k=-\infty}^{\infty}) - \phi(\{a_k\}_{k=-\infty}^{\infty})\| < \varepsilon_1.$$

Thus ϕ is continuous, as asserted.

Now, since Y is a compact metric space, it follows that $\phi : Y \to \phi(Y) \subset \mathbb{R}^n$ is a homeomorphism onto its image.

Finally since it is clear that

$$\alpha(\{\mathbf{w}_{k+1}\}_{k=-\infty}^{\infty}) = \{a_{k+1}\}_{k=-\infty}^{\infty} = \sigma(\{a_k\}_{k=-\infty}^{\infty}),$$

it follows that

$$\phi(\sigma(\{a_k\}_{k=-\infty}^{\infty})) = \mathbf{w}_1 = f(\mathbf{w}_0) = f(\phi(\{a_k\}_{k=-\infty}^{\infty})).$$

Hence

$$\phi \circ \sigma = f \circ \phi$$

and the proof of the theorem is complete.

Now we show that the dynamics on the set $\phi(Y)$ in the theorem is *chaotic* in the sense of the following definition (see Devaney [1989]).

Definition 5.2. Let X be a complete metric space with metric $d(\cdot,\cdot)$ and $f : X \to X$ a homeomorphism. Then f is said to be *chaotic* on X if

(i) f is *transitive* on X, that is, given nonempty open sets U and V in X and $N_0 > 0$, there is $m \geq N_0$ such that $f^m(U) \cap V$ is not empty;

(ii) the *periodic points* are dense in X;

(iii) f has *sensitive dependence on initial conditions*, that is, there exists a positive number δ such that for all $\varepsilon > 0$ and all $\mathbf{x} \in X$, there exist $\mathbf{y}$ with $d(\mathbf{y}, \mathbf{x}) < \varepsilon$ and a nonnegative integer N such that

$$d(f^N(\mathbf{y}), f^N(\mathbf{x})) \geq \delta.$$

Let us remark that Banks et al. [1992] (conf. also Silverman [1992]) showed that properties (i) and (ii) imply (iii).

Now we show that the f in Theorem 5.1 is chaotic on the set $\phi(Y)$. We start with the easiest property to verify, that is, (ii). Note, first of all, that $\{a_k\}_{k=-\infty}^{\infty}$ is a periodic point of σ if and only if there exists a positive integer m such that

$$a_{k+m} = a_k$$

for all $k \in \boldsymbol{Z}$. The set of all such sequences is dense in Y. For let the sequence $\{a_k\}_{k=-\infty}^{\infty}$ be in Y. Given any positive integer N, we can find positive integers n^+ and n^- exceeding N such that $a_k = 0$ if $n^+ - L < k \leq n^+$ or $-n^- \leq k < -n^- + L$. Then we define a sequence of period $n^+ + n^- + 1$ in Y by extending a_k periodically outside the segment $[-n^-, n^+]$. Thus the periodic points are dense in Y. Now the homeomorphism ϕ preserves the property of denseness and also of periodicity since $f^m = \phi \circ \sigma^m \circ \phi^{-1}$. So we conclude that $f : \phi(Y) \to \phi(Y)$ satisfies (ii) also.

To verify the transitivity, we need only consider $\sigma : Y \to Y$ since ϕ is a homeomorphism and $f^m = \phi \circ \sigma^m \circ \phi^{-1}$. So let U and V be open sets in Y and choose $\{a_k\}_{k=-\infty}^{\infty}$ in U and $\{b_k\}_{k=-\infty}^{\infty}$ in V. Next we set

$$a'_k = \begin{cases} a_k & \text{if } k \leq N, \\ b_{k-m} & \text{if } k > N, \end{cases}$$

where the positive integers N and m are chosen so that $\{a'_k\}_{k=-\infty}^{\infty}$ is in Y and N also so that $\{a'_k\}_{k=-\infty}^{\infty}$ is in U. Then we see that

$$\sigma^m(\{a'_k\}_{k=-\infty}^{\infty}) = \{b'_k\}_{k=-\infty}^{\infty},$$

where

$$b'_k = \begin{cases} a_{k+m} & \text{if } k \leq N - m, \\ b_k & \text{if } k > N - m. \end{cases}$$

Now if m is large compared to N, $\{b'_k\}_{k=-\infty}^{\infty}$ is in V, that is, $\sigma^m(\{a'_k\}_{k=-\infty}^{\infty}) \in V$ and, of course, we may choose m greater than N_0. This verifies the transitivity for $\sigma : Y \to Y$ and hence for $f : \phi(Y) \to \phi(Y)$.

Now, of course, the sensitive dependence is a consequence of the other two items in the definition of chaos. However, it is easy to give a direct proof in this situation. Let $\{a_k\}_{k=-\infty}^{\infty} \in Y$ and set $\mathbf{w}_0 = \phi(\{a_k\}_{k=-\infty}^{\infty})$. If $\{a_k\}_{k=-\infty}^{\infty}$ ends in an infinite sequence of zeros, we change J consecutive zeros to $\{1, 2, \ldots, J\}$. Otherwise, we change a segment $\{1, 2, \ldots, J\}$ to J zeros. In either case, the change can be made arbitrarily far along the sequence. So we obtain a new sequence $\{b_k\}_{k=-\infty}^{\infty}$ arbitrarily close to $\{a_k\}_{k=-\infty}^{\infty}$. Set $\mathbf{z}_0 = \phi(\{b_k\}_{k=-\infty}^{\infty})$. Then $\mathbf{z}_0$ is as close to $\mathbf{w}_0$ as we like. However, for some positive m, $f^m(\mathbf{z}_0) \in V_0$ and $f^m(\mathbf{w}_0) \in V_1$, or vice versa. So we have proved the sensitive dependence with $\delta = \operatorname{dist}(V_0, V_1)$.

6. HYPERBOLIC PERIODIC ORBITS OF ORDINARY DIFFERENTIAL EQUATIONS, STABLE AND UNSTABLE MANIFOLDS AND ASYMPTOTIC PHASE

We wish to develop a theory for autonomous systems of ordinary differential equations analogous to the theory we have developed for diffeomorphisms in Chapters 1 through 5. It turns out that the object analogous to the fixed point of a diffeomorphism is a periodic solution rather than an equilibrium point. To some extent, we can reduce the study of a periodic solution to that of the fixed point of a diffeomorphism by using the *Poincaré* map. However, first we begin by recalling a few elementary facts from the theory of ordinary differential equations.

Let U be an open subset of $\mathbb{R}^n$ and $F : U \to \mathbb{R}^n$ a C^1 vector field. Then for each pair $(\tau, \xi) \in \mathbb{R} \times U$ the initial value problem

$$\dot{\mathbf{x}} = F(\mathbf{x}), \ \ \mathbf{x}(\tau) = \xi \tag{1}$$

has a unique solution in the sense that any two solutions coincide on the intersection of their domains. This allows us to define a maximal interval of existence $I(\tau, \xi)$ for the solution of the initial value problem (1). (Note that if it is known that the solution of (1) lies in a compact subset of U for all t in $I(\tau, \xi)$, then $I(\tau, \xi) = (-\infty, \infty)$). Then the set

$$O = \{(t, \tau, \xi) : \tau \in \mathbb{R}, \ \xi \in U, \ t \in I(\tau, \xi)\}$$

is open and if we define $\Phi : O \to \mathbb{R}^n$ by

$$\Phi(t, \tau, \xi) = \mathbf{x}(t),$$

where $\mathbf{x}(t)$ is the solution of the initial value problem (1), then Φ is a C^1 function (C^r if F is C^r) and because of the uniqueness, the identity

$$\Phi(t, s, \Phi(s, \tau, \xi)) = \Phi(t, \tau, \xi)$$

and because of the time-independence of F, the identities

$$\Phi(t, \tau, \xi) = \Phi(t - \tau, 0, \xi), \ \ I(\tau, \xi) = \tau + I(0, \xi)$$

hold in the appropriate domains. Let us define the *flow*

$$\phi(t, \xi) = \phi^t(\xi) = \Phi(t, 0, \xi) \ (= \Phi(0, -t, \xi)).$$

For each fixed t, ϕ^t is a C^1 function (C^r if F is C^r) defined on the open set

$$U^t = \{\xi \in U : t \in I(0, \xi)\}.$$

Also we have the identity

$$\phi^t(\phi^{-t}(\xi)) = \phi(t,\phi(-t,\xi)) = \Phi(t,0,\Phi(0,t,\xi)) = \Phi(t,t,\xi) = \xi.$$

So ϕ^t is a C^1 diffeomorphism onto its range U^{-t}.

The solutions of the equation

$$\dot{\mathbf{x}} = F(\mathbf{x}) \tag{2}$$

can be divided into three classes:

(i) the constant or *equilibrium* solutions $\mathbf{x}(t) \equiv \mathbf{x}_0$, where $F(\mathbf{x}_0) = \mathbf{0}$;

(ii) solutions $\mathbf{x}(t)$ where $\mathbf{x}(t_1) \neq \mathbf{x}(t_2)$ if $t_1 \neq t_2$;

(iii) the *periodic solutions* $\mathbf{x}(t)$ for which there exists a positive number T, called the *minimal period*, such that $\mathbf{x}(t+T) \equiv \mathbf{x}(t)$ and $\mathbf{x}(t) \neq \mathbf{x}(s)$ if $0 < |t-s| < T$.

6.1 THE POINCARÉ MAP

Let $\mathbf{u}(t)$ be a periodic solution of the autonomous system (2) with minimal period $T > 0$. Set $\mathbf{x}_0 = \mathbf{u}(0)$ and consider the hyperplane

$$C = \{\mathbf{y} \in I\!R^n : \langle \mathbf{y} - \mathbf{x}_0,\ F(\mathbf{x}_0) \rangle = 0\}.$$

We consider solutions of Eq.(2) starting in C near $\mathbf{x}_0$ and look at the time these solutions take to hit C again near $\mathbf{x}_0$.

Theorem 6.1. *Let U be an open subset of $I\!R^n$, let $F : U \to I\!R^n$ be a $C^r\,(r \geq 1)$ vectorfield, let ϕ be the flow associated with Eq.(2) and let $\mathbf{u}(t) = \phi^t(\mathbf{x}_0)$ be a periodic solution of Eq.(2) with minimal period T. Then there exist $\Delta > 0$ and a C^r function $\tau : B(\mathbf{x}_0, \Delta) \cap C \to I\!R$ such that*

(i) $\phi^{\tau(\mathbf{x})}(\mathbf{x}) \in C$,

(ii) $\tau(\mathbf{x}_0) = T$ *and* $|\tau(\mathbf{x}) - T| \leq 4\,\|\phi^T(\mathbf{x}) - \mathbf{x}_0\| / \|F(\mathbf{x}_0)\|$,

$$\text{(iii) } \tau'(\mathbf{x})\mathbf{h} = -\langle F(\phi^{\tau(\mathbf{x})}(\mathbf{x})), F(\mathbf{x}_0) \rangle^{-1} \langle D\phi^{\tau(\mathbf{x})}(\mathbf{x})\mathbf{h}, F(\mathbf{x}_0) \rangle \tag{3}$$

if $\mathbf{h}$ is orthogonal to $F(\mathbf{x}_0)$, and

(iv) *there exists $\alpha > 0$ such that if $\mathbf{x}$ and $\phi^t(\mathbf{x})$ are in $B(\mathbf{x}_0, \Delta) \cap C$ with $-\alpha \leq t \leq T + \alpha$ then $t = 0$ or $\tau(\mathbf{x})$.*

In order to prove the theorem, we need the following lemma.

Lemma 6.2. *Let $F : U \to \mathbb{R}^n$ be a C^1 vectorfield on an open set U in $\mathbb{R}^n$. Let $\mathbf{x}$ and $\mathbf{y}$ be points in U and $\mathbf{v}$ a vector in $\mathbb{R}^n$ such that $\langle F(\mathbf{y}), \mathbf{v} \rangle \neq 0$ and such that*

$$\|F(\mathbf{x}) - F(\mathbf{y})\| \leq \frac{1}{2}|\langle F(\mathbf{y}), \mathbf{v} \rangle|/\|\mathbf{v}\|.$$

Suppose also that the solution $\phi^t(\mathbf{x})$ of Eq.(2) is defined on an interval $[-\alpha, \alpha]$ and that

$$\|DF(\phi^t(\mathbf{x}))F(\phi^t(\mathbf{x}))\| \leq |\langle F(\mathbf{y}), \mathbf{v} \rangle|/2\alpha\|\mathbf{v}\|$$

for $|t| \leq \alpha$. Then if

$$|\langle \mathbf{x} - \mathbf{y}, \mathbf{v} \rangle| \leq \alpha|\langle F(\mathbf{y}), \mathbf{v} \rangle|/4,$$

there exists t satisfying

$$|t| \leq 4|\langle \mathbf{x} - \mathbf{y}, \mathbf{v} \rangle|/|\langle F(\mathbf{y}), \mathbf{v} \rangle|$$

such that

$$\langle \phi^t(\mathbf{x}) - \mathbf{y}, \mathbf{v} \rangle = 0.$$

Moreover, this is the only t satisfying the last equation in $|t| \leq \alpha$.

Proof. Set

$$g(t) = \langle \phi^t(\mathbf{x}) - \mathbf{y}, \mathbf{v} \rangle.$$

Then, assuming without loss of generality that $\langle F(\mathbf{y}), \mathbf{v} \rangle$ is positive,

$$g'(0) = \langle F(\mathbf{x}), \mathbf{v} \rangle \geq \langle F(\mathbf{y}), \mathbf{v} \rangle - \|F(\mathbf{x}) - F(\mathbf{y})\|\|\mathbf{v}\| \geq \frac{1}{2}\langle F(\mathbf{y}), \mathbf{v} \rangle.$$

Also we see that when $|t| \leq \alpha$,

$$|g''(t)| = |\langle DF(\phi^t(\mathbf{x}))F(\phi^t(\mathbf{x})), \mathbf{v} \rangle| \leq \langle F(\mathbf{y}), \mathbf{v}, \rangle/2\alpha.$$

Now, by Taylor's theorem, there exists θ between 0 and t such that

$$g(t) = g(0) + g'(0)t + \frac{1}{2}g''(\theta)t^2.$$

Then if $t = 4|g(0)|/\langle F(\mathbf{y}), \mathbf{v} \rangle$,

$$g(t) \geq -|g(0)| + \frac{1}{2}\langle F(\mathbf{y}), \mathbf{v} \rangle t \; - \langle F(\mathbf{y}), \mathbf{v} \rangle t^2/4\alpha \geq 0$$

and if $t = -4|g(0)|/\langle F(\mathbf{y}), \mathbf{v}\rangle$,

$$g(t) \leq |g(0)| + \frac{1}{2}\langle F(\mathbf{y}), \mathbf{v}\rangle t + \langle F(\mathbf{y}), \mathbf{v}\rangle t^2/4\alpha \leq 0.$$

The existence of a t follows. Its uniqueness follows from the fact that

$$g'(t) = g'(0) + \int_0^t g''(s)ds \geq \frac{1}{2}\langle F(\mathbf{y}), \mathbf{v}\rangle - \langle F(\mathbf{y}), \mathbf{v}\rangle |t|/2\alpha > 0$$

if $|t| < \alpha$. This completes the proof of the lemma.

Proof of Theorem 6.1. Assume $\Delta > 0$ is so small that when $\|\mathbf{x} - \mathbf{x}_0\| \leq \Delta$, $\phi^t(\mathbf{x})$ is defined for $|t| \leq T + 1$. Let U_1 be a compact set in U containing the set

$$\{\phi^t(\mathbf{x}) : 0 \leq t \leq T+1, \|\mathbf{x} - \mathbf{x}_0\| \leq \Delta\}$$

in its interior. Then set

$$\alpha = \min\{\|F(\mathbf{x}_0)\|/2M_0\overline{M}_1, 1\},$$

where

$$M_0 = \sup_{\mathbf{x}\in U_1} \|F(\mathbf{x})\|, \; \overline{M}_1 = \sup\left\{\frac{\|F(\mathbf{x}) - F(\mathbf{y})\|}{\|\mathbf{x} - \mathbf{y}\|} : \mathbf{x}, \mathbf{y} \in U_1, \; \mathbf{x} \neq \mathbf{y}\right\}. \quad (4)$$

Let $\mathbf{x} \in B(\mathbf{x}_0, \Delta)$. Then, by Gronwall's lemma,

$$\|\phi^T(\mathbf{x}) - \mathbf{x}_0\| = \|\phi^T(\mathbf{x}) - \phi^T(\mathbf{x}_0)\| \leq e^{\overline{M}_1 T}\|\mathbf{x} - \mathbf{x}_0\|.$$

Hence we may apply Lemma 6.2 with $\phi^T(\mathbf{x})$ as $\mathbf{x}$, $\mathbf{x}_0$ as $\mathbf{y}$, $F(\mathbf{x}_0)/\|F(\mathbf{x}_0)\|$ as $\mathbf{v}$ and α as above to deduce that if

$$\Delta < e^{-\overline{M}_1 T} \min\{\|F(\mathbf{x}_0)\|/2\overline{M}_1, \; \|F(\mathbf{x}_0)\|^2/8M_0\overline{M}_1, \; \|F(\mathbf{x}_0)\|/4\},$$

then there exists t satisfying

$$|t - T| \leq 4\|\phi^T(\mathbf{x}) - \mathbf{x}_0\|/\|F(\mathbf{x}_0)\| \quad (5)$$

such that $\phi^t(\mathbf{x})$ is in C. Also this t is the unique t such that $\phi^t(\mathbf{x}) \in C$ and $|t - T| \leq \alpha$.

We write the t just found as $\tau(\mathbf{x})$. By uniqueness $\tau(\mathbf{x}_0) = T$. Next we show that the function τ is C^r. Observe that $t = \tau(\mathbf{x})$ solves the equation

$$g(t, \mathbf{x}) = \langle \phi^t(\mathbf{x}) - \mathbf{x}_0, \; F(\mathbf{x}_0)\rangle = 0$$

and that $g(t, \mathbf{x})$ is a C^r function of $(t, \mathbf{x})$ in its domain. Also

$$\begin{aligned}
\frac{\partial g}{\partial t}(t, \mathbf{x}) &= \langle F(\phi^t(\mathbf{x})), F(\mathbf{x}_0) \rangle \\
&\geq \|F(\mathbf{x}_0)\|^2 - \|F(\phi^t(\mathbf{x})) - F(\mathbf{x}_0)\| \|F(\mathbf{x}_0)\| \\
&\geq \|F(\mathbf{x}_0)\| \left[\|F(\mathbf{x}_0)\| - \overline{M}_1 \|\phi^t(\mathbf{x}) - \mathbf{x}_0\| \right].
\end{aligned}$$

However if $t = \tau(\mathbf{x})$, where $\mathbf{x} \in B(\mathbf{x}_0, \Delta) \cap C$,

$$\begin{aligned}
\|\phi^t(\mathbf{x}) - \mathbf{x}_0\| &\leq \|\phi^t(\mathbf{x}) - \phi^T(\mathbf{x})\| + \|\phi^T(\mathbf{x}) - \mathbf{x}_0\| \\
&\leq M_0|t - T| + \|\phi^T(\mathbf{x}) - \mathbf{x}_0\| \\
&\leq 4M_0\|\phi^T(\mathbf{x}) - \mathbf{x}_0\|/\|F(\mathbf{x}_0)\| + \|\phi^T(\mathbf{x}) - \mathbf{x}_0\| \quad \text{by Eq.(5)} \\
&= (1 + 4M_0/\|F(\mathbf{x}_0)\|)\|\phi^T(\mathbf{x}) - \mathbf{x}_0\| \\
&\leq e^{\overline{M}_1 T}\|\mathbf{x} - \mathbf{x}_0\|(1 + 4M_0/\|F(\mathbf{x}_0)\|) \\
&< \Delta e^{\overline{M}_1 T}(\|F(\mathbf{x}_0)\| + 4M_0)/\|F(\mathbf{x}_0)\| \\
&\leq \|F(\mathbf{x}_0)\|/\overline{M}_1.
\end{aligned}$$

So if $\mathbf{x} \in B(\mathbf{x}_0, \Delta) \cap C$,

$$\frac{\partial g}{\partial t}(\tau(\mathbf{x}), \mathbf{x}) > 0$$

and therefore it follows from the implicit function theorem and the uniqueness of $\tau(\mathbf{x})$ that $\tau(\mathbf{x})$ is C^r in $\mathbf{x}$.

Next we calculate $\tau'(\mathbf{x})$. Differentiating the equation

$$g(\tau(\mathbf{x}), \mathbf{x}) = \langle \phi^{\tau(\mathbf{x})}(\mathbf{x}) - \mathbf{x}_0, F(\mathbf{x}_0) \rangle = 0$$

with respect to $\mathbf{x}$, we find that for $\mathbf{h}$ orthogonal to $F(\mathbf{x}_0)$

$$\langle F(\phi^{\tau(\mathbf{x})}(\mathbf{x})), F(\mathbf{x}_0) \rangle \tau'(\mathbf{x})\mathbf{h} \; + \langle D\phi^{\tau(\mathbf{x})}(\mathbf{x})\mathbf{h}, F(\mathbf{x}_0) \rangle = 0.$$

Since

$$\langle F(\phi^{\tau(\mathbf{x})}(\mathbf{x})), F(\mathbf{x}_0) \rangle = \frac{\partial g}{\partial t}(\tau(\mathbf{x}), \mathbf{x}) > 0,$$

it follows that

$$\tau'(\mathbf{x})\mathbf{h} = -\langle F(\phi^{\tau(\mathbf{x})}(\mathbf{x})), F(\mathbf{x}_0) \rangle^{-1} \langle D\phi^{\tau(\mathbf{x})}(\mathbf{x})\mathbf{h}, F(\mathbf{x}_0) \rangle.$$

All that remains to prove is (iv). We already know that if $\mathbf{x} \in C$ and $\|\mathbf{x} - \mathbf{x}_0\| \leq \Delta$, where Δ satisfies all the conditions imposed above, and if $\phi^t(\mathbf{x}) \in C$ with $|t - T| \leq \alpha$ then $t = \tau(\mathbf{x})$. If we apply Lemma 6.2 again, with $\mathbf{x}_0$ as $\mathbf{y}$, $F(\mathbf{x}_0)/\|F(\mathbf{x}_0)\|$ as $\mathbf{v}$ and α as defined at the beginning of the proof, it follows that if $\mathbf{x} \in C$ and $\|\mathbf{x} - \mathbf{x}_0\| \leq \Delta$, then 0 is the unique t in $|t| \leq \alpha$ such that $\phi^t(\mathbf{x}) \in C$. Now $\{\mathbf{x}_0\}$ and $\{\phi^t(\mathbf{x}_0) : \alpha \leq t \leq T - \alpha\}$ are disjoint compact sets. So if the positive number Δ is sufficiently small and $\|\mathbf{x} - \mathbf{x}_0\| \leq \Delta$, then $\|\phi^t(\mathbf{x}) - \mathbf{x}_0\| \geq \Delta$ for $\alpha \leq t \leq T - \alpha$. Hence, making Δ smaller if necessary, (iv) follows and the proof of Theorem 6.1 is complete.

The time $\tau(\mathbf{x})$ found in Theorem 6.1 is called the *first return time.* We now define the *Poincaré map* $P : B(\mathbf{x}_0, \Delta) \cap C \to C$ by

$$P(\mathbf{x}) = \phi^{\tau(\mathbf{x})}(\mathbf{x}).$$

It follows from Theorem 6.1 that P is a C^r map with $P(\mathbf{x}_0) = \mathbf{x}_0$ and

$$DP(\mathbf{x})\mathbf{h} = F(P(\mathbf{x}))\tau'(\mathbf{x})\mathbf{h} + D\phi^{\tau(\mathbf{x})}(\mathbf{x})\mathbf{h} \tag{6}$$

if $\mathbf{h}$ is orthogonal to $F(\mathbf{x}_0)$. Now it follows from Eq.(3) that

$$\tau'(\mathbf{x}_0)\mathbf{h} = -\|F(\mathbf{x}_0)\|^{-2}\langle D\phi^T(\mathbf{x}_0)\mathbf{h}, F(\mathbf{x}_0)\rangle.$$

Hence

$$DP(\mathbf{x}_0)\mathbf{h} = -\|F(\mathbf{x}_0)\|^{-2}\langle D\phi^T(\mathbf{x}_0)\mathbf{h}, F(\mathbf{x}_0)\rangle F(\mathbf{x}_0) + D\phi^T(\mathbf{x}_0)\mathbf{h}, \tag{7}$$

which is just the orthogonal projection of $D\phi^T(\mathbf{x}_0)\mathbf{h}$ onto $\text{span}\{F(\mathbf{x}_0)\}^\perp$, where $\perp$ denotes the orthogonal complement. Since $D\phi^T(\mathbf{x}_0)$ is invertible and $D\phi^T(\mathbf{x}_0)F(\mathbf{x}_0) = F(\mathbf{x}_0)$, it follows that $DP(\mathbf{x}_0)$ is an invertible mapping of $\text{span}\{F(\mathbf{x}_0)\}^\perp$ onto itself. So we deduce from the inverse function theorem that if Δ is sufficiently small, then *P is a C^r diffeomorphism of $B(\mathbf{x}_0, \Delta) \cap C$ onto some open neighbourhood in C of $\mathbf{x}_0$.*

6.2 HYPERBOLIC PERIODIC ORBITS AND THEIR STABLE AND UNSTABLE MANIFOLDS

We begin with the definition of a hyperbolic periodic orbit.

Definition 6.3. Let $F : U \to I\!R^n$ be a C^r vectorfield ($r \geq 1$) with associated flow ϕ, and let $\mathbf{u}(t)$ be a periodic solution of Eq.(2) with minimal period $T > 0$. Then $\mathbf{u}(t)$ is said to be *hyperbolic* if all but one of the eigenvalues of $D\phi^T(\mathbf{u}(0))$ lie off the unit circle. (Of course, these eigenvalues are just the Floquet multipliers of the variational equation

$$\dot{\mathbf{x}} = DF(\mathbf{u}(t))\mathbf{x}.)$$

Now we know that with $\mathbf{x}_0 = \mathbf{u}(0)$

$$D\phi^T(\mathbf{x}_0)F(\mathbf{x}_0) = F(\mathbf{x}_0).$$

Since $F(\mathbf{x}_0) \neq \mathbf{0}$, $D\phi^T(\mathbf{x}_0)$ has 1 as an eigenvalue. Next, in view of Eq.(7), $D\phi^T(\mathbf{x}_0)$ has the upper triangular form

$$\begin{bmatrix} 1 & \cdots\cdots \\ 0 & \\ \vdots & DP(\mathbf{x}_0) \\ 0 & \end{bmatrix}$$

when $\mathbb{R}^n$ is regarded as $\text{span}\{F(\mathbf{x}_0)\} \oplus \text{span}\{F(\mathbf{x}_0)\}^\perp$. So the eigenvalues of $DP(\mathbf{x}_0)$ are those of $D\phi^T(\mathbf{x}_0)$ excluding the 1 associated with the eigenvector $F(\mathbf{x}_0)$. Hence we conclude that $\mathbf{u}(t)$ *is hyperbolic if and only if* $\mathbf{x}_0 = \mathbf{u}(0)$ *is a hyperbolic fixed point of the Poincaré map* P.

Now we define the stable and unstable manifolds of a hyperbolic periodic orbit.

Definition 6.4. If $\mathbf{u}(t)$ is a hyperbolic periodic orbit with period T of Eq.(2), which has associated flow ϕ, we define its *stable manifold* as the set

$$W^s(\mathbf{u}) = \left\{\mathbf{x} \in U : \text{dist}(\phi^t(\mathbf{x}), \mathbf{u}) = \min_{0 \le s \le T} \|\phi^t(\mathbf{x}) - \mathbf{u}(s)\| \to 0 \text{ as } t \to \infty\right\}$$

and its *unstable manifold* as the set

$$W^u(\mathbf{u}) = \{\mathbf{x} \in U : \text{dist}(\phi^t(\mathbf{x}), \mathbf{u}) \to 0 \text{ as } t \to -\infty\}.$$

In the next proposition, we relate these sets to $W^s(\mathbf{x}_0)$ and $W^u(\mathbf{x}_0)$, the stable and unstable manifolds of $\mathbf{x}_0 = \mathbf{u}(0)$ considered as a fixed point of the Poincaré map P.

Proposition 6.5. *Let* $\mathbf{u}(t)$ *be a hyperbolic periodic orbit of Eq.*(2), *which has associated flow* ϕ. *Then the stable and unstable manifolds are given by*

$$W^s(\mathbf{u}) = \bigcup_{t<0} \phi^t(W^s(\mathbf{x}_0)) = \bigcup_{t<0} \phi^t(W^{s,\varepsilon}(\mathbf{x}_0)) \tag{8}$$

and

$$W^u(\mathbf{u}) = \bigcup_{t>0} \phi^t(W^u(\mathbf{x}_0)) = \bigcup_{t>0} \phi^t(W^{u,\varepsilon}(\mathbf{x}_0)), \tag{9}$$

where $\mathbf{x}_0 = \mathbf{u}(0)$ *and* ε *is any positive number.*

Proof. Let $\mathbf{x} \in W^s(\mathbf{x}_0)$. Then $\|P^k(\mathbf{x}) - \mathbf{x}_0\| \to 0$ as $k \to \infty$. This means in the notation of Theorem 6.1 that for $k \geq 0$

$$P^{k+1}(\mathbf{x}) = \phi^{\tau(P^k(\mathbf{x}))}(P^k(\mathbf{x}))$$

with $\tau(P^k(\mathbf{x})) \to \tau(\mathbf{x}_0) = T$ as $k \to \infty$. So

$$P^k(\mathbf{x}) = \phi^{t_k}(\mathbf{x})$$

with $t_{k+1} - t_k = \tau(P^k(\mathbf{x})) \to T$ as $k \to \infty$. Now if $t_k \leq t \leq t_{k+1}$, it follows from Gronwall's lemma that for k sufficiently large

$$\|\phi^t(\mathbf{x}) - \phi^{t-t_k}(\mathbf{x}_0)\| = \|\phi^{t-t_k}(P^k(\mathbf{x})) - \phi^{t-t_k}(\mathbf{x}_0)\| \leq e^{\overline{M}_1(t-t_k)}\|P^k(\mathbf{x}) - \mathbf{x}_0\|,$$

where $\overline{M}_1$ is as in the proof of Theorem 6.1 (conf. Eq.(4)). So if k is sufficiently large and $t_k \leq t \leq t_{k+1}$,

$$\|\phi^t(\mathbf{x}) - \phi^{t-t_k}(\mathbf{x}_0)\| \leq e^{\overline{M}_1 \max_k(t_{k+1}-t_k)}\|P^k(\mathbf{x}) - \mathbf{x}_0\|.$$

Hence $\mathrm{dist}(\phi^t(\mathbf{x}), \mathbf{u}) \to 0$ as $t \to \infty$ and thus

$$W^s(\mathbf{x}_0) \subset W^s(\mathbf{u}). \tag{10}$$

Conversely, let $\mathbf{x} \in W^s(\mathbf{u})$. So $\mathrm{dist}(\phi^t(\mathbf{x}), \mathbf{u}) \to 0$ as $t \to \infty$. Let $\varepsilon_0 > 0$ be a positive number with the properties that $\overline{B(\mathbf{x}_0, \varepsilon_0)} \cap C$ is in the domain of the Poincaré map P and if $\|\mathbf{y} - \mathbf{x}_0\| \leq \varepsilon_0$, the first return time $\tau(\mathbf{y})$ satisfies

$$|\tau(\mathbf{y}) - T| \leq T/2$$

and, moreover, $\|\phi^{\tau(\mathbf{y})}(\mathbf{y}) - \mathbf{x}_0\|$ is the minimum distance from the point $\phi^{\tau(\mathbf{y})}(\mathbf{y})$ to the set $\{\mathbf{u}(t) : 0 \leq t \leq T\}$.

First we prove there exists t_0 such that $\phi^{t_0}(\mathbf{x}) \in C$ and $\|\phi^{t_0}(\mathbf{x}) - \mathbf{x}_0\| \leq \varepsilon_0$. To this end, choose ε_1 so that $\phi^t(\mathbf{y})$ is defined for $|t| \leq 2T$ when $\mathrm{dist}(\mathbf{y}, \mathbf{u}) = \min\{\|\mathbf{y} - \mathbf{u}(t)\| : 0 \leq t \leq T\} \leq \varepsilon_1$. Take U_1 as a compact set in U containing the compact set

$$\{\phi^t(\mathbf{y}) : |t| \leq 2T,\ \mathrm{dist}(\mathbf{y}, \mathbf{u}) \leq \varepsilon_1\}$$

in its interior and define M_0, $\overline{M}_1$ as in Eq.(4). Next assume, in addition, that ε_1 does not exceed the quantity

$$e^{-\overline{M}_1 T}\|F(\mathbf{x}_0)\| \min\left\{\varepsilon_0/(4M_0 + \|F(\mathbf{x}_0)\|),\ 1/2\overline{M}_1,\ T/8,\ \|F(\mathbf{x}_0)\|/8M_0\overline{M}_1\right\}.$$

Now there exists τ such that

$$\mathrm{dist}(\phi^t(\mathbf{x}), \mathbf{u}) < \varepsilon_1 \quad \text{for } t \geq \tau.$$

So for some s in $[0, T)$

$$\|\phi^{\tau+T}(\mathbf{x}) - \mathbf{u}(s)\| \le \varepsilon_1.$$

Then, by Gronwall's lemma,

$$\|\phi^{\tau+T-s}(\mathbf{x}) - \mathbf{x}_0\| \le e^{\overline{M}_1 T}\varepsilon_1.$$

Applying Lemma 6.2 with $\phi^{\tau+T-s}(\mathbf{x})$ as $\mathbf{x}, \mathbf{x}_0$ as $\mathbf{y}$, $F(\mathbf{x}_0)/\|F(\mathbf{x}_0)\|$ as $\mathbf{v}$ and $\alpha = 4e^{\overline{M}_1 T}\varepsilon_1/\|F(\mathbf{x}_0)\|$, we conclude there is t in

$$|t| \le 4e^{\overline{M}_1 T}\varepsilon_1/\|F(\mathbf{x}_0)\| \le T/2$$

such that $\phi^{\tau+T-s+t}(\mathbf{x}) \in C$. Moreover,

$$\begin{aligned}
\|\phi^{\tau+T-s+t}(\mathbf{x}) - \mathbf{x}_0\| &\le \|\phi^t(\phi^{\tau+T-s}(\mathbf{x})) - \phi^{\tau+T-s}(\mathbf{x})\| + \|\phi^{\tau+T-s}(\mathbf{x}) - \mathbf{x}_0\| \\
&\le M_0|t| + e^{\overline{M}_1 T}\varepsilon_1 \\
&\le \left(\frac{4M_0}{\|F(\mathbf{x}_0)\|} + 1\right) e^{\overline{M}_1 T}\varepsilon_1 \\
&\le \varepsilon_0.
\end{aligned}$$

Thus if we define

$$t_0 = \tau + T - s + t,$$

we have $\|\phi^{t_0}(\mathbf{x}) - \mathbf{x}_0\| \le \varepsilon_0$ and $\phi^{t_0}(\mathbf{x}) \in C$. Note also that $t_0 \ge \tau - T/2$.

Next from the definition of ε_0, if we let τ_1 be the first return time $\tau(\phi^{t_0}(\mathbf{x}))$, then $|\tau_1 - T| \le T/2$ and $\phi^{t_0+\tau_1}(\mathbf{x}) \in C$, $\phi^{t_0+\tau_1}(\mathbf{x}) = P(\phi^{t_0}(\mathbf{x}))$. Note that $t_0 + \tau_1 \ge \tau$ and so $\mathrm{dist}(\phi^{t_0+\tau_1}(\mathbf{x}), \mathbf{u}) < \varepsilon_1 \le \varepsilon_0$. Also it follows from the definition of ε_0 that $\|\phi^{t_0+\tau_1}(\mathbf{x}) - \mathbf{x}_0\| = \mathrm{dist}(\phi^{t_0+\tau_1}(\mathbf{x}), \mathbf{u}) < \varepsilon_0$. Then we define $t_1 = t_0 + \tau_1$ and note that $t_1 \ge \tau$, $\|\phi^{t_1}(\mathbf{x}) - \mathbf{x}_0\| = \mathrm{dist}(\phi^{t_1}(\mathbf{x}), \mathbf{u}) < \varepsilon_0$, $\phi^{t_1}(\mathbf{x}) \in C$ and $\phi^{t_1}(\mathbf{x}) = P(\phi^{t_0}(\mathbf{x}))$. Repeating the argument we find a sequence $\{t_k\}_{k=1}^{\infty}$ such that $t_{k+1} - t_k = \tau_k \ge T/2$ with $\phi^{t_k}(\mathbf{x}) \in C$ and

$$\phi^{t_{k+1}}(\mathbf{x}) = P(\phi^{t_k}(\mathbf{x})),\ \|\phi^{t_k}(\mathbf{x}) - \mathbf{x}_0\| = \mathrm{dist}(\phi^{t_k}(\mathbf{x}), \mathbf{u}) < \varepsilon_0.$$

Thus $t_k \to \infty$ as $k \to \infty$ and

$$\|P^k(\phi^{t_0}(\mathbf{x})) - \mathbf{x}_0\| = \|\phi^{t_k}(\mathbf{x}) - \mathbf{x}_0\| \to 0 \text{ as } k \to \infty.$$

This means that $\phi^{t_k}(\mathbf{x}) \in W^s(\mathbf{x}_0)$ for $k \ge 0$. Hence we have proved that

$$W^s(\mathbf{u}) \subset \bigcup_{t<0} \phi^t(W^s(\mathbf{x}_0)). \tag{11}$$

Then the first equality in assertion (8) of the proposition follows from the relations in Eqs.(10) and (11), the second equality being trivial. Assertion (9) can be proved similarly to complete the proof of the proposition.

How do we define a "piece" of the stable manifold $W^s(\mathbf{u})$ through $\mathbf{x} \in W^s(\mathbf{u})$? Consider first $W^{s,\varepsilon}(\mathbf{x}_0)$, where ε is so small that $W^{s,\varepsilon}(\mathbf{x}_0)$ is a manifold. Also assume σ and ε are so small that $F(\phi^t(\mathbf{y}))$ and $D\phi^t(\mathbf{y})\mathbf{h}$ are linearly independent when $\|\mathbf{y}-\mathbf{x}_0\| < \varepsilon$, $|t| < \sigma$ and $\mathbf{h}$ is orthogonal to $F(\mathbf{x}_0)$, and also so that the mapping $(t, \mathbf{y}) \to \phi^t(\mathbf{y})$ with $|t| < \sigma$, $\mathbf{y} \in C$, $\|\mathbf{y}-\mathbf{x}_0\| < \varepsilon$ is one to one. Then the set

$$W^{s,\varepsilon,\sigma}(\mathbf{u}) = \{\phi^t(W^{s,\varepsilon}(\mathbf{x}_0)) :| t |< \sigma\} \subset W^s(\mathbf{u})$$

is a C^r submanifold of dimension one greater than that of $W^{s,\varepsilon}(\mathbf{x}_0)$. Now let $\mathbf{x} \in W^s(\mathbf{u})$. Then $\mathbf{x} = \phi^{t_0}(\mathbf{z})$ where $\mathbf{z} \in W^{s,\varepsilon}(\mathbf{x}_0)$, and the set

$$W^{s,\varepsilon,\sigma,t_0}(\mathbf{u}) = \phi^{t_0}(W^{s,\varepsilon,\sigma}(\mathbf{u})) = \{\phi^t(W^{s,\varepsilon}(\mathbf{x}_0)) : |t - t_0| < \sigma\}$$

is a smooth piece of $W^s(\mathbf{u})$ containing $\mathbf{x}$ and with tangent space

$$T_{\mathbf{x}}W^{s,\varepsilon,\sigma,t_0}(\mathbf{u}) = \text{span}\{F(\mathbf{x})\} \oplus D\phi^{t_0}(\mathbf{z})(T_{\mathbf{z}}W^{s,\varepsilon}(\mathbf{x}_0))$$

at $\mathbf{x}$.

We show that this definition of tangent space is independent of the choice of ε and t_0. For suppose we also have $\mathbf{x} = \phi^{s_0}(\mathbf{w})$ where $\mathbf{w} \in W^{s,\bar{\varepsilon}}(\mathbf{x}_0)$, assuming without loss of generality that $t_0 > s_0$. If we define

$$\mathbf{z}_k = P^k(\mathbf{z}) \quad \text{for} \quad k \geq 0,$$

then $\mathbf{z}_k \in W^{s,\varepsilon}(\mathbf{x}_0)$ for all k and, since $\mathbf{w}$ is in the domain of P, it follows from Theorem 6.1 that there exists $\ell > 0$ such that

$$\mathbf{w} = \phi^{t_0 - s_0}(\mathbf{z}) = P^\ell(\mathbf{z}).$$

In particular this means that $\mathbf{w} \in W^{s,\varepsilon}(\mathbf{x}_0)$ also. So, noting that

$$\phi^{t_k}(\mathbf{z}_k) = \mathbf{x},$$

where

$$t_k = t_0 - \tau(\mathbf{z}_0) - \ldots - \tau(\mathbf{z}_{k-1}),$$

and that

$$t_\ell = s_0,$$

we conclude that it suffices to show that the subspace

$$E_k = \text{span}\{F(\mathbf{x})\} \oplus D\phi^{t_k}(\mathbf{z}_k)(T_{\mathbf{z}_k}W^{s,\varepsilon}(\mathbf{x}_0))$$

is independent of k.

To establish this first observe, by the invariance of $W^{s,\varepsilon}(\mathbf{x}_0)$ under P, that

$$T_{\mathbf{z}_{k+1}}(W^{s,\varepsilon}(\mathbf{x}_0)) = DP(\mathbf{z}_k)(T_{\mathbf{z}_k}W^{s,\varepsilon}(\mathbf{x}_0)).$$

However, from Eq.(6),

$$DP(\mathbf{z}_k)\mathbf{h} = F(\mathbf{z}_{k+1})\tau'(\mathbf{z}_k)\mathbf{h} + D\phi^{\tau(\mathbf{z}_k)}(\mathbf{z}_k)\mathbf{h}$$

and so

$$T_{\mathbf{z}_{k+1}}(W^{s,\varepsilon}(\mathbf{x}_0)) \subset \text{span}\{F(\mathbf{z}_{k+1})\} \oplus D\phi^{\tau(\mathbf{z}_k)}(\mathbf{z}_k)(T_{\mathbf{z}_k}W^{s,\varepsilon}(\mathbf{x}_0)).$$

This implies that

$$\begin{aligned} &D\phi^{t_{k+1}}(\mathbf{z}_{k+1})(T_{\mathbf{z}_{k+1}}W^{s,\varepsilon}(\mathbf{x}_0)) \\ &\qquad \subset \text{span}\{F(\phi^{t_{k+1}}(\mathbf{z}_{k+1}))\} \oplus D\phi^{t_{k+1}+\tau(\mathbf{z}_k)}(\mathbf{z}_k)(T_{\mathbf{z}_k}W^{s,\varepsilon}(\mathbf{x}_0)) \\ &\qquad = \text{span}\{F(\mathbf{x})\} \oplus D\phi^{t_k}(\mathbf{z}_k)(T_{\mathbf{z}_k}W^{s,\varepsilon}(\mathbf{x}_0)) \\ &\qquad = E_k \end{aligned}$$

and hence that

$$E_{k+1} \subset E_k.$$

Since $\dim E_k = \dim E_{k+1}$, it follows that $E_k = E_{k+1}$. Hence the definition of $T_{\mathbf{x}}W^{s,\varepsilon,\sigma,t_0}(\mathbf{u})$ is independent of ε and t_0 and so we take it as the definition of $T_{\mathbf{x}}W^s(\mathbf{u})$. With this definition of $T_{\mathbf{x}}W^s(\mathbf{u})$, it is also clear that the invariance property

$$D\phi^t(\mathbf{x})(T_{\mathbf{x}}W^s(\mathbf{u})) = T_{\phi^t(\mathbf{x})}W^s(\mathbf{u})$$

holds.

Similar considerations can be carried out for the unstable manifold.

6.3 ASYMPTOTIC PHASE AND THE STABLE AND UNSTABLE FOLIATIONS

First of all, we define the term "asymptotic phase".

Definition 6.6. Let $\mathbf{u}(t)$ be a hyperbolic periodic orbit of Eq.(2). Then a point $\mathbf{x}$ in $W^s(\mathbf{u})$ is said to have ***asymptotic phase*** γ if

$$\|\phi^t(\mathbf{x}) - \mathbf{u}(t+\gamma)\| \to 0 \text{ as } t \to \infty.$$

First we examine the extent to which an asymptotic phase is unique. Let $\mathbf{u}(t)$ have minimal period $T > 0$ and suppose $\mathbf{x}$ has asymptotic phases γ_1 and γ_2. Then

$$\|\mathbf{u}(t+\gamma_1) - \mathbf{u}(t+\gamma_2)\| \to 0 \quad \text{as } t \to \infty. \tag{12}$$

Let

$$\gamma_1 = \overline{\gamma}_1 + k_1 T, \quad \gamma_2 = \overline{\gamma}_2 + k_2 T$$

where $0 \le \overline{\gamma}_1, \overline{\gamma}_2 < T$ and k_1, k_2 are integers. Then it follows from Eq.(12) that

$$\|\mathbf{u}(\overline{\gamma}_1) - \mathbf{u}(\overline{\gamma}_2)\| = \|\mathbf{u}((k-k_1)T + \gamma_1) - \mathbf{u}((k-k_2)T + \gamma_2)\| \to 0 \quad \text{as } k \to \infty.$$

That is, $\mathbf{u}(\overline{\gamma}_1) = \mathbf{u}(\overline{\gamma}_2)$ and so $\overline{\gamma}_1 = \overline{\gamma}_2$. Hence if γ_1, γ_2 are two asymptotic phases, $\gamma_1 - \gamma_2$ is a multiple of T. Conversely, it is easy to see that if γ is an asymptotic phase, then so also is $\gamma + kT$ for any integer k.

Now we show the existence of an asymptotic phase.

Theorem 6.7. *Let $\mathbf{u}(t)$ be a hyperbolic periodic orbit of Eq.(2). Then if $\mathbf{x} \in W^s(\mathbf{u})$, $\mathbf{x}$ has an asymptotic phase.*

Proof. Consider first the case $\mathbf{y} \in W^{s,\varepsilon}(\mathbf{x}_0)$, where $\mathbf{x}_0 = \mathbf{u}(0)$ (conf. Proposition 6.5). Then

$$\|P^k(\mathbf{y}) - \mathbf{x}_0\| \le \varepsilon \quad \text{for all } k \ge 0$$

and, if ε is sufficiently small, it follows from Proposition 1.8 that there are positive constants L and β with $\beta < 1$ such that

$$\|P^k(\mathbf{y}) - \mathbf{x}_0\| \le L\|\mathbf{y} - \mathbf{x}_0\|\beta^k \quad \text{for } k \ge 0. \tag{13}$$

Let us write

$$\mathbf{y}_k = P^k(\mathbf{y}) = \phi^{t_k}(\mathbf{y}).$$

Then, from the definition of P,

$$t_{k+1} - t_k = \tau(\mathbf{y}_k).$$

We show that

$$s_k = t_k - kT$$

is a convergent sequence. Note that

$$s_{k+1} - s_k = \tau(\mathbf{y}_k) - T.$$

It follows from (ii) in Theorem 6.1 that

$$|s_{k+1} - s_k| \leq 4\|\phi^T(\mathbf{y}_k) - \mathbf{x}_0\|/\|F(\mathbf{x}_0)\|$$

and hence, by Gronwall's lemma, that

$$|s_{k+1} - s_k| \leq 4e^{\overline{M}_1 T}\|\mathbf{y}_k - \mathbf{x}_0\|/\|F(\mathbf{x}_0)\|,$$

where $\overline{M}_1$ is as in Eq.(4). Next, using Eq.(13), it follows that

$$|s_{k+1} - s_k| \leq (4e^{\overline{M}_1 T}L\|\mathbf{y} - \mathbf{x}_0\|/\|F(\mathbf{x}_0)\|)\beta^k.$$

Thus $s_k = t_k - kT$ is indeed a Cauchy sequence and so there exists a real number γ such that $t_k - kT \to -\gamma$ as $k \to \infty$. This implies, in particular, that $t_{k+1} - t_k \to T$ as $k \to \infty$. Also

$$\begin{aligned}\|\phi^{t_k}(\mathbf{y}) - \mathbf{u}(t_k + \gamma)\| &\leq \|\phi^{t_k}(\mathbf{y}) - \mathbf{x}_0\| + \|\mathbf{u}(0) - \mathbf{u}(t_k + \gamma - kT)\| \\ &\to 0 \text{ as } k \to \infty.\end{aligned}$$

Now, by Gronwall's lemma again,

$$\|\phi^t(\mathbf{y}) - \mathbf{u}(t + \gamma)\| \leq e^{\overline{M}_1(t-t_k)}\|\phi^{t_k}(\mathbf{y}) - \mathbf{u}(t_k + \gamma)\|$$

if $t_k \leq t \leq t_{k+1}$. So there is a constant L_1 such that for $k \geq 0$

$$\|\phi^t(\mathbf{y}) - \mathbf{u}(t + \gamma)\| \leq L_1\|\phi^{t_k}(\mathbf{y}) - \mathbf{u}(t_k + \gamma)\|$$

if $t_k \leq t \leq t_{k+1}$. Hence we may conclude that

$$\|\phi^t(\mathbf{y}) - \mathbf{u}(t + \gamma)\| \to 0 \text{ as } t \to \infty$$

so that γ is an asymptotic phase for $\mathbf{y}$.

Now let $\mathbf{x} \in W^s(\mathbf{u})$. Then by Proposition 6.5 there exists t_0 such that $\mathbf{x} = \phi^{t_0}(\mathbf{y})$ where $\mathbf{y} \in W^{s,\varepsilon}(\mathbf{x}_0)$. Then if γ is an asymptotic phase for $\mathbf{y}$, we see that

$$\|\phi^t(\mathbf{x}) - \mathbf{u}(t + \gamma + t_0)\| = \|\phi^{t+t_0}(\mathbf{y}) - \mathbf{u}(t + t_0 + \gamma)\| \to 0 \text{ as } t \to \infty.$$

That is, $\gamma + t_0$ is an asymptotic phase for $\mathbf{x}$. So the proof of Theorem 6.7 is complete.

Next we show that the asymptotic phase is a smooth function defined on the stable manifold. Here "smooth" means smooth when restricted to a smooth piece of the stable manifold through a given point (see the end of the previous section).

Theorem 6.8. *Let* $\mathbf{u}(t)$ *be a hyperbolic periodic orbit of Eq.(2). Denote by* $\gamma(\mathbf{x})$ *the asymptotic phase of* $\mathbf{x} \in W^s(\mathbf{u})$. *Then* $\gamma(\mathbf{x})$ *is well-defined up to a multiple of* T *and* $\mathbf{x} \to e^{\frac{i2\pi\gamma(\mathbf{x})}{T}}$ *is a* C^1 *function on* $W^s(\mathbf{u})$.

Proof. The fact that $\gamma(\mathbf{x})$ is well-defined follows from Theorem 6.7 and the remarks before it.

To show the smoothness, suppose first that $\mathbf{y} \in W^{s,\varepsilon}(\mathbf{x}_0)$ where $\mathbf{x}_0 = \mathbf{u}(0)$. Then, as in the proof of Theorem 6.7, if ε is sufficiently small, we can write

$$\mathbf{y}_k = P^k(\mathbf{y}) = \phi^{t_k}(\mathbf{y}),$$

where $t_k = t_k(\mathbf{y})$ with $t_0(\mathbf{y}) = 0$ and $t_{k+1}(\mathbf{y}) - t_k(\mathbf{y}) = \tau(\mathbf{y}_k)$. We found in the proof of Theorem 6.7 that

$$\gamma(\mathbf{y}) = \lim_{k\to\infty} - s_k(\mathbf{y}) = \lim_{k\to\infty} [-t_k(\mathbf{y}) + kT].$$

Note that

$$s_{k+1}(\mathbf{y}) - s_k(\mathbf{y}) = \tau(\mathbf{y}_k) - T = \tau(P^k(\mathbf{y})) - T.$$

So if $\mathbf{h} \in T_{\mathbf{y}}W^{s,\varepsilon}(\mathbf{x}_0)$ (which is a subspace of $\text{span}\{F(\mathbf{x}_0)\}^{\perp}$),

$$[s'_{k+1}(\mathbf{y}) - s'_k(\mathbf{y})]\mathbf{h} = \tau'(P^k(\mathbf{y}))DP^k(\mathbf{y})\mathbf{h} = \tau'(\mathbf{y}_k)DP^k(\mathbf{y})\mathbf{h}. \tag{14}$$

Now, since $\tau'(\cdot)$ is continuous and $\|\mathbf{y}_k - \mathbf{x}_0\| < \varepsilon$ for all k, there is a constant L_1, not depending on k or $\mathbf{y} \in W^{s,\varepsilon}(\mathbf{x}_0)$, such that

$$\|\tau'(\mathbf{y}_k)\| \le L_1. \tag{15}$$

Also it follows from Proposition 3.3 that if ε is sufficiently small, there exist positive constants L_2 and β with $\beta < 1$ such that

$$\|DP^k(\mathbf{y})\mathbf{h}\| \le L_2\|\mathbf{h}\|\beta^k \tag{16}$$

for $k \ge 0$, $\mathbf{y} \in W^{s,\varepsilon}(\mathbf{x}_0)$ and $\mathbf{h} \in T_{\mathbf{y}}W^{s,\varepsilon}(\mathbf{x}_0)$. Then it follows from Eqs.(14), (15) and (16) that

$$\|[s'_{k+1}(\mathbf{y}) - s'_k(\mathbf{y})]\mathbf{h}\| \le L_1L_2\|\mathbf{h}\|\beta^k$$

for $k \ge 0$, $\mathbf{y} \in W^{s,\varepsilon}(\mathbf{x}_0)$ and $\mathbf{h} \in T_{\mathbf{y}}W^{s,\varepsilon}(\mathbf{x}_0)$. This inequality implies that $s'_k(\mathbf{y})$ converges uniformly for $\mathbf{y} \in W^{s,\varepsilon}(\mathbf{x}_0)$. Hence $\gamma'(\mathbf{y})$ exists and is continuous for $\mathbf{y} \in W^{s,\varepsilon}(\mathbf{x}_0)$.

Now, as we saw at the end of the previous section, each point in $W^s(\mathbf{u})$ is contained in a smooth piece of the stable manifold of the form

$$\{\phi^t(W^{s,\varepsilon}(\mathbf{x}_0)) : |t - t_0| < \sigma\}$$

and on this set the asymptotic phase is $\gamma(\mathbf{y}) + t$ with $\mathbf{y} \in W^{s,\varepsilon}(\mathbf{x}_0)$. Clearly this is a C^1 function of $(t, \mathbf{y})$. So the theorem follows.

Now we discuss the *stable foliation* induced by the asymptotic phase on the stable manifold $W^s(\mathbf{u})$. If $\mathbf{x} \in W^s(\mathbf{u})$, denote by $\gamma(\mathbf{x})$ its asymptotic phase which is uniquely defined up to a multiple of T. Given a real number γ, we define

$$W^{s,\gamma}(\mathbf{u}) = \{\mathbf{x} \in W^s(\mathbf{u}) : \gamma(\mathbf{x}) \equiv \gamma \,(\mathrm{mod}\, T)\}.$$

We show that $W^{s,\gamma}(\mathbf{u})$ is a manifold in a certain sense. Recall from the end of the previous section that each point in $W^s(\mathbf{u})$ is contained in a smooth piece of $W^s(\mathbf{u})$ of the form

$$W^{s,\varepsilon,\sigma,t_0}(\mathbf{u}) = \{\phi^\tau(W^{s,\varepsilon}(\mathbf{x}_0)) : |\tau - t_0| < \sigma\}$$

on which $\gamma(\cdot)$ can be defined smoothly as $\gamma(\phi^\tau(\mathbf{y})) = \gamma(\mathbf{y}) + \tau$, where $\gamma(\mathbf{y})$ is a smooth function on $W^{s,\varepsilon}(\mathbf{x}_0)$. So if $\mathbf{x} = \phi^\tau(\mathbf{y}) \in W^{s,\varepsilon,\sigma,t_0}(\mathbf{u})$ and t is small,

$$\gamma(\phi^t(\mathbf{x})) = \gamma(\phi^{t+\tau}(\mathbf{y})) = \gamma(\mathbf{y}) + t + \tau$$

from which it follows that

$$\left.\frac{d}{dt}\gamma(\phi^t(\mathbf{x}))\right|_{t=0} = 1.$$

On the other hand, if $\mathbf{x} \in W^{s,\varepsilon,\sigma,t_0}(\mathbf{u})$,

$$\left.\frac{d}{dt}\gamma(\phi^t(\mathbf{x}))\right|_{t=0} = \left.\gamma'(\phi^t(\mathbf{x}))F(\phi^t(\mathbf{x}))\right|_{t=0} = \gamma'(\mathbf{x})F(\mathbf{x}).$$

Then

$$\gamma'(\mathbf{x})F(\mathbf{x}) = 1.$$

In particular, this means that $\gamma'(\mathbf{x}) \neq \mathbf{0}$ for all $\mathbf{x}$ in $W^{s,\varepsilon,\sigma,t_0}(\mathbf{u})$ and from this it follows that $W^{s,\gamma}(\mathbf{u}) \cap W^{s,\varepsilon,\sigma,t_0}(\mathbf{u})$ is a manifold.

Next if $\mathbf{x} \in W^{s,\gamma}(\mathbf{u})$, we define its tangent space $T_{\mathbf{x}}W^{s,\gamma}(\mathbf{u})$. Now for suitable $\sigma > 0$,

$$W^{s,\varepsilon,\sigma,t_0}(\mathbf{u}) = \{\phi^t(W^{s,\varepsilon}(\mathbf{x}_0)) : |t - t_0| < \sigma\}$$

is a smooth piece of $W^s(\mathbf{u})$ containing $\mathbf{x}$ and

$$W^{s,\gamma}(\mathbf{u}) \cap W^{s,\varepsilon,\sigma,t_0}(\mathbf{u}) = \{\mathbf{z} \in W^{s,\varepsilon,\sigma,t_0}(\mathbf{u}) : \gamma(\mathbf{z}) \equiv \gamma(\mathrm{mod}\, T)\},$$

where $\gamma(\mathbf{z})$ can be chosen as a C^1 function with $\gamma'(\mathbf{z})F(\mathbf{z}) = 1$. Then we take

$$T_{\mathbf{x}}W^{s,\gamma}(\mathbf{u}) = T_{\mathbf{x}}[W^{s,\gamma}(\mathbf{u}) \cap W^{s,\varepsilon,\sigma,t_0}(\mathbf{u})] = \{\mathbf{h} \in T_{\mathbf{x}}W^s(\mathbf{u}) : \gamma'(\mathbf{x})\mathbf{h} = 0\}.$$

Recall from the end of the previous section that

$$T_{\mathbf{x}}W^s(\mathbf{u}) = T_{\mathbf{x}}W^{s,\varepsilon,\sigma,t_0}(\mathbf{u}) = \{sF(\mathbf{x}) + D\phi^{t_0}(\mathbf{y})\mathbf{g} : s \in \mathbb{R},\ \mathbf{g} \in T_{\mathbf{y}}W^s(\mathbf{x}_0)\},$$

where $\mathbf{y} \in W^{s,\varepsilon}(\mathbf{x}_0)$ and $\mathbf{x} = \phi^{t_0}(\mathbf{y})$. So

$$\begin{aligned} &T_{\mathbf{x}}W^{s,\gamma}(\mathbf{u}) \\ &= \{sF(\mathbf{x}) + D\phi^{t_0}(\mathbf{y})\mathbf{g} : s \in \mathbb{R},\ \mathbf{g} \in T_{\mathbf{y}}W^s(\mathbf{x}_0),\ \gamma'(\mathbf{x})(sF(\mathbf{x}) + D\phi^{t_0}(\mathbf{y})\mathbf{g}) = 0\} \\ &= \{sF(\mathbf{x}) + D\phi^{t_0}(\mathbf{y})\mathbf{g} : s \in \mathbb{R},\ \mathbf{g} \in T_{\mathbf{y}}W^s(\mathbf{x}_0),\ s + \gamma'(\mathbf{x})D\phi^{t_0}(\mathbf{y})\mathbf{g} = 0\} \end{aligned}$$

and hence

$$T_{\mathbf{x}}W^{s,\gamma}(\mathbf{u}) = \{-(\gamma'(\mathbf{x})D\phi^{t_0}(\mathbf{y})\mathbf{g})F(\mathbf{x}) + D\phi^{t_0}(\mathbf{y})\mathbf{g} : \mathbf{g} \in T_{\mathbf{y}}W^s(\mathbf{x}_0)\}.$$

In particular, this means that

$$T_{\mathbf{x}}W^s(\mathbf{u}) = \operatorname{span}\{F(\mathbf{x})\} \oplus T_{\mathbf{x}}W^{s,\gamma}(\mathbf{u}).$$

Also note that when $\mathbf{y} \in W^{s,\varepsilon}(\mathbf{x}_0)$,

$$T_{\mathbf{y}}W^s(\mathbf{u}) = \operatorname{span}\{F(\mathbf{y})\} \oplus T_{\mathbf{y}}W^{s,\varepsilon}(\mathbf{x}_0). \tag{17}$$

We see that $W^{s,\gamma+T}(\mathbf{u}) = W^{s,\gamma}(\mathbf{u})$ for all γ and that $W^s(\mathbf{u})$ is the disjoint union of $W^{s,\gamma}(\mathbf{u})$, $0 \le \gamma < T$. Moreover the foliation has the invariance property that

$$\phi^t(W^{s,\gamma}(\mathbf{u})) \subset W^{s,\gamma+t}(\mathbf{u})$$

since $\gamma(\phi^t(\mathbf{x})) \equiv \gamma(\mathbf{x}) + t \pmod{T}$. The family of manifolds $W^{s,\gamma}(\mathbf{u})$, $-\infty < \gamma < \infty$, is referred to as the *stable foliation.*

Similarly, there is a "backward" asymptotic phase which induces an *unstable foliation* on the unstable manifold $W^u(\mathbf{u})$. If $\mathbf{x} \in W^u(\mathbf{u})$ we can prove the existence of a number γ such that

$$\|\phi^t(\mathbf{x}) - \mathbf{u}(t+\gamma)\| \to 0 \ \text{ as } t \to -\infty.$$

This number $\gamma = \gamma(\mathbf{x})$ is unique up to a multiple of T and $\mathbf{x} \to e^{\frac{i2\pi\gamma(\mathbf{x})}{T}}$ is a smooth function on $W^u(\mathbf{u})$. Then the *unstable foliation* is defined by

$$W^{u,\gamma}(\mathbf{u}) = \{\mathbf{x} \in W^u(\mathbf{u}) : \gamma(\mathbf{x}) \equiv \gamma \pmod{T}\}$$

and has similar properties to those enjoyed by the stable foliation.

7. HYPERBOLIC SETS OF ORDINARY DIFFERENTIAL EQUATIONS

In this chapter we develop the theory of hyperbolic sets for flows. First we show that the continuity of the splitting into stable and unstable bundles follows from the other items in the definition. Next we develop the theory of exponential dichotomies for linear differential equations, paying special attention to the roughness theorem. We use the latter to prove that hyperbolic sets are expansive both in a "continuous" way and a "discrete" way. Finally we show that hyperbolic sets are robust under perturbation, our major tool here being Lemma 2.17.

7.1 DEFINITION OF HYPERBOLIC SET

We consider the autonomous system of ordinary differential equations

$$\dot{\mathbf{x}} = F(\mathbf{x}), \tag{1}$$

where $F : U \to I\!R^n$ is a C^1 vector field, denoting by ϕ the corresponding flow. In order to lead up to the definition of hyperbolicity given below, we look at hyperbolic periodic orbits a little more closely.

So let $\mathbf{u}(t)$ be a hyperbolic periodic orbit of Eq.(1) with minimal period T. This means that the Poincaré map P has $\mathbf{x}_0 = \mathbf{u}(0)$ as a hyperbolic fixed point. Also if we denote by W the orthogonal complement of span$\{F(\mathbf{x}_0)\}$, it follows from Eq.(6) in Chapter 6 that

$$D\phi^T(\mathbf{x}_0)\mathbf{w} = (\beta^*\mathbf{w})F(\mathbf{x}_0) + DP(\mathbf{x}_0)\mathbf{w} \quad \text{for } \mathbf{w} \in W,$$

where β is the vector in W such that

$$\beta^*\mathbf{w} = -\tau'(\mathbf{x}_0)\mathbf{w} \quad \text{for} \quad \mathbf{w} \in W.$$

Let $E_P^s(\mathbf{x}_0), E_P^u(\mathbf{x}_0) \subset W$ be the stable and unstable subspaces for $DP(\mathbf{x}_0)$. Since 1 is not an eigenvalue for $DP(\mathbf{x}_0)$, it is not an eigenvalue for $DP(\mathbf{x}_0)^*$ and so we may define

$$\alpha = -(I - DP(\mathbf{x}_0)^*)^{-1}\beta.$$

It is easy to show using the equation for $D\phi^T(\mathbf{x}_0)\mathbf{w}$ above and the fact

$$D\phi^T(\mathbf{x}_0)F(\mathbf{x}_0) = F(\mathbf{x}_0),$$

that for all integers k and $\mathbf{w} \in W$

$$[D\phi^T(\mathbf{x}_0)]^k((\alpha^*\mathbf{w})F(\mathbf{x}_0) + \mathbf{w}) = (\alpha^*[DP(\mathbf{x}_0)]^k\mathbf{w})F(\mathbf{x}_0) + [DP(\mathbf{x}_0)]^k\mathbf{w}. \tag{2}$$

Then, using this equation with $k = 1$, it is easy to verify that

$$E^s(\mathbf{x}_0) = \{(\alpha^*\mathbf{w})F(\mathbf{x}_0) + \mathbf{w} : \mathbf{w} \in E^s_P(\mathbf{x}_0)\}$$

and

$$E^u(\mathbf{x}_0) = \{(\alpha^*\mathbf{w})F(\mathbf{x}_0) + \mathbf{w} : \mathbf{w} \in E^u_P(\mathbf{x}_0)\}$$

are invariant subspaces for $D\phi^T(\mathbf{x}_0)$. It is also clear that they have the same dimensions as $E^s_P(\mathbf{x}_0)$ and $E^u_P(\mathbf{x}_0)$ respectively and that

$$\mathbb{R}^n = \text{span}\{F(\mathbf{x}_0)\} \oplus E^s(\mathbf{x}_0) \oplus E^u(\mathbf{x}_0).$$

Now we know there are positive constants $K_7, K_8, \lambda_1, \lambda_2$ with $\lambda_1 < 1, \lambda_2 < 1$ such that for $k \geq 0$

$$\|[DP(\mathbf{x}_0)]^k\xi\| \leq K_7\lambda_1^k\|\xi\| \quad \text{for } \xi \in E^s_P(\mathbf{x}_0)$$

and

$$\|[DP(\mathbf{x}_0)]^{-k}\xi\| \leq K_8\lambda_2^k\|\xi\| \quad \text{for } \xi \in E^u_P(\mathbf{x}_0).$$

Hence, using Eq.(2), if $k \geq 0$ and $\xi = (\alpha^*\mathbf{w})F(\mathbf{x}_0) + \mathbf{w} \in E^s(\mathbf{x}_0)$,

$$\|D\phi^{kT}(\mathbf{x}_0)\xi\| \leq (\|\alpha\|\ \|F(\mathbf{x}_0)\| + 1)K_7\lambda_1^k\|\mathbf{w}\| \leq K_5\lambda_1^k\|\xi\|,$$

where

$$K_5 = (\|\alpha\|\ \|F(\mathbf{x}_0)\| + 1)K_7$$

and we have made use of the fact that $\mathbf{w}$ is the orthogonal projection of ξ onto W. Now if $t \geq 0$, there exists a nonnegative integer k such that

$$kT \leq t < (k+1)T.$$

Then, if $\xi \in E^s(\mathbf{x}_0)$,

$$\|D\phi^t(\mathbf{x}_0)\xi\| = \|D\phi^{t-kT}(\mathbf{x}_0)D\phi^{kT}(\mathbf{x}_0)\xi\| \leq e^{M_1T}\|D\phi^{kT}(\mathbf{x}_0)\xi\|,$$

using Gronwall's lemma with

$$M_1 = \sup_{0\leq t\leq T}\|DF(\mathbf{u}(t))\|. \tag{3}$$

Hence if $t \geq 0$ and $\xi \in E^s(\mathbf{x}_0)$,

$$\|D\phi^t(\mathbf{x}_0)\xi\| \leq e^{M_1T}K_5\lambda_1^k\|\xi\| \leq e^{M_1T}K_5\lambda_1^{\frac{t}{T}-1}\|\xi\| \leq K_3e^{-\alpha_1t}\|\xi\|, \tag{4}$$

where

$$K_3 = e^{M_1 T} K_5 \lambda_1^{-1}, \quad \alpha_1 = -\log \lambda_1 / T.$$

Now we define

$$E^s(\mathbf{u}(t)) = D\phi^t(\mathbf{x}_0)(E^s(\mathbf{x}_0)), \;\; E^u(\mathbf{u}(t)) = D\phi^t(\mathbf{x}_0)(E^u(\mathbf{x}_0)).$$

It is clear that these subspaces have the invariance property

$$D\phi^t(\mathbf{u}(\tau))(E^{s,u}(\mathbf{u}(\tau))) = E^{s,u}(\phi^t(\mathbf{u}(\tau))$$

and that

$$\mathbb{R}^n = \text{span}\{F(\mathbf{u}(t))\} \oplus E^s(\mathbf{u}(t)) \oplus E^u(\mathbf{u}(t))$$

for all t. Also if $t \geq 0$ and $\xi = D\phi^\tau(\mathbf{x}_0)\eta \in E^s(\mathbf{u}(\tau))$ with $0 \leq \tau \leq T$, it follows from Eq.(4) that

$$\|D\phi^t(\mathbf{u}(\tau))\xi\| = \|D\phi^{t+\tau}(\mathbf{x}_0)\eta\| \leq K_3 e^{-\alpha_1(t+\tau)}\|\eta\| \leq K_3 e^{-\alpha_1(t+\tau)} e^{M_1\tau}\|\xi\|.$$

Hence if $t \geq 0$ and $\xi \in E^s(\mathbf{u}(\tau))$,

$$\|D\phi^t(\mathbf{u}(\tau))\xi\| \leq K_1 e^{-\alpha_1 t}\|\xi\|,$$

where

$$K_1 = K_3 e^{M_1 T}.$$

Similarly we prove that if $t \geq 0$ and $\xi \in E^u(\mathbf{u}(\tau))$, then

$$\|D\phi^{-t}(\mathbf{u}(\tau))\xi\| \leq K_2 e^{-\alpha_2 t}\|\xi\|,$$

where

$$K_2 = e^{2M_1 T}\lambda_2^{-1}(\|\alpha\| \, \|F(\mathbf{x}_0)\| + 1)K_8, \quad \alpha_2 = -\log \lambda_2 / T.$$

Thus we are led to make the following definition.

Definition 7.1. A compact set $S \subset U$ is said to be *hyperbolic* for Eq.(1) if

(i) $F(\mathbf{x}) \neq 0$ for all $\mathbf{x}$ in S;

(ii) S is invariant, that is, $\phi^t(S) = S$ for all t;

(iii) there is a continuous splitting

$$\mathbb{R}^n = E^0(\mathbf{x}) \oplus E^s(\mathbf{x}) \oplus E^u(\mathbf{x}) \quad \text{for } \mathbf{x} \in S \tag{5}$$

with $E^0(\mathbf{x}) = \text{span}\{F(\mathbf{x})\}$ and $\dim E^s(\mathbf{x})$, $\dim E^u(\mathbf{x})$ constant, such that for all t and $\mathbf{x}$ in S

$$D\phi^t(\mathbf{x})(E^s(\mathbf{x})) = E^s(\phi^t(\mathbf{x})), \quad D\phi^t(\mathbf{x})(E^u(\mathbf{x})) = E^u(\phi^t(\mathbf{x})),$$

and such that there are positive constants $K_1, K_2, \alpha_1, \alpha_2$ with the property that for all $t \geq 0$ and $\mathbf{x}$ in S

$$\|D\phi^t(\mathbf{x})\xi\| \leq K_1 e^{-\alpha_1 t}\|\xi\| \quad \text{for } \xi \in E^s(\mathbf{x}), \tag{6}$$

$$\|D\phi^{-t}(\mathbf{x})\xi\| \leq K_2 e^{-\alpha_2 t}\|\xi\| \quad \text{for } \xi \in E^u(\mathbf{x}). \tag{7}$$

In the following two sections we show that the continuity of the splitting follows from the other conditions. First we show that the associated projections are bounded and then in the following section that they are continuous.

7.2 BOUNDEDNESS OF THE PROJECTIONS

We associate with the splitting (5) the projections $\mathcal{P}^0(\mathbf{x})$, $\mathcal{P}^s(\mathbf{x})$ and $\mathcal{P}^u(\mathbf{x})$. In this and the next section we show that the other conditions in Definition 7.1 imply that the splitting (5) is continuous, that is, the projections are continuous. In this section we show that they are bounded and in the next that they are continuous.

To show the projections are bounded, let $\mathbf{v} \in E^0(\mathbf{x})$, $\xi \in E^s(\mathbf{x})$, $\eta \in E^u(\mathbf{x})$ be nonzero vectors and define

$$\mathbf{x}^0(t) = D\phi^t(\mathbf{x})\mathbf{v}, \quad \mathbf{x}^s(t) = D\phi^t(\mathbf{x})\xi, \quad \mathbf{x}^u(t) = D\phi^t(\mathbf{x})\eta$$

where $\mathbf{x} \in S$. Note that $\mathbf{v} = \alpha F(\mathbf{x})$ for some real number α and so

$$\mathbf{x}^0(t) = \alpha F(\phi^t(\mathbf{x})).$$

Hence for all t

$$M_0^{-1}\Delta\|\mathbf{v}\| \leq \|\mathbf{x}^0(t)\| \leq M_0\Delta^{-1}\|\mathbf{v}\|,$$

where

$$M_0 = \sup_{\mathbf{x}\in S}\|F(\mathbf{x})\|, \quad \Delta = \inf_{\mathbf{x}\in S}\|F(\mathbf{x})\|. \tag{8}$$

Now choose a positive number T so that

$$\sigma = \Delta M_0^{-1} \min \left\{ K_1^{-1} e^{\alpha_1 T},\ K_2^{-1} e^{\alpha_2 T} \right\} > 1.$$

Then

$$\begin{aligned}
\left\| \frac{\mathbf{x}^0(T)}{\|\mathbf{v}\|} + \frac{\mathbf{x}^s(T)}{\|\xi\|} \right\| &= \frac{\|\mathbf{x}^s(T)\|}{\|\xi\|} \left\| \frac{\|\xi\|}{\|\mathbf{x}^s(T)\|} \cdot \frac{\mathbf{x}^0(T)}{\|\mathbf{v}\|} + \frac{\mathbf{x}^s(T)}{\|\mathbf{x}^s(T)\|} \right\| \\
&\geq \frac{\|\mathbf{x}^s(T)\|}{\|\xi\|} \left[\frac{\|\mathbf{x}^0(T)\|}{\|\mathbf{v}\|} \cdot \frac{\|\xi\|}{\|\mathbf{x}^s(T)\|} - 1 \right] \\
&\geq e^{-M_1 T} [\Delta M_0^{-1} K_1^{-1} e^{\alpha_1 T} - 1] \\
&\geq (\sigma - 1) e^{-M_1 T},
\end{aligned}$$

where

$$M_1 = \sup_{\mathbf{x} \in S} \|DF(\mathbf{x})\|.$$

On the other hand,

$$\left\| \frac{\mathbf{x}^0(T)}{\|\mathbf{v}\|} + \frac{\mathbf{x}^s(T)}{\|\xi\|} \right\| = \left\| D\phi^T(\mathbf{x}) \left[\frac{\mathbf{v}}{\|\mathbf{v}\|} + \frac{\xi}{\|\xi\|} \right] \right\| \leq e^{M_1 T} \left\| \frac{\mathbf{v}}{\|\mathbf{v}\|} + \frac{\xi}{\|\xi\|} \right\|.$$

Hence

$$\left\| \frac{\mathbf{v}}{\|\mathbf{v}\|} + \frac{\xi}{\|\xi\|} \right\| \geq (\sigma - 1) e^{-2M_1 T}.$$

By inequality (8) in Chapter 2 it follows that

$$\max\{\|\mathbf{v}\|, \|\xi\|\} \leq 2(\sigma - 1)^{-1} e^{2M_1 T} \|\mathbf{v} + \xi\|. \tag{9}$$

Similarly, since

$$\frac{\|\mathbf{v}\|}{\|\mathbf{x}^0(T)\|} \cdot \frac{\|\mathbf{x}^u(T)\|}{\|\eta\|} \geq \Delta M_0^{-1} K_2^{-1} e^{\alpha_2 T}$$

we deduce that

$$\max\{\|\mathbf{v}\|, \|\eta\|\} \leq 2(\sigma - 1)^{-1} e^{2M_1 T} \|\mathbf{v} + \eta\|.$$

By invariance, this implies for all t that

$$\max\{\|\mathbf{x}^0(t)\|, \|\mathbf{x}^u(t)\|\} \leq 2(\sigma - 1)^{-1} e^{2M_1 T} \|\mathbf{x}^0(t) + \mathbf{x}^u(t)\|. \tag{10}$$

Next we observe that for $t \geq 0$

$$\begin{aligned}\frac{\|\xi\|}{\|\mathbf{x}^s(t)\|}\frac{\|\mathbf{x}^0(t)+\mathbf{x}^u(t)\|}{\|v+\eta\|} &\geq K_1^{-1}e^{\alpha_1 t}\frac{\|\mathbf{x}^0(t)+\mathbf{x}^u(t)\|}{\|\mathbf{v}\|+\|\eta\|}\\ &\geq K_1^{-1}e^{\alpha_1 t}\frac{\|\mathbf{x}^0(t)+\mathbf{x}^u(t)\|}{M_0\Delta^{-1}\|\mathbf{x}^0(t)\|+K_2e^{-\alpha_2 t}\|\mathbf{x}^u(t)\|}\\ &\geq K_1^{-1}e^{\alpha_1 t}\frac{1}{(M_0\Delta^{-1}+K_2e^{-\alpha_2 t})2(\sigma-1)^{-1}e^{2M_1T}},\end{aligned}$$

using Eq.(10). Now choose the positive number T_1 so that

$$\sigma_1 = K_1^{-1}e^{\alpha_1 T_1}\frac{1}{(M_0\Delta^{-1}+K_2e^{-\alpha_2 T_1})2(\sigma-1)^{-1}e^{2M_1T}} > 1.$$

Then, by considering $\left\|\frac{\mathbf{x}^s(T_1)}{\|\xi\|}+\frac{\mathbf{x}^0(T_1)+\mathbf{x}^u(T_1)}{\|\mathbf{v}+\eta\|}\right\|$ and reasoning as above, we deduce that

$$\|\xi\| \leq 2(\sigma_1-1)^{-1}e^{2M_1T_1}\|\mathbf{v}+\xi+\eta\|.$$

Thus

$$\|\mathcal{P}^s(\mathbf{x})\| \leq 2(\sigma_1-1)^{-1}e^{2M_1T_1}.$$

Again observe that for $t \geq 0$

$$\begin{aligned}\frac{\|\mathbf{x}^u(t)\|}{\|\eta\|}\frac{\|\mathbf{v}+\xi\|}{\|\mathbf{x}^0(t)+\mathbf{x}^s(t)\|} &\geq K_2^{-1}e^{\alpha_2 t}\frac{\|\mathbf{v}+\xi\|}{\|\mathbf{x}^0(t)+\mathbf{x}^s(t)\|}\\ &\geq K_2^{-1}e^{\alpha_2 t}\frac{\|\mathbf{v}+\xi\|}{M_0\Delta^{-1}\|\mathbf{v}\|+K_1e^{-\alpha_1 t}\|\xi\|}\\ &\geq K_2^{-1}e^{\alpha_2 t}\frac{1}{(M_0\Delta^{-1}+K_1e^{-\alpha_1 t})2(\sigma-1)^{-1}e^{2M_1T}},\end{aligned}$$

using Eq.(9). Now choose $T_2 > 0$ so that

$$\sigma_2 = K_2^{-1}e^{\alpha_2 T_2}\frac{1}{(M_0\Delta^{-1}+K_1e^{-\alpha_1 T_2})2(\sigma-1)^{-1}e^{2M_1T}} > 1.$$

Then, by considering $\left\|\frac{\mathbf{x}^0(T_2)+\mathbf{x}^s(T_2)}{\|\mathbf{v}+\xi\|}+\frac{\mathbf{x}^u(T_2)}{\|\eta\|}\right\|$, we deduce that

$$\|\mathcal{P}^u(\mathbf{x})\| \leq 2(\sigma_2-1)^{-1}e^{2M_1T_2}.$$

So $\mathcal{P}^s(\mathbf{x}), \mathcal{P}^u(\mathbf{x})$ are both bounded and since $\mathcal{P}^0(\mathbf{x}) = I - \mathcal{P}^s(\mathbf{x}) - \mathcal{P}^u(\mathbf{x})$, the projection $\mathcal{P}^0(\mathbf{x})$ is bounded also.

7.3 CONTINUITY OF THE PROJECTIONS

In the previous section we showed there are constants M^0, M^s, M^u such that for all $\mathbf{x}$ in S

$$\|\mathcal{P}^0(\mathbf{x})\| \leq M^0, \quad \|\mathcal{P}^s(\mathbf{x})\| \leq M^s, \quad \|\mathcal{P}^u(\mathbf{x})\| \leq M^u.$$

From Eqs.(6) and (7) in the definition of hyperbolicity and also the fact that

$$D\phi^t(\mathbf{x})F(\mathbf{x}) = F(\phi^t(\mathbf{x})),$$

it follows that for $\mathbf{x}$ in S

$$\|D\phi^t(\mathbf{x})\mathcal{P}^0(\mathbf{x})\| \leq M_0 M^0 \Delta^{-1} \quad \text{for all } t, \tag{11}$$

$$\|D\phi^t(\mathbf{x})\mathcal{P}^s(\mathbf{x})\| \leq K_1 M^s e^{-\alpha_1 t} \quad \text{for } t \geq 0 \tag{12}$$

and

$$\|D\phi^{-t}(\mathbf{x})\mathcal{P}^u(\mathbf{x})\| \leq K_2 M^u e^{-\alpha_2 t} \quad \text{for } t \geq 0, \tag{13}$$

where M_0 and Δ are as in Eq.(8). Note also that the invariance of the splitting (5) implies that the identity

$$D\phi^t(\mathbf{x})\mathcal{P}(\mathbf{x}) = \mathcal{P}(\phi^t(\mathbf{x}))D\phi^t(\mathbf{x}) \tag{14}$$

holds with $\mathcal{P}(\mathbf{x}) = \mathcal{P}^0(\mathbf{x}),\ \mathcal{P}^s(\mathbf{x})$ or $\mathcal{P}^u(\mathbf{x})$.

In order to prove that the projections are continuous, we consider two solutions $\mathbf{x}(t), \mathbf{y}(t)$ of Eq.(1) in S such that for some positive numbers δ and T

$$\|DF(\mathbf{y}(t)) - DF(\mathbf{x}(t))\| \leq \delta \tag{15}$$

for $0 \leq t \leq T$. Note that the matrix function

$$U(t) = D\phi^{t-T}(\mathbf{y}(T))(\mathcal{P}^0(\mathbf{y}(T)) + \mathcal{P}^u(\mathbf{y}(T)))$$

is a solution of

$$\dot{U} = DF(\mathbf{x}(t))U + B(t)U,$$

where

$$B(t) = DF(\mathbf{y}(t)) - DF(\mathbf{x}(t)). \tag{16}$$

Hence, by variation of constants,

$$U(t) = D\phi^t(\mathbf{x}(0))U(0) + \int_0^t D\phi^{t-s}(\mathbf{x}(s))B(s)U(s)ds.$$

Taking $t = T$, we find that

$$\begin{aligned}\mathcal{P}^0(\mathbf{y}(T)) + P^u(\mathbf{y}(T)) &= D\phi^T(\mathbf{x}(0))D\phi^{-T}(\mathbf{y}(T))[\mathcal{P}^0(\mathbf{y}(T)) + \mathcal{P}^u(\mathbf{y}(T))] \\ &+ \int_0^T D\phi^{T-t}(\mathbf{x}(t))B(t)D\phi^{t-T}(\mathbf{y}(T))[\mathcal{P}^0(\mathbf{y}(T)) + \mathcal{P}^u(\mathbf{y}(T))]dt.\end{aligned}$$

Multiplying by $\mathcal{P}^s(\mathbf{x}(T))$ and using the invariance (14), we obtain

$$\begin{aligned}&\mathcal{P}^s(\mathbf{x}(T))[\mathcal{P}^0(\mathbf{y}(T)) + \mathcal{P}^u(\mathbf{y}(T))] \\ &\quad = D\phi^T(\mathbf{x}(0))\mathcal{P}^s(\mathbf{x}(0))D\phi^{-T}(\mathbf{y}(T))[\mathcal{P}^0(\mathbf{y}(T)) + \mathcal{P}^u(\mathbf{y}(T))] \\ &+ \int_0^T D\phi^{T-t}(\mathbf{x}(t))\mathcal{P}^s(\mathbf{x}(t))B(t)D\phi^{t-T}(\mathbf{y}(T))[\mathcal{P}^0(\mathbf{y}(T)) + \mathcal{P}^u(\mathbf{y}(T))]dt.\end{aligned}$$

Hence, using Eqs.(11), (12) and (13),

$$\begin{aligned}\|\mathcal{P}^s(\mathbf{x}(T))[I - \mathcal{P}^s(\mathbf{y}(T))]\| &\le K_1M^s e^{-\alpha_1 T}[M_0M^0\Delta^{-1} + K_2M^u] \\ &+ \int_0^T K_1M^s e^{-\alpha_1(T-t)}\delta[M_0M^0\Delta^{-1} + K_2M^u]dt.\end{aligned}$$

Thus, if inequality (15) holds for $0 \le t \le T$,

$$\begin{aligned}&\|\mathcal{P}^s(\mathbf{x}(T))[I - \mathcal{P}^s(\mathbf{y}(T))]\| \\ &\qquad \le K_1M^s(M_0M^0\Delta^{-1} + K_2M^u)(e^{-\alpha_1 T} + \alpha_1^{-1}\delta).\end{aligned} \tag{17}$$

Next if we set

$$U(t) = D\phi^t(\mathbf{y}(0))\mathcal{P}^s(\mathbf{y}(0)),$$

it follows by variation of constants again that

$$U(0) = D\phi^{-T}(\mathbf{x}(T))U(T) - \int_0^T D\phi^{-t}(\mathbf{x}(t))B(t)U(t)dt,$$

where $B(t)$ is as in Eq.(16). Hence

$$\begin{aligned}\mathcal{P}^s(\mathbf{y}(0)) &= D\phi^{-T}(\mathbf{x}(T))D\phi^T(\mathbf{y}(0))\mathcal{P}^s(\mathbf{y}(0)) \\ &\quad - \int_0^T D\phi^{-t}(\mathbf{x}(t))B(t)D\phi^t(\mathbf{y}(0))\mathcal{P}^s(\mathbf{y}(0))dt.\end{aligned}$$

Thus, using the invariance (14),

$$\begin{aligned}&[\mathcal{P}^0(\mathbf{x}(0)) + \mathcal{P}^u(\mathbf{x}(0))]\mathcal{P}^s(\mathbf{y}(0)) \\ &\qquad = D\phi^{-T}(\mathbf{x}(T))[\mathcal{P}^0(\mathbf{x}(T)) + \mathcal{P}^u(\mathbf{x}(T))]D\phi^T(\mathbf{y}(0))\mathcal{P}^s(\mathbf{y}(0)) \\ &- \int_0^T D\phi^{-t}(\mathbf{x}(t))[\mathcal{P}^0(\mathbf{x}(t)) + P^u(\mathbf{x}(t))]B(t)D\phi^t(\mathbf{y}(0))\mathcal{P}^s(\mathbf{y}(0))dt\end{aligned}$$

and so, using Eqs.(11), (12) and (13),

$$\begin{aligned}\|[I - \mathcal{P}^s(\mathbf{x}(0))]\mathcal{P}^s(\mathbf{y}(0))\| &\leq (M_0M^0\Delta^{-1} + K_2M^u)K_1M^se^{-\alpha_1T} \\ &\quad + \int_0^T (M_0M^0\Delta^{-1} + K_2M^u)\delta K_1M^se^{-\alpha_1t}dt.\end{aligned}$$

Therefore, if inequality (15) holds for $0 \leq t \leq T$,

$$\|[I - \mathcal{P}^s(\mathbf{x}(0))]\mathcal{P}^s(\mathbf{y}(0))\| \leq K_1M^s(M_0M^0\Delta^{-1} + K_2M^u)(e^{-\alpha_1T} + \alpha_1^{-1}\delta). \quad (18)$$

Then it follows from Eqs.(17) and (18) that if $\mathbf{x}(t)$ and $\mathbf{y}(t)$ are two solutions in S satisfying inequality (15) for $-T \leq t \leq T$,

$$\|\mathcal{P}^s(\mathbf{y}(0)) - \mathcal{P}^s(\mathbf{x}(0))\| \leq 2K_1M^s(M_0M^0\Delta^{-1} + K_2M^u)(e^{-\alpha_1T} + \alpha_1^{-1}\delta), \quad (19)$$

where we apply inequality (17) to the interval $[-T, 0]$ and (18) to the interval $[0, T]$.

Next we derive a similar inequality for $\mathcal{P}^u$. We consider two solutions $\mathbf{x}(t), \mathbf{y}(t)$ in S satisfying inequality (15) for $0 \leq t \leq T$. First we examine

$$U(t) = D\phi^t(\mathbf{y}(0))[\mathcal{P}^0(\mathbf{y}(0)) + \mathcal{P}^s(\mathbf{y}(0))].$$

By variation of constants,

$$U(0) = D\phi^{-T}(\mathbf{x}(T))U(T) - \int_0^T D\phi^{-t}(\mathbf{x}(t))B(t)U(t)dt,$$

where $B(t)$ is as in Eq.(16). Multiplying by $\mathcal{P}^u(\mathbf{x}(0))$ and using Eq.(14), we obtain

$$\begin{aligned}
&\mathcal{P}^u(\mathbf{x}(0))[\mathcal{P}^0(\mathbf{y}(0)) + \mathcal{P}^s(\mathbf{y}(0))] \\
&\qquad = D\phi^{-T}(\mathbf{x}(T))\mathcal{P}^u(\mathbf{x}(T))D\phi^T(\mathbf{y}(0))[\mathcal{P}^0(\mathbf{y}(0)) + \mathcal{P}^s(\mathbf{y}(0))] \\
&\qquad - \int_0^T D\phi^{-t}(\mathbf{x}(t))\mathcal{P}^u(\mathbf{x}(t))B(t)D\phi^t(\mathbf{y}(0))[\mathcal{P}^0(\mathbf{y}(0)) + \mathcal{P}^s(\mathbf{y}(0))]dt.
\end{aligned}$$

Hence

$$\begin{aligned}
&\|\mathcal{P}^u(\mathbf{x}(0))[I - \mathcal{P}^u(\mathbf{y}(0))]\| \\
&\qquad \leq K_2 M^u e^{-\alpha_2 T}[M_0 M^0 \Delta^{-1} + K_1 M^s] \\
&\qquad\quad + \int_0^T K_2 M^u e^{-\alpha_2 t}\delta[M_0 M^0 \Delta^{-1} + K_1 M^s]dt \\
&\qquad \leq K_2 M^u [M_0 M^0 \Delta^{-1} + K_1 M^s](e^{-\alpha_2 T} + \alpha_2^{-1}\delta).
\end{aligned} \tag{20}$$

Next set

$$U(t) = D\phi^{t-T}(\mathbf{y}(T))\mathcal{P}^u(\mathbf{y}(T)).$$

By variation of constants,

$$U(T) = D\phi^T(\mathbf{x}(0))U(0) + \int_0^T D\phi^{T-t}(\mathbf{x}(t))B(t)U(t)dt.$$

Multiplying by $\mathcal{P}^0(\mathbf{x}(T)) + \mathcal{P}^s(\mathbf{x}(T))$ and using Eq.(14), we obtain

$$\begin{aligned}
&[\mathcal{P}^0(\mathbf{x}(T)) + \mathcal{P}^s(\mathbf{x}(T))]\mathcal{P}^u(\mathbf{y}(T)) \\
&\qquad = D\phi^T(\mathbf{x}(0))[\mathcal{P}^0(\mathbf{x}(0)) + \mathcal{P}^s(\mathbf{x}(0))]D\phi^{-T}(\mathbf{y}(T))\mathcal{P}^u(\mathbf{y}(T)) \\
&\qquad + \int_0^T D\phi^{T-t}(\mathbf{x}(t))[\mathcal{P}^0(\mathbf{x}(t)) + \mathcal{P}^s(\mathbf{x}(t))]B(t)D\phi^{t-T}(\mathbf{y}(T))\mathcal{P}^u(\mathbf{y}(T))dt.
\end{aligned}$$

Hence

$$\begin{aligned}
&\|[I - \mathcal{P}^u(\mathbf{x}(T))]\mathcal{P}^u(\mathbf{y}(T))\| \\
&\qquad \le [M_0 M^0 \Delta^{-1} + K_1 M^s] K_2 M^u e^{-\alpha_2 T} \\
&\qquad\qquad + \int_0^T [M_0 M^0 \Delta^{-1} + K_1 M^s] \delta K_2 M^u e^{-\alpha_2 (T-t)} dt \qquad (21) \\
&\qquad \le K_2 M^u (M_0 M^0 \Delta^{-1} + K_1 M^s)(e^{-\alpha_2 T} + \alpha_2^{-1}\delta).
\end{aligned}$$

Now let $\mathbf{x}(t)$ and $\mathbf{y}(t)$ be two solutions in S satisfying inequality (15) for $-T \le t \le T$. Then, applying Eq.(20) on $[0, T]$ and (21) on $[-T, 0]$, we deduce that

$$\|\mathcal{P}^u(\mathbf{y}(0)) - \mathcal{P}^u(\mathbf{x}(0))\| \le 2K_2 M^u (M_0 M^0 \Delta^{-1} + K_1 M^s)(e^{-\alpha_2 T} + \alpha_2^{-1}\delta). \quad (22)$$

To finish the proof of the continuity of $\mathcal{P}^s$ and $\mathcal{P}^u$, fix a positive number ε. Let $T > 0$ satisfy

$$4K_1 M^s (M_0 M^0 \Delta^{-1} + K_2 M^u) e^{-\alpha_1 T} \le \varepsilon$$

and

$$4K_2 M^u (M_0 M^0 \Delta^{-1} + K_1 M^s) e^{-\alpha_2 T} \le \varepsilon,$$

and choose $\delta > 0$ so that

$$4K_1 M^s (M_0 M^0 \Delta^{-1} + K_2 M^u) \alpha_1^{-1} \delta \le \varepsilon$$

and

$$4K_2 M^u (M_0 M^0 \Delta^{-1} + K_1 M^s) \alpha_2^{-1} \delta \le \varepsilon.$$

Then choose $\delta_1 > 0$ so that

$$\|DF(\phi^t(\mathbf{y})) - DF(\phi^t(\mathbf{x}))\| \le \delta$$

if $\mathbf{x}, \mathbf{y} \in S$, $-T \le t \le T$ and $\|\mathbf{x} - \mathbf{y}\| \le \delta_1$.

Then if $\mathbf{x}, \mathbf{y} \in S$ and $\|\mathbf{y} - \mathbf{x}\| \le \delta_1$ it follows from Eqs.(19) and (22) with $\mathbf{x}(t) = \phi^t(\mathbf{x})$ and $\mathbf{y}(t) = \phi^t(\mathbf{y})$ that

$$\|\mathcal{P}^s(\mathbf{y}) - \mathcal{P}^s(\mathbf{x})\| \le \varepsilon, \quad \|\mathcal{P}^u(\mathbf{y}) - \mathcal{P}^u(\mathbf{x})\| \le \varepsilon.$$

So $\mathcal{P}^s, \mathcal{P}^u$ are both continuous on S and the continuity of $\mathcal{P}^0$ follows from the fact that $\mathcal{P}^0 = I - \mathcal{P}^s - \mathcal{P}^u$.

7.4 EXPONENTIAL DICHOTOMIES FOR DIFFERENTIAL EQUATIONS

In order to assist us in proving additional properties of hyperbolic sets, we need to develop the perturbation theory of exponential dichotomies for linear differential equations (conf. Massera and Schäffer [1966], Coppel [1978], Daleckii and Krein [1974]). We also show how exponential dichotomy is related to hyperbolicity. Lastly, for completeness, we demonstrate how hyperbolicity can be defined using linear skew product flows (conf. Sacker and Sell [1974]).

Definition 7.2. Let $A(t)$ be a piecewise continuous $n \times n$ matrix valued function defined on an interval J. The linear differential equation

$$\dot{\mathbf{x}} = A(t)\mathbf{x} \tag{23}$$

is said to have an *exponential dichotomy* on J if there are projections $P(t)$, $t \in J$, and positive constants $K_1, K_2, \alpha_1, \alpha_2$ such that

$$X(t)X^{-1}(s)P(s) = P(t)X(t)X^{-1}(s) \quad \text{for} \quad t, s \in J, \tag{24}$$

$$\|X(t)X^{-1}(s)P(s)\| \le K_1 e^{-\alpha_1(t-s)} \quad \text{for} \quad t, s, \in J \quad \text{with} \quad t \ge s \tag{25}$$

and

$$\|X(t)X^{-1}(s)\,(I - P(s))\,\| \le K_2 e^{-\alpha_2(s-t)} \quad \text{for} \quad t, s \in J \quad \text{with} \quad t \le s. \tag{26}$$

Here $X(t)$ is any fundamental matrix for Eq.(23). Note that K_1, K_2 are called *constants* and α_1, α_2 *exponents* associated with the dichotomy.

We note four facts about exponential dichotomies in the following proposition.

Proposition 7.3. *Let* Eq.(23) *have a fundamental matrix* $X(t)$ *and an exponential dichotomy on an interval* J *with projection* $P(t)$, *constants* K_1, K_2 *and exponents* α_1, α_2. *Then*

(i) *the adjoint equation*

$$\dot{\mathbf{x}} = -A^*(t)\mathbf{x}$$

has an exponential dichotomy on J with projection $I - P^*(t)$, constants K_2, K_1 and exponents α_2, α_1;

(ii) *if $J = (-\infty, \infty)$, then*

$$\begin{aligned} \mathcal{R}(P(t)) &= \{\xi : X(\tau)X^{-1}(t)\xi \to 0 \quad \text{as} \quad \tau \to \infty\} \\ &= \{\xi : \sup_{\tau \geq t} \|X(\tau)X^{-1}(t)\xi\| < \infty\} \end{aligned}$$

and

$$\begin{aligned} \mathcal{N}(P(t)) &= \{\xi : X(\tau)X^{-1}(t)\xi \to 0 \quad \text{as} \quad \tau \to -\infty\} \\ &= \{\xi : \sup_{\tau \leq t} \|X(\tau)X^{-1}(t)\xi\| < \infty\}; \end{aligned}$$

(iii) *if $J = (-\infty, \infty)$ and $f(t)$ is a bounded continuous $\mathbb{R}^n$-valued function defined on $(-\infty, \infty)$, then the equation*

$$\dot{\mathbf{x}} = A(t)\mathbf{x} + f(t) \tag{27}$$

has a unique bounded solution $\mathbf{x}(t)$. Moreover

$$\|\mathbf{x}(t)\| \leq \left(K_1\alpha_1^{-1} + K_2\alpha_2^{-1}\right) \|f\|_\infty$$

for all t, where

$$\|f\|_\infty = \sup_{-\infty < t < \infty} \|f(t)\|;$$

(iv) *if Eq.(23) has an exponential dichotomy on $(-\infty, \infty)$, the difference equation*

$$\mathbf{u}_{k+1} = X(k+1)X^{-1}(k)\mathbf{u}_k \tag{28}$$

has an exponential dichotomy on $(-\infty, \infty)$. The converse is true if $A(t)$ is bounded.

Proof. (i) It follows by taking adjoints in Eq.(24) that for all t, s in J

$$X^{*-1}(s)X^*(t)P^*(t) = P^*(s)X^{*-1}(s)X^*(t)$$

and hence that

$$X^{*-1}(t)X^*(s)(I - P^*(s)) = (I - P^*(t))X^{*-1}(t)X^*(s). \tag{29}$$

Also taking adjoints in inequalities (25) and (26), we find (since we are using the Euclidean norm in $\mathbb{R}^n$) that

$$\|P^*(s)X^{*-1}(s)X^*(t))\| \leq K_1 e^{-\alpha_1(t-s)} \text{ for } t \geq s \text{ in } J$$

and

$$\|(I - P^*(s))X^{*-1}(s)X^*(t)\| \leq K_2 e^{-\alpha_2(s-t)} \text{ for } t \leq s \text{ in } J.$$

Using the invariance (29), it follows that

$$\|X^{*-1}(t)X^*(s)(I-P^*(s))\| \le K_2 e^{-\alpha_2(t-s)} \text{ for } t \ge s \text{ in } J$$

and

$$\|X^{*-1}(t)X^*(s)P^*(s)\| \le K_1 e^{-\alpha_1(s-t)} \text{ for } t \le s \text{ in } J.$$

(ii) It is clear that

$$\begin{aligned}\mathcal{R}(P(t)) &\subset \{\xi : X(\tau)X^{-1}(t)\xi \to \mathbf{0} \quad \text{as} \quad \tau \to \infty\} \\ &\subset \{\xi : \sup_{\tau\ge t} \|X(\tau)X^{-1}(t)\xi\| < \infty\}.\end{aligned}$$

To show these subspaces all coincide, suppose $\sup_{\tau\ge t}\|X(\tau)X^{-1}(t)\xi\| < \infty$. Then for $\tau \ge t$,

$$\begin{aligned}\|(I-P(t))\xi\| &= \|(I-P(t))X(t)X^{-1}(\tau)X(\tau)X^{-1}(t)\xi\| \\ &= \|X(t)X^{-1}(\tau)(I-P(\tau))X(\tau)X^{-1}(t)\xi\| \quad \text{by invariance} \\ &\le K_2 e^{-\alpha_2(\tau-t)}\|X(\tau)X^{-1}(t)\xi\| \\ &\to 0 \quad \text{as} \quad \tau \to \infty.\end{aligned}$$

Hence $(I-P(t))\xi = \mathbf{0}$ and so $\xi \in \mathcal{R}(P(t))$. This proves the assertion about $\mathcal{R}(P(t))$ and the assertion about $\mathcal{N}(P(t))$ is similarly proved.

(iii) The difference between two bounded solutions of Eq.(27) would be a bounded solution $\mathbf{x}(t)$ of Eq.(23). Then if $t \le 0$

$$\begin{aligned}\|P(0)\mathbf{x}(0)\| &= \|P(0)X(0)X^{-1}(t)\mathbf{x}(t)\| \\ &= \|X(0)X^{-1}(t)P(t)\mathbf{x}(t)\| \\ &\le K_1 e^{\alpha_1 t}\|\mathbf{x}(t)\|.\end{aligned}$$

So $P(0)\mathbf{x}(0) = 0$. Similarly, $(I-P(0))\,\mathbf{x}(0) = 0$ and so $\mathbf{x}(0) = 0$. Therefore $\mathbf{x}(t) \equiv 0$ and the bounded solution of Eq.(27) is unique.

Next one verifies easily that

$$\mathbf{x}(t) = \int_{-\infty}^{t} X(t)X^{-1}(s)P(s)f(s)ds - \int_{t}^{\infty} X(t)X^{-1}(s)(I-P(s))f(s)ds$$

is well-defined and a solution of Eq.(27). Moreover,

$$\begin{aligned}\|\mathbf{x}(t)\| &\le \left\{\int_{-\infty}^{t} K_1 e^{-\alpha_1(t-s)}ds + \int_{t}^{\infty} K_2 e^{-\alpha_2(s-t)}ds\right\}\|f\|_\infty \\ &= \left(K_1\alpha_1^{-1} + K_2\alpha_2^{-1}\right)\|f\|_\infty.\end{aligned}$$

(iv) Suppose Eq.(23) has an exponential dichotomy on $(-\infty, \infty)$ as in Eqs. (24), (25) and (26). Then if we define

$$A_k = X(k+1)X^{-1}(k),\ P_k = P(k),$$

it follows that

$$A_k P_k = P_{k+1} A_k, \tag{30}$$

$$\|\Phi(k,m)P_m\| \le K_1(e^{-\alpha_1})^{k-m} \quad \text{for} \quad k \ge m \tag{31}$$

and

$$\|\Phi(k,m)(I - P_m)\| \le K_2(e^{-\alpha_2})^{m-k} \quad \text{for} \quad k \le m. \tag{32}$$

Conversely, suppose (30), (31) and (32) hold and that

$$\|A(t)\| \le M$$

for all t. By Gronwall's lemma, this implies that

$$\|X(t)X^{-1}(s)\| \le e^{M|t-s|}$$

for all t, s, where we assume without loss of generality that $X(0) = I$. If we define

$$P(t) = X(t)P_0X^{-1}(t),$$

it is clear that the invariance condition (24) is satisfied. Suppose $t \ge s$. Then there are integers $m \le k$ such that $m \le s \le m+1$ and $k \le t \le k+1$ and so

$$\begin{aligned}
&\|X(t)X^{-1}(s)P(s)\| \\
&\quad = \|X(t)X^{-1}(k)X(k)X^{-1}(s)X(s)P_0X^{-1}(s)\| \\
&\quad = \|X(t)X^{-1}(k)X(k)X^{-1}(s)X(s)X^{-1}(m)P_mX(m)X^{-1}(s)\| \\
&\quad = \|X(t)X^{-1}(k)X(k)X^{-1}(m)P_mX(m)X^{-1}(s)\| \\
&\quad \le e^M K_1(e^{-\alpha_1})^{k-m}e^M \\
&\quad \le e^{2M} K_1(e^{-\alpha_1})^{t-s-1} \\
&\quad = K_1 e^{2M+\alpha_1} e^{-\alpha_1(t-s)}.
\end{aligned}$$

Similarly, we show when $t \le s$ that

$$\|X(t)X^{-1}(s)(I - P(s))\| \le K_2 e^{2M+\alpha_2} e^{-\alpha_2(s-t)}.$$

So the proof of the proposition is complete.

Now, for use in the sequel, we indicate a way in which exponential dichotomies fit into the theory of hyperbolic sets for flows. Indeed let S be a compact hyperbolic set for Eq.(1) as in Definition 7.1. For fixed $\mathbf{x} \in S$, we triangularise the variational system

$$\dot{\mathbf{y}} = DF(\phi^t(\mathbf{x}))\mathbf{y}.$$

Choose an $n \times n$ orthogonal matrix

$$T_0 = \left[\frac{F(\mathbf{x})}{\|F(\mathbf{x})\|} \middle| S_0 \right].$$

We apply the Gram-Schmidt procedure to $D\phi^t(\mathbf{x})T_0$ to obtain an orthogonal matrix

$$\left[\frac{F(\phi^t(\mathbf{x}))}{\|F(\phi^t(\mathbf{x}))\|} \middle| S(t) \right]$$

and an upper triangular matrix

$$Y(t) = S^*(t)D\phi^t(\mathbf{x})S_0$$

with positive diagonal entries such that

$$D\phi^t(\mathbf{x})T_0 = \left[\frac{F(\phi^t(\mathbf{x}))}{\|F(\phi^t(\mathbf{x}))\|} \middle| S(t) \right] \begin{bmatrix} \frac{\|F(\phi^t(\mathbf{x}))\|}{\|F(\mathbf{x})\|} & *** \\ 0 & Y(t) \end{bmatrix}. \tag{33}$$

Note that

$$S^*(t)S(t) = I, \; S(t)S^*(t) = I - \frac{F(\phi^t(\mathbf{x}))F(\phi^t(\mathbf{x}))^*}{\|F(\phi^t(\mathbf{x}))\|^2}$$

and also that

$$\dot{Y}(t) = A(t)Y(t),$$

where

$$A(t) = S^*(t)[DF(\phi^t(\mathbf{x}))S(t) - \dot{S}(t)].$$

We define

$$Q(t) = S^*(t)\mathcal{P}^s(\phi^t(\mathbf{x}))S(t),$$

where $\mathcal{P}^s(\mathbf{x})$ is the projection onto $E^s(\mathbf{x})$ along $E^0(\mathbf{x}) \oplus E^u(\mathbf{x})$ as in Sections

7.2 and 7.3. $Q(t)$ is a projection since

$$\begin{aligned} Q(t)^2 &= S^*(t)\mathcal{P}^s(\phi^t(\mathbf{x}))S(t)S^*(t)\mathcal{P}^s(\phi^t(\mathbf{x}))S(t) \\ &= S^*(t)\mathcal{P}^s(\phi^t(\mathbf{x}))\left(I - \frac{F(\phi^t(\mathbf{x}))F(\phi^t(\mathbf{x}))^*}{\|F(\phi^t(\mathbf{x}))\|^2}\right)\mathcal{P}^s(\phi^t(\mathbf{x}))S(t) \\ &= S(t)\mathcal{P}^s(\phi^t(\mathbf{x}))\mathcal{P}^s(\phi^t(\mathbf{x}))S(t) \\ &= Q(t), \end{aligned}$$

where we used the identity $\mathcal{P}^s(\mathbf{x})F(\mathbf{x}) = \mathbf{0}$. Also

$$\begin{aligned} &Y(t)Y^{-1}(s)Q(s) \\ &\quad = S^*(t)D\phi^t(\mathbf{x})S_0S_0^*D\phi^{-s}(\phi^s(\mathbf{x}))S(s)S^*(s)\mathcal{P}^s(\phi^s(\mathbf{x}))S(s) \\ &\quad = S^*(t)D\phi^t(\mathbf{x})\left(I - \frac{F(\mathbf{x})F(\mathbf{x})^*}{\|F(\mathbf{x})\|^2}\right)D\phi^{-s}(\phi^s(\mathbf{x}))S(s)S^*(s)\mathcal{P}^s(\phi^s(\mathbf{x}))S(s) \\ &\quad = S^*(t)\left(D\phi^t(\mathbf{x}) - \frac{F(\phi^t(\mathbf{x}))F(\mathbf{x})^*}{\|F(\mathbf{x})\|^2}\right)D\phi^{-s}(\phi^s(\mathbf{x}))S(s)S^*(s)\mathcal{P}^s(\phi^s(\mathbf{x}))S(s) \\ &\qquad\qquad \text{since } D\phi^t(\mathbf{x})F(\mathbf{x}) = F(\phi^t(\mathbf{x})) \\ &\quad = S^*(t)D\phi^t(\mathbf{x})D\phi^{-s}(\phi^s(\mathbf{x}))S(s)S^*(s)\mathcal{P}^s(\phi^s(\mathbf{x}))S(s) \\ &\qquad\qquad \text{by orthogonality} \\ &\quad = S^*(t)D\phi^{t-s}(\phi^s(\mathbf{x}))\left(I - \frac{F(\phi^s(\mathbf{x}))F(\phi^s(\mathbf{x}))^*}{\|F(\phi^s(\mathbf{x}))\|^2}\right)\mathcal{P}^s(\phi^s(\mathbf{x}))S(s) \\ &\quad = S^*(t)D\phi^{t-s}(\phi^s(\mathbf{x}))\mathcal{P}^s(\phi^s(\mathbf{x}))S(s) \quad \text{by orthogonality} \\ &\quad = S^*(t)\mathcal{P}^s(\phi^t(\mathbf{x}))D\phi^{t-s}(\phi^s(\mathbf{x}))S(s) \quad \text{by invariance of } \mathcal{P}^s \\ &\quad = S^*(t)\mathcal{P}^s(\phi^t(\mathbf{x}))S(t)S^*(t)D\phi^t(\mathbf{x})S_0S_0^*D\phi^{-s}(\phi^s(\mathbf{x}))S(s) \\ &\qquad\qquad \text{reversing the above arguments} \\ &\quad = Q(t)Y(t)Y^{-1}(s) \end{aligned}$$

and so we have the required invariance property. Moreover, using one of the equations just derived,

$$\begin{aligned}\|Y(t)Y^{-1}(s)Q(s)\| &= \|S^*(t)D\phi^{t-s}(\phi^s(\mathbf{x}))\mathcal{P}^s(\phi^s(\mathbf{x}))S(s)\| \\ &\leq \|D\phi^{t-s}(\phi^s(\mathbf{x}))\mathcal{P}^s(\phi^s(\mathbf{x}))\| \\ &\leq K_1 M^s e^{-\alpha_1(t-s)}\end{aligned}$$

for $t \geq s$, with M^s the bound on $\|\mathcal{P}^s(\cdot)\|$. Similarly, if $t \leq s$,

$$\|Y(t)Y^{-1}(s)(I - Q(s))\| \leq K_2 M^u e^{-\alpha_2(s-t)},$$

with M^u the bound on $\|\mathcal{P}^u(\cdot)\|$. *Thus the equation*

$$\dot{\mathbf{y}} = A(t)\mathbf{y} = S^*(t)[DF(\phi^t(\mathbf{x}))S(t) - \dot{S}(t)]\mathbf{y},$$

which has fundamental matrix $Y(t)$, *has an exponential dichotomy on* $(-\infty, \infty)$ *with projection* $Q(t)$, *constants* $K_1 M^s$, $K_2 M^u$ *and exponents* α_1, α_2.

There is a second way in which exponential dichotomies fit into the theory of hyperbolic sets. Again let S be a compact hyperbolic set for Eq.(1) as in Definition 7.1 and let $\mathcal{P}^0(\mathbf{x})$, $\mathcal{P}^s(\mathbf{x})$, $\mathcal{P}^u(\mathbf{x})$ be the projections associated with the splitting (5). For given $\mathbf{x}$ in S, we examine the equation

$$\dot{\mathbf{y}} = DF(\phi^t(\mathbf{x}))\mathbf{y}.$$

If $\mathbf{y}(t)$ is a solution of this equation in $E^u(\phi^t(\mathbf{x}))$, then for $t \leq s$

$$\|\mathbf{y}(t)\| \leq K_2 e^{-\alpha_2(s-t)}\|\mathbf{y}(s)\|. \tag{34}$$

On the other hand, if $\mathbf{y}(t)$ is a solution in $E^s(\phi^t(\mathbf{x}))$ and α is a real number, then for $t \geq s$

$$\begin{aligned}\|\mathbf{y}(t) + \alpha F(\phi^t(\mathbf{x}))\| &\leq \|\mathbf{y}(t)\| + |\alpha|\|F(\phi^t(\mathbf{x}))\| \\ &\leq K_1 e^{-\alpha_1(t-s)}\|\mathbf{y}(s)\| + |\alpha| M_0 \Delta^{-1}\|F(\phi^s(\mathbf{x}))\| \\ &\leq \max\{K_1, M_0\Delta^{-1}\}[\|\mathbf{y}(s)\| + |\alpha|\|F(\phi^s(\mathbf{x}))\|],\end{aligned}$$

where

$$M_0 = \sup_{\mathbf{x}\in S}\|F(\mathbf{x})\| \quad \text{and} \quad \Delta = \inf_{\mathbf{x}\in S}\|F(\mathbf{x})\|.$$

Notice that

$$\mathbf{y}(s) = \mathcal{P}^s(\phi^s(\mathbf{x}))[\mathbf{y}(s) + \alpha F(\phi^s(\mathbf{x}))]$$

and

$$\alpha F(\phi^s(\mathbf{x})) = \mathcal{P}^0(\phi^s(\mathbf{x}))[\mathbf{y}(s) + \alpha F(\phi^s(\mathbf{x}))].$$

Hence

$$\|\mathbf{y}(s)\| + |\alpha|\|F(\phi^s(\mathbf{x}))\| \le \max\{M^0, M^s\}\|\mathbf{y}(s) + \alpha F(\phi^s(\mathbf{x}))\|,$$

where

$$M^0 = \sup_{\mathbf{x}\in S} \|\mathcal{P}^0(\mathbf{x})\|,\ M^s = \sup_{\mathbf{x}\in S} \|\mathcal{P}^s(\mathbf{x})\|,\ M^u = \sup_{\mathbf{x}\in S} \|\mathcal{P}^u(\mathbf{x})\|.$$

Thus, if $\mathbf{y}(t)$ is a solution in $E^s(\phi^t(\mathbf{x}))$ and α is a real number, then for $t \ge s$,

$$\begin{aligned}&\|\mathbf{y}(t) + \alpha F(\phi^t(\mathbf{x}))\| \\ &\qquad \le \max\{K_1, M_0\Delta^{-1}\}\max\{M^0, M^s\}\|\mathbf{y}(s) + \alpha F(\phi^s(\mathbf{x}))\|.\end{aligned} \tag{35}$$

It follows from inequalities (34) and (35) that, *when* $0 < \lambda < \alpha_2$, *the equation*

$$\dot{\mathbf{y}} = [DF(\phi^t(\mathbf{x})) - \lambda]\mathbf{y}$$

has an exponential dichotomy on $(-\infty, \infty)$ *with projections* $\mathcal{P}^0(\phi^t(\mathbf{x}))+ \mathcal{P}^s(\phi^t(\mathbf{x}))$, *exponents* λ, $\alpha_2 - \lambda$ *and constants*

$$(M^0 + M^s)\max\{K_1, M_0\Delta^{-1}\}\max\{M^0, M^s\} \quad \text{and} \quad K_2M^u.$$

Similarly, *when* $0 < \lambda < \alpha_1$, *the equation*

$$\dot{\mathbf{y}} = [DF(\phi^t(\mathbf{x})) + \lambda]\mathbf{y}$$

has an exponential dichotomy on $(-\infty, \infty)$ *with projections* $\mathcal{P}^s(\phi^t(\mathbf{x}))$, *exponents* $\alpha_1 - \lambda$, λ *and constants*

$$K_1M^s \quad \text{and} \quad (M^0 + M^u)\max\{K_2, M_0\Delta^{-1}\}\max\{M^0, M^u\}.$$

So this is the second way exponential dichotomies fit into the theory of hyperbolicity.

Now we prove a roughness result, analogous to Lemma 2.8 for difference equations.

Lemma 7.4. *Let the differential equation* (23) *have an exponential dichotomy on an interval* $J = [a, b]$ *(interpreted as* $[a, \infty)$ *when* $b = \infty$*, etc.) with projections* $P(t)$, *constants* K_1, K_2 *and exponents* α_1, α_2 *and suppose* β_1 *and* β_2 *are numbers satisfying*

$$0 < \beta_1 < \alpha_1,\ 0 < \beta_2 < \alpha_2.$$

Then there exists a positive number $\delta_0 = \delta_0(K_1, K_2, \alpha_1, \alpha_2, \beta_1, \beta_2)$ such that if $B(t)$ is a piecewise continuous $n \times n$ matrix valued function with

$$\|B(t)\| \leq \delta \leq \delta_0$$

for $t \in J$, the perturbed system

$$\dot{\mathbf{x}} = [A(t) + B(t)]\mathbf{x} \tag{36}$$

has an exponential dichotomy on J with constants L_1, L_2 exponents β_1, β_2 and projections $Q(t)$ satisfying

$$\|Q(t) - P(t)\| \leq N\delta,$$

where L_1, L_2, N are constants depending only on $K_1, K_2, \alpha_1, \alpha_2$.

Proof. We assume that δ_0 is the least positive number satisfying the inequalities

$$2(K_1\alpha_1^{-1} + K_2\alpha_2^{-1})\delta_0 \leq 1,$$
$$2K_1\delta_0 \leq \alpha_1 - \beta_1, \; 2K_2\delta_0 \leq \alpha_2 - \beta_2$$

and

$$4K_1K_2(\alpha_1^{-1} + \alpha_2^{-1})\delta_0 \leq 1.$$

First consider the case where $J = [a, b]$ is a finite interval. If $\mathbf{u}(t)$ is a solution of Eq.(36), then by variation of constants

$$\mathbf{u}(t) = X(t)X^{-1}(a)\mathbf{u}(a) + \int_a^t X(t)X^{-1}(s)B(s)\mathbf{u}(s)ds \tag{37}$$

or

$$\mathbf{u}(t) = X(t)X^{-1}(b)\mathbf{u}(b) - \int_t^b X(t)X^{-1}(s)B(s)\mathbf{u}(s)ds. \tag{38}$$

Multiplying Eq.(37) by $P(t)$, Eq.(38) by $I - P(t)$, using the invariance and adding, we obtain

$$\begin{aligned} \mathbf{u}(t) &= X(t)X^{-1}(a)P(a)\mathbf{u}(a) + X(t)X^{-1}(b)(I - P(b))\mathbf{u}(b) \\ &\quad + \int_a^t X(t)X^{-1}(s)P(s)B(s)\mathbf{u}(s)ds \\ &\quad - \int_t^b X(t)X^{-1}(s)(I - P(s))B(s)\mathbf{u}(s)ds, \end{aligned} \tag{39}$$

a representation of the solution in terms of its boundary values $P(a)\mathbf{u}(a)$ and $(I - P(b))\mathbf{u}(b)$.

We consider those solutions $\mathbf{u}(t)$ of Eq.(36) for which $(I - P(b))\mathbf{u}(b) = 0$ and show that they satisfy certain exponential estimates. By Eq.(39) with s instead of a, we have

$$\begin{aligned}\mathbf{u}(t) &= X(t)X^{-1}(s)P(s)\mathbf{u}(s) + \int_s^t X(t)X^{-1}(\tau)P(\tau)B(\tau)\mathbf{u}(\tau)d\tau \\ &\quad - \int_t^b X(t)X^{-1}(\tau)\left(I - P(\tau)\right)B(\tau)\mathbf{u}(\tau)d\tau\end{aligned} \tag{40}$$

for $a \le s \le t \le b$. Hence

$$\begin{aligned}\|\mathbf{u}(t)\| &\le K_1 e^{-\alpha_1(t-s)}\|\mathbf{u}(s)\| + K_1\delta\int_s^t e^{-\alpha_1(t-\tau)}\|\mathbf{u}(\tau)\|d\tau \\ &\quad + K_2\delta\int_t^b e^{-\alpha_2(\tau-t)}\|\mathbf{u}(\tau)\|d\tau\end{aligned}$$

for $a \le s \le t \le b$. Then, since $\sigma\delta \le \sigma\delta_0 \le \frac{1}{2}$, where

$$\sigma = K_1\alpha_1^{-1} + K_2\alpha_2^{-1},$$

it follows from the argument used in the proof of Theorem 13 in Coppel [1965, p. 80] that

$$\|\mathbf{u}(t)\| \le K_1(1-\sigma\delta)^{-1}e^{-(\alpha_1 - K_1(1-\sigma\delta)^{-1}\delta)(t-s)}\|\mathbf{u}(s)\| \tag{41}$$

for $a \le s \le t \le b$.

Now we consider those solutions $\mathbf{u}(t)$ of Eq.(36) for which $P(a)\mathbf{u}(a) = 0$. According to Eq.(39), with s instead of b,

$$\begin{aligned}\mathbf{u}(t) &= X(t)X^{-1}(s)\left(I - P(s)\right)\mathbf{u}(s) \\ &\quad + \int_a^t X(t)X^{-1}(\tau)P(\tau)B(\tau)\mathbf{u}(\tau)d\tau \\ &\quad - \int_t^s X(t)X^{-1}(\tau)\left(I - P(\tau)\right)B(\tau)\mathbf{u}(\tau)d\tau\end{aligned} \tag{42}$$

for $a \le t \le s \le b$. Hence

$$\begin{aligned}\|\mathbf{u}(t)\| &\le K_2 e^{-\alpha_2(s-t)}\|\mathbf{u}(s)\| + K_1\delta\int_a^t e^{-\alpha_1(t-\tau)}\|\mathbf{u}(\tau)\|d\tau \\ &\quad + K_2\delta\int_t^s e^{-\alpha_2(\tau-t)}\|\mathbf{u}(\tau)\|d\tau\end{aligned}$$

for $a \leq t \leq s \leq b$. By an argument analogous to that used above, we find that

$$\|\mathbf{u}(t)\| \leq K_2(1-\sigma\delta)^{-1}e^{-(\alpha_2 - K_2(1-\sigma\delta)^{-1}\delta)(s-t)}\|\mathbf{u}(s)\| \tag{43}$$

for $a \leq t \leq s \leq b$.

Next we show given $\xi \in \mathcal{R}(P(a))$ and $\eta \in \mathcal{N}(P(b))$, there is a unique solution $\mathbf{u}(t)$ of Eq.(36) such that

$$P(a)\mathbf{u}(a) = \xi, \ (I - P(b))\,\mathbf{u}(b) = \eta. \tag{44}$$

To show this, let X be the Banach space of continuous functions $\mathbf{u} = \mathbf{u}(t)$, $a \leq t \leq b$, with norm

$$\|\mathbf{u}\| = \sup_{a \leq t \leq b} \|\mathbf{u}(t)\|.$$

For $a \leq t \leq b$, define the operator

$$\begin{aligned}(T\mathbf{u})(t) \ &= X(t)X^{-1}(a)\xi + X(t)X^{-1}(b)\eta + \int_a^t X(t)X^{-1}(s)P(s)B(s)\mathbf{u}(s)ds \\ &\quad - \int_t^b X(t)X^{-1}(s)\,(I - P(s))\,B(s)\mathbf{u}(s)ds.\end{aligned}$$

T maps X into itself and if $\mathbf{u} = \mathbf{u}(t), \mathbf{v} = \mathbf{v}(t)$ are in X

$$\begin{aligned}\|(T\mathbf{u})(t) - (T\mathbf{v})(t)\| \ &= \ \Big\| \int_a^t X(t)X^{-1}(s)P(s)B(s)[\mathbf{u}(s) - \mathbf{v}(s)]ds \\ &\quad - \int_t^b X(t)X^{-1}(s)\,(I - P(s))\,B(s)[\mathbf{u}(s) - \mathbf{v}(s)]ds \Big\| \\ &\leq \ \int_a^t K_1 e^{-\alpha_1(t-s)}\delta\|\mathbf{u}(s) - \mathbf{v}(s)\|ds \\ &\quad + \int_t^b K_2 e^{-\alpha_2(s-t)}\delta\|\mathbf{u}(s) - \mathbf{v}(s)\|ds.\end{aligned}$$

Hence

$$\|T\mathbf{u} - T\mathbf{v}\| \leq \sigma\delta\|\mathbf{u} - \mathbf{v}\|.$$

Since $\sigma\delta < 1$, it follows that T is a contraction and hence has a unique fixed point $\mathbf{u} = \mathbf{u}(t)$ which satisfies

$$\begin{aligned}\mathbf{u}(t) &= X(t)X^{-1}(a)\xi + X(t)X^{-1}(b)\eta \\ &+ \int_a^t X(t)X^{-1}(s)P(s)B(s)\mathbf{u}(s)ds \\ &- \int_t^b X(t)X^{-1}(s)\left(I - P(s)\right)B(s)\mathbf{u}(s)ds\end{aligned} \tag{45}$$

for $a \le t \le b$. Clearly Eq.(44) holds and for $a \le t \le b$

$$\dot{\mathbf{u}}(t) = A(t)\mathbf{u}(t) + P(t)B(t)\mathbf{u}(t) + \left(I - P(t)\right)B(t)\mathbf{u}(t) = A(t)\mathbf{u}(t) + B(t)\mathbf{u}(t).$$

So $\mathbf{u}(t)$ is indeed a solution of Eq.(36) satisfying (44). On the other hand, it follows from the discussion at the beginning of the proof that any such solution satisfies (39) with $P(a)\mathbf{u}(a) = \xi$ and $\left(I - P(b)\right)\mathbf{u}(b) = \eta$ and so must be a fixed point of T. Thus the uniqueness is established.

Now denote by E_t^s the subspace of $\mathbb{R}^n$ consisting of the values $\mathbf{u}(t)$ of the solutions of Eq.(36) satisfying (44) with $\eta = \mathbf{0}$ and by E_t^u those satisfying (44) with $\xi = \mathbf{0}$. Clearly $E_t^s \cap E_t^u = \{\mathbf{0}\}$ since a solution in the intersection would satisfy (44) with $\xi = \mathbf{0}$, $\eta = \mathbf{0}$ and hence, by uniqueness, must be zero. Also, by definition, $E_t^s = Y(t)Y^{-1}(a)(E_a^s)$ and $E_t^u = Y(t)Y^{-1}(b)(E_b^u)$, where $Y(t)$ is a fundamental matrix for Eq.(36). Moreover, by the existence and uniqueness of solutions of Eq.(36) satisfying (44), $\dim E_t^s = \dim E_a^s = \operatorname{rank} P(a)$, $\dim E_t^u = \dim E_b^u = n - \operatorname{rank} P(a)$ and so $\mathbb{R}^n = E_t^s \oplus E_t^u$. Therefore, if we let $Q(t)$ be the projection of $\mathbb{R}^n$ onto E_t^s along E_t^u, we see that $\operatorname{rank} Q(t) = \operatorname{rank} P(t)$ and that $Q(t)$ has the invariance property

$$Q(t)Y(t)Y^{-1}(s) = Y(t)Y^{-1}(s)Q(s).$$

Also it follows from Eqs.(41) and (43) that

$$\|Y(t)Y^{-1}(s)Q(s)\| \le 2K_1 e^{-\beta_1(t-s)}\|Q(s)\| \tag{46}$$

for $a \le s \le t \le b$ and

$$\|Y(t)Y^{-1}(s)(I - Q(s))\| \le 2K_2 e^{-\beta_2(s-t)}\|I - Q(s)\| \tag{47}$$

for $a \le t \le s \le b$.

To complete the proof for the case $J = [a, b]$, we need to estimate

$$\|Q(t) - P(t)\|.$$

We follow an argument from Coppel [1978, pp.32-33]. If $\mathbf{u}(t)$ is a solution of Eq.(36) in E_t^s, then it satisfies Eq.(40) and so, using Eq.(41) also,

$$\begin{aligned}
\|(I-P(t))\mathbf{u}(t)\| &= \left\| \int_t^b X(t)X^{-1}(\tau)\,(I-P(\tau))\,B(\tau)\mathbf{u}(\tau)d\tau \right\| \\
&\leq \int_t^b K_2 e^{-\alpha_2(\tau-t)}\delta\|\mathbf{u}(\tau)\|d\tau \\
&\leq \int_t^b K_2 e^{-\alpha_2(\tau-t)}\delta\cdot 2K_1\|\mathbf{u}(t)\|d\tau \\
&\leq a_1\delta\|\mathbf{u}(t)\|,
\end{aligned}$$

where

$$a_1 = 2K_1K_2\alpha_2^{-1}.$$

Hence, for all ξ in $\mathbb{R}^n$,

$$\|\,(I-P(t))\,Y(t)Y^{-1}(a)Q(a)\xi\| \leq a_1\delta\|Y(t)Y^{-1}(a)Q(a)\xi\|.$$

Replacing ξ by $Y(a)Y^{-1}(t)\xi$, we deduce that

$$\|\,(I-P(t))\,Q(t)\| \leq a_1\delta\|Q(t)\|. \tag{48}$$

Next if $\mathbf{u}(t)$ is a solution of Eq.(36) in E_t^u, then it satisfies Eq.(42) and so, using Eq.(43) also,

$$\begin{aligned}
\|P(t)\mathbf{u}(t)\| &= \left\| \int_a^t X(t)X^{-1}(s)P(s)B(s)\mathbf{u}(s)ds \right\| \\
&\leq \int_a^t K_1 e^{-\alpha_1(t-s)}\delta\|\mathbf{u}(s)\|ds \\
&\leq \int_a^t K_1 e^{-\alpha_1(t-s)}\delta\cdot 2K_2\|\mathbf{u}(t)\|ds \\
&\leq a_2\delta\|\mathbf{u}(t)\|,
\end{aligned}$$

where

$$a_2 = 2K_1K_2\alpha_1^{-1}.$$

Hence for all η in $\mathbb{R}^n$,

$$\|P(t)Y(t)Y^{-1}(b)\,(I-Q(b))\,\eta\| \leq a_2\delta\|Y(t)Y^{-1}(b)\,(I-Q(b))\,\eta\|.$$

Replacing η by $Y(b)Y^{-1}(t)\eta$ we conclude that

$$\|P(t)\,(I-Q(t))\,\| \le a_2\delta\|I-Q(t)\|. \tag{49}$$

Then it follows from Eqs.(48) and (49) that

$$\begin{aligned}
\|Q(t)-P(t)\| &= \|\,(I-P(t))\,Q(t)-P(t)\,(I-Q(t))\,\| \\
&\le a_1\delta\|Q(t)\|+a_2\delta\|I-Q(t)\| \\
&\le a_1\delta\,[\|P(t)\|+\|Q(t)-P(t)\|] \\
&\qquad + a_2\delta\,[\|I-P(t)\|+\|Q(t)-P(t)\|] \\
&\le a_1\delta\,[K_1+\|Q(t)-P(t)\|]+a_2\delta\,[K_2+\|Q(t)-P(t)\|]\,.
\end{aligned}$$

Hence

$$\begin{aligned}
\|Q(t)-P(t)\| &\le [1-(a_1+a_2)\delta]^{-1}(K_1a_1+K_2a_2)\delta \\
&\le 4K_1K_2(K_2\alpha_1^{-1}+K_1\alpha_2^{-1})\delta \qquad (50)\\
&\le K_1+K_2.
\end{aligned}$$

This also means that

$$\|Q(t)\| \le 2K_1+K_2,\ \|I-Q(t)\| \le 2K_2+K_1$$

and so referring to Eqs.(46), (47) and (50) we see that in the case $J=[a,b]$ the lemma has been established with

$$L_1 = 2K_1(2K_1+K_2),\ \ L_2 = 2K_2(2K_2+K_1)$$

and

$$N = 4K_1K_2(K_2\alpha_1^{-1}+K_1\alpha_2^{-1}).$$

Next we consider the case $J=[a,\infty)$. In this case, the solutions $\mathbf{u}(t)$ in E_t^s are those for which

$$\mu = \sup_{t\ge a}\|\mathbf{u}(t)\| < \infty.$$

If $\mathbf{u}(t)$ is such a solution, we multiply Eq.(38) by $I-P(t)$ to obtain for $a\le t\le b$

$$\begin{aligned}
(I-P(t))\mathbf{u}(t) &= X(t)X^{-1}(b)\,(I-P(b))\,\mathbf{u}(b) \\
&\quad -\int_t^b X(t)X^{-1}(s)\,(I-P(s))\,B(s)\mathbf{u}(s)ds.
\end{aligned} \tag{51}$$

Note that

$$\|X(t)X^{-1}(b)\,(I-P(b))\,\mathbf{u}(b)\| \le K_2 e^{-\alpha_2(b-t)}\mu$$

and

$$\int_t^\infty \|X(t)X^{-1}(s)\,(I-P(s))\,B(s)\mathbf{u}(s)\|ds \le \int_t^\infty K_2 e^{-\alpha_2(s-t)}\delta\mu ds = K_2\alpha_2^{-1}\delta\mu$$

and so we may let $b \to \infty$ in Eq.(51) to obtain

$$(I-P(t))\,\mathbf{u}(t) = -\int_t^\infty X(t)X^{-1}(s)\,(I-P(s))\,B(s)\mathbf{u}(s)ds. \qquad (52)$$

When we add this to the equation obtained by multiplying Eq.(37) by $P(t)$, we obtain

$$\begin{aligned}\mathbf{u}(t) = \; & X(t)X^{-1}(a)P(a)\mathbf{u}(a) + \int_a^t X(t)X^{-1}(s)P(s)B(s)\mathbf{u}(s)ds \\ & -\int_t^\infty X(t)X^{-1}(s)(I-P(s))B(s)\mathbf{u}(s)ds.\end{aligned}$$

The same reasoning shows that for $a \le s \le t$

$$\begin{aligned}\mathbf{u}(t) \; & = X(t)X^{-1}(s)P(s)\mathbf{u}(s) + \int_s^t X(t)X^{-1}(\tau)P(\tau)B(\tau)\mathbf{u}(\tau)d\tau \\ & -\int_t^\infty X(t)X^{-1}(\tau)(I-P(\tau))B(\tau)\mathbf{u}(\tau)d\tau.\end{aligned}$$

It follows that for $a \le s \le t$

$$\begin{aligned}\|\mathbf{u}(t)\| \; & \le K_1 e^{-\alpha_1(t-s)}\|\mathbf{u}(s)\| + \int_s^t K_1 e^{-\alpha_1(t-\tau)}\delta\|\mathbf{u}(\tau)\|d\tau \\ & + \int_t^\infty K_2 e^{-\alpha_2(\tau-t)}\delta\|\mathbf{u}(\tau)\|d\tau.\end{aligned}$$

Then it follows from the argument used in the proof of Theorem 13 in Coppel [1965, p.80] that Eq.(41) holds for $a \le s \le t$.

Next we show that for each $\xi \in \mathbb{R}^n$ there is a unique bounded solution $\mathbf{u}(t)$ of Eq.(36) on $t \ge a$ such that

$$P(a)\mathbf{u}(a) = P(a)\xi.$$

Reasoning as above, such a solution must satisfy

$$\begin{aligned}\mathbf{u}(t) \; & = X(t)X^{-1}(a)P(a)\xi + \int_a^t X(t)X^{-1}(s)P(s)B(s)\mathbf{u}(s)ds \\ & -\int_t^\infty X(t)X^{-1}(s)(I-P(s))B(s)\mathbf{u}(s)ds.\end{aligned} \qquad (53)$$

Now let X be the Banach space of bounded continuous functions $\mathbf{u}(t)$, $t \geq a$, with norm

$$\|\mathbf{u}\| = \sup_{t \geq a} \|\mathbf{u}(t)\|.$$

Define an operator T on X by taking $(T\mathbf{u})(t)$ as the right hand side of Eq.(53). Then

$$\begin{aligned} &\|(T\mathbf{u})(t)\| \\ &\leq K_1 e^{-\alpha_1(t-a)}\|\xi\| + \left(\int_a^t K_1 e^{-\alpha_1(t-s)}\delta ds + \int_t^\infty K_2 e^{-\alpha_2(s-t)}\delta ds\right)\|\mathbf{u}\| \\ &\leq K_1\|\xi\| + (K_1\alpha_1^{-1} + K_2\alpha_2^{-1})\delta\|\mathbf{u}\|. \end{aligned}$$

So T maps X into itself. The fact that T is a contraction follows as in the case $J = [a, b]$.

Summarising so far for the case $J = [a, \infty)$, what we have shown is that if $\mathbf{u}(t)$ is a solution of Eq.(36) bounded on $t \geq a$, then for $a \leq s \leq t$

$$\|\mathbf{u}(t)\| \leq 2K_1\, e^{-\beta_1(t-s)}\|\mathbf{u}(s)\|.$$

Moreover, for all ξ in $\mathcal{R}(P(a))$, there is a unique such solution satisfying

$$P(a)\mathbf{u}(a) = \xi.$$

The solutions $\mathbf{u}(t)$ in E_t^u consist of those for which $P(a)\mathbf{u}(a) = \mathbf{0}$, that is, $\mathbf{u}(a) \in \mathcal{N}(P(a))$. Then we can apply the reasoning for the case $J = [a, b]$ on any interval $[a, b]$ to deduce that Eq.(43) holds for $a \leq s \leq t$. Also since $\mathbf{u}(t) \equiv \mathbf{0}$ is the unique solution bounded in $t \geq a$ with $P(a)\mathbf{u}(a) = \mathbf{0}$, $E_t^s \cap E_t^u = \{\mathbf{0}\}$. $Q(t)$ is defined as in the first case and inequalities (46), (47) hold for $a \leq s \leq t$. The remaining reasoning for the case $J = [a, \infty)$ is the same as for the first case with obvious modifications.

The case $J = (-\infty, b]$ is treated similarly.

Finally we come to the case $J = (-\infty, \infty)$. Then the solutions in E_t^s are those for which

$$\sup_{t \geq 0} \|\mathbf{u}(t)\| < \infty$$

and we can apply the reasoning from the case $J = [a, \infty)$ in any interval $[a, \infty)$ to deduce that inequality (41) holds for $s \leq t$. Also for any $\xi \in \mathcal{R}(P(0))$ there is a unique such solution $\mathbf{u}(t)$ satisfying

$$P(0)\mathbf{u}(0) = \xi.$$

Similarly, we take the solutions in E_t^u to be those for which

$$\sup_{t\le 0}\|\mathbf{u}(t)\| < \infty.$$

Then inequality (43) holds for $t \le s$ and for any $\eta \in \mathcal{N}(P(0))$ there is a unique such solution satisfying

$$(I - P(0))\mathbf{u}(0) = \eta.$$

Next if $\mathbf{u}(t)$ is a solution in $E_t^s \cap E_t^u$, then it is bounded on $(-\infty, \infty)$ and we may let $a \to -\infty$ and $b \to \infty$ in Eq.(39) to obtain

$$\begin{aligned}\mathbf{u}(t) &= \int_{-\infty}^{t} X(t)X^{-1}(s)P(s)B(s)\mathbf{u}(s)ds \\ &\quad - \int_{t}^{\infty} X(t)X^{-1}(s)(I - P(s))B(s)\mathbf{u}(s)ds.\end{aligned}$$

If

$$\|\mathbf{u}\| = \sup_{-\infty<t<\infty} \|\mathbf{u}(t)\|,$$

it follows that

$$\|\mathbf{u}\| \le \left(K_1\alpha_1^{-1} + K_2\alpha_2^{-1}\right)\delta\|\mathbf{u}\| \le \frac{1}{2}\|\mathbf{u}\|.$$

So $\|\mathbf{u}\| = 0$ and $E_t^s \cap E_t^u = \{\mathbf{0}\}$.

Now if $\mathbf{u}(t)$ is in E_t^s it follows as in the case $J = [a, \infty)$ that Eq.(52) holds and if $\mathbf{u}(t)$ is in E_t^u it follows similarly that

$$P(t)\mathbf{u}(t) = \int_{-\infty}^{t} X(t)X^{-1}(s)P(s)B(s)\mathbf{u}(s)ds.$$

Then if we take the projection $Q(t)$ as before, the proof in this case can be completed as in the previous cases.

Thus the proof of Lemma 7.4 is complete.

Lastly in this section, we give the definition of a linear skew product flow and go on to characterise hyperbolicity in terms of such flows. Note that this part is included for completeness only and is not used in the sequel.

Definition 7.5. Let V be a vector bundle with base space S, a complete metric space. A flow Φ on V is said to be a ***skew product flow*** if there is a flow ϕ on S such that

$$p(\Phi(t, \mathbf{v})) = \phi(t, p(\mathbf{v})) = \phi^t(p(\mathbf{v})),$$

where $p : V \to S$ is the canonical projection. Φ is a *linear skew product flow* if, in addition,

$$\Phi(t, \lambda\mathbf{v} + \mathbf{w}) = \lambda\Phi(t, \mathbf{v}) + \Phi(t, \mathbf{w}).$$

Thus, if Φ is a linear skew product flow, there are invertible linear transformations $\Phi_{\mathbf{x}}^t : V_{\mathbf{x}} \to V_{\phi^t(\mathbf{x})}$ $\big(\mathbf{x} \in S,\ V_{\mathbf{x}} = p^{-1}(\mathbf{x})\big)$ given by $\Phi_{\mathbf{x}}^t(\mathbf{v}) = \Phi(t, \mathbf{v})$ satisfying the *cocycle* property

$$\Phi_{\mathbf{x}}^{t+s} = \Phi_{\phi^s(\mathbf{x})}^t \circ \Phi_{\mathbf{x}}^s.$$

Definition 7.6. Let Φ be a linear skew product flow on the vector bundle $p : V \to S$ as in Definition 7.5. Φ is said to have an *exponential dichotomy* if there are projections $P(\mathbf{x}) : V_{\mathbf{x}} \to V_{\mathbf{x}}$ $(\mathbf{x} \in S)$ of constant rank and positive constants $K_1, K_2, \alpha_1, \alpha_2$ such that for $\mathbf{x} \in S$

$$\Phi_{\mathbf{x}}^t P(\mathbf{x}) = P(\phi^t(\mathbf{x}))\Phi_{\mathbf{x}}^t \quad \text{for all } t,$$

and for $t \geq 0$

$$\begin{aligned} \|\Phi_{\mathbf{x}}^t P(\mathbf{x})\| &\leq K_1 e^{-\alpha_1 t}, \\ \|\Phi_{\mathbf{x}}^{-t}(I - P(\mathbf{x}))\| &\leq K_2 e^{-\alpha_2 t}. \end{aligned}$$

We now characterize hyperbolic sets in terms of dichotomies of linear skew product flows. Let S be a compact, invariant set for Eq.(1) with $F(\mathbf{x}) \neq 0$ for all $\mathbf{x}$ in S. We define the *normal bundle*

$$N = \{(\mathbf{x}, \mathbf{v}) \in S \times \mathbb{R}^n : \langle \mathbf{v}, F(\mathbf{x}) \rangle = 0\},$$

with canonical projection $p : N \to S$ given by $p(\mathbf{x}, \mathbf{v}) = \mathbf{x}$. Let $R(\mathbf{x})$ be the orthogonal projection on $\mathbb{R}^n$ with nullspace $\text{span}\{F(\mathbf{x})\}$ and range

$$N_{\mathbf{x}} = \{\mathbf{v} \in \mathbb{R}^n : (\mathbf{x}, \mathbf{v}) \in N\} = \{\mathbf{v} \in \mathbb{R}^n : \langle \mathbf{v}, F(\mathbf{x}) \rangle = 0\}.$$

Note that since $D\phi^t(\mathbf{x})F(\mathbf{x}) = F(\phi^t(\mathbf{x}))$,

$$R(\phi^t(\mathbf{x}))D\phi^t(\mathbf{x}) = R(\phi^t(\mathbf{x}))D\phi^t(\mathbf{x})R(\mathbf{x}).$$

It is easy to verify that

$$(t, (\mathbf{x}, \mathbf{v})) \to (\phi^t(\mathbf{x}), R(\phi^t(\mathbf{x}))D\phi^t(\mathbf{x})\mathbf{v})$$

defines a linear skew product flow on the normal bundle N. Then we have the following proposition.

Proposition 7.7. *Let S be a compact, invariant set for Eq.(1) with $F(\mathbf{x}) \neq 0$ for all $\mathbf{x}$ in S. Then S is hyperbolic if and only if the induced linear skew product flow on the normal bundle has an exponential dichotomy.*

Proof. Suppose first that S is hyperbolic. Let $\mathcal{P}^0(\mathbf{x}), \mathcal{P}^s(\mathbf{x}), \mathcal{P}^u(\mathbf{x})$ be the supplementary projections associated with the splitting in Eq.(5). Define $Q(\mathbf{x}) : N_{\mathbf{x}} \to N_{\mathbf{x}}$ by

$$Q(\mathbf{x})\mathbf{v} = R(\mathbf{x})\mathcal{P}^s(\mathbf{x})\mathbf{v}.$$

Since $\mathcal{P}^s(\mathbf{x})(I - R(\mathbf{x})) = 0$, $Q(\mathbf{x})$ is a projection and it has rank equal to that of $\mathcal{P}^s(\mathbf{x})$. Also if I is the identity on $N_{\mathbf{x}}$, we see that

$$I - Q(\mathbf{x}) = R(\mathbf{x})\mathcal{P}^u(\mathbf{x})$$

since $R(\mathbf{x})\mathcal{P}^0(\mathbf{x}) = 0$.

Now the linear skew product flow on the normal bundle N is given by

$$(t, (\mathbf{x}, \mathbf{v})) \to \Phi^t(\mathbf{x}, \mathbf{v}) = (\phi^t(\mathbf{x}), \Psi^t_{\mathbf{x}}\mathbf{v}),$$

where

$$\Psi^t_{\mathbf{x}} = R(\phi^t(\mathbf{x}))D\phi^t(\mathbf{x}).$$

The projections $Q(\mathbf{x})$ are invariant with respect to this flow since if $\mathbf{v} \in N_{\mathbf{x}}$

$$\begin{aligned}
\Psi^t_{\mathbf{x}}Q(\mathbf{x})\mathbf{v} &= R(\phi^t(\mathbf{x}))D\phi^t(\mathbf{x})R(\mathbf{x})\mathcal{P}^s(\mathbf{x})\mathbf{v} \\
&= R(\phi^t(\mathbf{x}))D\phi^t(\mathbf{x})\mathcal{P}^s(\mathbf{x})\mathbf{v} \\
&= R(\phi^t(\mathbf{x}))\mathcal{P}^s(\phi^t(\mathbf{x}))D\phi^t(\mathbf{x})\mathbf{v} \quad \text{by invariance of } \mathcal{P}^s \\
&= R(\phi^t(\mathbf{x}))\mathcal{P}^s(\phi^t(\mathbf{x}))R(\phi^t(\mathbf{x}))D\phi^t(\mathbf{x})\mathbf{v} \\
&\qquad \text{since } \mathcal{P}^s(\phi^t(\mathbf{x}))F(\phi^t(\mathbf{x})) = \mathbf{0} \\
&= Q(\phi^t(\mathbf{x}))\Psi^t_{\mathbf{x}}\mathbf{v}.
\end{aligned}$$

Moreover if $t \geq 0$,

$$\begin{aligned}
\|\Psi^t_{\mathbf{x}}Q(\mathbf{x})\| &= \|R(\phi^t(\mathbf{x}))D\phi^t(\mathbf{x})R(\mathbf{x})\mathcal{P}^s(\mathbf{x})\| \\
&= \|R(\phi^t(\mathbf{x}))D\phi^t(\mathbf{x})\mathcal{P}^s(\mathbf{x})\| \\
&\leq \|D\phi^t(\mathbf{x})\mathcal{P}^s(\mathbf{x})\| \\
&\leq K_1 e^{-\alpha_1 t}.
\end{aligned}$$

Similarly if $t \geq 0$,

$$\|\Psi^{-t}_{\mathbf{x}}(I - Q(\mathbf{x}))\| \leq K_2 e^{-\alpha_2 t}.$$

Hence the induced linear skew product flow on the normal bundle does indeed have an exponential dichotomy.

Next suppose, conversely, that the induced linear skew product flow on the normal bundle has an exponential dichotomy with projections $Q(\mathbf{x})$, constants K_1, K_2 and exponents α_1, α_2. For fixed $\mathbf{x} \in S$, we triangularize the variational system

$$\dot{\mathbf{y}} = DF(\phi^t(\mathbf{x}))\mathbf{y}, \tag{54}$$

as above, using the matrix $S(t)$ constructed in Eq.(33). Define the projection $Q(t) : \mathbb{R}^{n-1} \to \mathbb{R}^{n-1}$ by

$$Q(t) = S^*(t)Q(\phi^t(\mathbf{x}))S(t).$$

We show that $Y(t) = S^*(t)D\phi^t(\mathbf{x})S_0$ (conf. Eq.(33)) is the fundamental matrix of a linear system having an exponential dichotomy with projections $Q(t)$. First note that

$$R(\phi^t(\mathbf{x})) = S(t)S^*(t), \; Y(t)Y^{-1}(s) = S^*(t)\Psi^{t-s}_{\phi^s(\mathbf{x})}S(s)$$

and hence, using the invariance of $Q(\mathbf{x})$ and the fact that $R(\mathbf{x})$ is the identity on $N_{\mathbf{x}}$,

$$\begin{aligned}
Y(t)Y^{-1}(s)Q(s) &= S^*(t)\Psi^{t-s}_{\phi^s(\mathbf{x})}S(s)S^*(s)Q(\phi^s(\mathbf{x}))S(s) \\
&= S^*(t)\Psi^{t-s}_{\phi^s(\mathbf{x})}R(\phi^s(\mathbf{x}))Q(\phi^s(\mathbf{x}))S(s) \\
&= S^*(t)\Psi^{t-s}_{\phi^s(\mathbf{x})}Q(\phi^s(\mathbf{x}))S(s) \\
&= S^*(t)Q(\phi^t(\mathbf{x}))\Psi^{t-s}_{\phi^s(\mathbf{x})}S(s) \\
&= S^*(t)Q(\phi^t(\mathbf{x}))R(\phi^t(\mathbf{x}))\Psi^{t-s}_{\phi^s(\mathbf{x})}S(s) \\
&= S^*(t)Q(\phi^t(\mathbf{x}))S(t)S^*(t)\Psi^{t-s}_{\phi^s(\mathbf{x})}S(s) \\
&= Q(t)Y(t)Y^{-1}(s).
\end{aligned}$$

Thus we have the invariance property

$$Y(t)Y^{-1}(s)Q(s) = Q(t)Y(t)Y^{-1}(s).$$

Next if $t \geq s$,

$$\begin{aligned}
\|Y(t)Y^{-1}(s)Q(s)\| &= \|S^*(t)\Psi^{t-s}_{\phi^s(\mathbf{x})}Q(\phi^s(\mathbf{x}))S(s)\|, \text{ as above} \\
&\leq \|\Psi^{t-s}_{\phi^s(\mathbf{x})}Q(\phi^s(\mathbf{x}))\| \\
&\leq K_1 e^{-\alpha_1(t-s)}.
\end{aligned}$$

Similarly, if $t \leq s$,

$$\|Y(t)Y^{-1}(s)(I - Q(s))\| \leq K_2 e^{-\alpha_2(s-t)}.$$

Thus $Y(t)$ is indeed the fundamental matrix of a linear system having an exponential dichotomy on $(-\infty, \infty)$ with projection $Q(t)$.

Next notice that the transformation

$$\mathbf{y} = T(\mathbf{t})\mathbf{z} = [F(\phi^t(\mathbf{x}))\ S(t)]\mathbf{z}$$

takes Eq.(54) into the triangularised equation

$$\dot{\mathbf{z}} = \begin{bmatrix} 0 & \mathbf{b}^*(t) \\ 0 & A(t) \end{bmatrix} \mathbf{z}, \tag{55}$$

where

$$\mathbf{b}^*(t) = \frac{F(\phi^t(\mathbf{x}))^*}{\|F(\phi^t(\mathbf{x}))\|^2}[DF(\phi^t(\mathbf{x}))S(t) - \dot{S}(t)],\ A(t) = \dot{Y}(t)Y^{-1}(t)$$

and both have bounds depending only on

$$M_1 = \sup_{\mathbf{x} \in S} \|DF(\mathbf{x})\| \quad \text{and} \quad \Delta = \inf_{\mathbf{x} \in S} \|F(\mathbf{x})\|.$$

(Here we are using the fact from Coppel [1965, p.130] that if

$$T(t) = \left[\frac{F(\phi^t(\mathbf{x}))}{\|F(\phi^t(\mathbf{x}))\|} \middle| S(t) \right],$$

then $A(t)$ is unchanged but

$$\mathbf{b}^*(t) = \frac{F(\phi^t(\mathbf{x}))^*}{\|F(\phi^t(\mathbf{x}))\|}[DF(\phi^t(\mathbf{x}))S(t) - \dot{S}(t)]$$

and the norms of this $\mathbf{b}(t)$ and $A(t)$ have bounds depending only on M_1.)

Now the equation

$$\dot{\mathbf{w}} = -A^*(t)\mathbf{w} \tag{56}$$

is adjoint to the equation

$$\dot{\mathbf{w}} = A(t)\mathbf{w}$$

which has fundamental matrix $Y(t)$ and so, as shown above, has an exponential dichotomy on $(-\infty, \infty)$ with projections $Q(t)$, constants K_1, K_2 and exponents α_1, α_2. Hence, by Proposition 7.3, Eq.(56) also has an exponential dichotomy on $(-\infty, \infty)$ with projections $I - Q^*(t)$, constants K_2, K_1 and exponents α_2, α_1. So, by Proposition 7.3 again, the inhomogeneous equation

$$\dot{\mathbf{w}} = -A^*(t)\mathbf{w} + \mathbf{b}(t)$$

has a unique bounded solution $\mathbf{d}(t)$, with bound a constant depending only on M_1, K_1, K_2, α_1, α_2 and Δ. Let

$$H(t) = \begin{bmatrix} 1 & \mathbf{d}^*(t) \\ 0 & I \end{bmatrix}. \tag{57}$$

Then

$$H(t)\begin{bmatrix} 0 & 0 \\ 0 & A(t) \end{bmatrix} - \begin{bmatrix} 0 & \mathbf{b}^*(t) \\ 0 & A(t) \end{bmatrix} H(t) + \dot{H}(t)$$

$$= \begin{bmatrix} 0 & \mathbf{d}^*(t)A(t) - \mathbf{b}^*(t) + \dot{\mathbf{d}}^*(t) \\ 0 & 0 \end{bmatrix}$$

$$= \begin{bmatrix} 0 & 0 \\ 0 & 0 \end{bmatrix}.$$

So the transformation $\mathbf{z} = H(t)\mathbf{w}$ takes Eq.(55) into the block diagonal system

$$\dot{\mathbf{w}} = \begin{bmatrix} 0 & 0 \\ 0 & A(t) \end{bmatrix} \mathbf{w}. \tag{58}$$

We define

$$E^s(\mathbf{x}) = \left\{ T(0)H(0) \begin{bmatrix} 0 \\ \eta \end{bmatrix} : \eta \in \mathcal{R}(Q(0)) \right\}$$

and

$$E^u(\mathbf{x}) = \left\{ T(0)H(0) \begin{bmatrix} 0 \\ \eta \end{bmatrix} : \eta \in \mathcal{N}(Q(0)) \right\}.$$

Then $E^s(\mathbf{x}) \cap E^u(\mathbf{x}) = \{\mathbf{0}\}$ and the codimension of $E^s(\mathbf{x}) \oplus E^u(\mathbf{x})$ is 1. Notice also that if

$$\xi = T(0)H(0) \begin{bmatrix} 0 \\ \eta \end{bmatrix}$$

is in $E^s(\mathbf{x})$, then

$$T(t)H(t) \begin{bmatrix} 0 \\ Y(t)\eta \end{bmatrix}$$

is a solution of Eq.(54) which equals ξ at $t = 0$. So

$$D\phi^t(\mathbf{x})\xi = T(t)H(t) \begin{bmatrix} 0 \\ Y(t)\eta \end{bmatrix}$$

and hence if $t \geq 0$

$$\begin{aligned} \|D\phi^t(\mathbf{x})\xi\| &\leq \|T\|_\infty \|H\|_\infty K_1 e^{-\alpha_1 t} \|\eta\| \\ &\leq \|T\|_\infty \|H\|_\infty K_1 e^{-\alpha_1 t} \|[T(0)H(0)]^{-1}\| \, \|\xi\|, \end{aligned}$$

where $\|\cdot\|_\infty$ denotes the supremum norm. So, if $\xi \in E^s(\mathbf{x})$,

$$\|D\phi^t(\mathbf{x})\xi\| \le \overline{K}_1 e^{-\alpha_1 t}\|\xi\| \quad \text{for } t \ge 0, \tag{59}$$

where $\overline{K}_1$ depends only on M_0, M_1, K_1, K_2, α_1, α_2 and Δ with

$$M_0 = \sup_{\mathbf{x}\in S} \|F(\mathbf{x})\|.$$

Similarly, if $\xi \in E^u(\mathbf{x})$,

$$\|D\phi^{-t}(\mathbf{x})\xi\| \le \overline{K}_2 e^{-\alpha_2 t}\|\xi\| \text{ for } \ t \ge 0, \tag{60}$$

where $\overline{K}_2$ depends only on M_0, M_1, K_1, K_2, α_1, α_2 and Δ.

In virtue of the transformations we have carried out and the fact that $D\phi^t(\mathbf{x})$ is the fundamental matrix for Eq.(54) which equals the identity at $t = 0$, we have

$$D\phi^t(\mathbf{x}) = T(t)H(t)\begin{bmatrix} 1 & 0 \\ 0 & Y(t) \end{bmatrix} H^{-1}(0)T^{-1}(0).$$

It follows that

$$D\phi^t(\mathbf{x})\xi \to \mathbf{0} \quad \text{as} \quad t \to \infty$$

if and only if

$$\begin{bmatrix} 1 & 0 \\ 0 & Y(t) \end{bmatrix} \overline{\xi} \to \mathbf{0} \,\text{as}\, t \to \infty,$$

where $\overline{\xi} = H^{-1}(0)T^{-1}(0)\xi$. However, by Proposition 7.3 (ii), this happens if and only if

$$\overline{\xi} = \begin{bmatrix} 0 \\ \eta \end{bmatrix}$$

where $\eta \in \mathcal{R}(Q(0))$, that is, $\xi \in E^s(\mathbf{x})$. Hence

$$E^s(\mathbf{x}) = \{\xi \in I\!R^n : D\phi^t(\mathbf{x})\xi \to \mathbf{0} \text{ as } t \to \infty\}$$

and similarly,

$$E^u(\mathbf{x}) = \{\xi \in I\!R^n : D\phi^t(\mathbf{x})\xi \to \mathbf{0} \text{ as } t \to -\infty\}.$$

From this the invariance of $E^s(\mathbf{x})$ and $E^u(\mathbf{x})$, that is,

$$D\phi^t(\mathbf{x})(E^{s,u}(\mathbf{x})) = E^{s,u}(\phi^t(\mathbf{x}))$$

is easily seen to follow. Moreover, we know $\dim E^s(\mathbf{x}) = \dim \mathcal{R}(Q(\mathbf{x}))$ and $\dim E^u(\mathbf{x}) = \dim \mathcal{N}(Q(\mathbf{x}))$ and also that $E^s(\mathbf{x}) \cap E^u(\mathbf{x}) = \{\mathbf{0}\}$ so that the codimension of $E^s(\mathbf{x}) \oplus E^u(\mathbf{x})$ is 1. Since we also have the exponential estimates (59) and (60), all that remains to be established is that $F(\mathbf{x}) \notin E^s(\mathbf{x}) \oplus E^u(\mathbf{x})$. But this clearly follows from

$$F(\mathbf{x}) = T(0)H(0)\begin{bmatrix} 1 \\ 0 \end{bmatrix}.$$

Thus the proof of Proposition 7.7 is complete.

7.5 EXPANSIVITY PROPERTY OF HYPERBOLIC SETS

Let $U \subset \mathbb{R}^n$ be a convex open set, let $F : U \to \mathbb{R}^n$ be a C^1 vector field and let ϕ be the flow associated with Eq.(1). Let S be a compact hyperbolic set for Eq.(1) as in Definition 7.1. Our aim in this section is to show that the flow ϕ is *expansive* on S. First we present the definition of this concept.

Definition 7.8. Let $F : U \to \mathbb{R}^n$ be a C^1 vector field, let ϕ be the flow associated with Eq.(1) and let S be *invariant*, that is, $\phi^t(S) = S$ for all real t. Then Eq.(1) is said to be *expansive* on S with *expansivity constant* $d > 0$ if whenever $\mathbf{x}, \mathbf{y} \in S$ and there exists a continuous real valued function $\alpha(t)$ such that

$$\|\phi^{\alpha(t)}(\mathbf{y}) - \phi^t(\mathbf{x})\| \le d \quad \text{for all real } t,$$

then $\mathbf{y} = \phi^\tau(\mathbf{x})$ for some real τ.

Definition 7.8 is slightly different from the definition in Bowen and Walters [1972] which requires that for all $\varepsilon > 0$, there exists $d > 0$ such that if $\mathbf{x}$, $\mathbf{y}$ are in S and

$$\|\phi^{\alpha(t)}(\mathbf{y}) - \phi^t(\mathbf{x})\| \le d$$

for all t, where $\alpha(t)$ is continuous with $\alpha(0) = 0$, then $\mathbf{y} = \phi^\tau(\mathbf{x})$ with $|\tau| < \varepsilon$.

In order to prove the expansiveness of a hyperbolic set, we use the following proposition. Before proving it, we introduce some notation. If S is a compact hyperbolic set for Eq.(1) as in Definition 7.1, denote by $\mathcal{P}^0(\mathbf{x}), \mathcal{P}^s(\mathbf{x}), \mathcal{P}^u(\mathbf{x})$ the projections associated with the splitting in Eq.(5) and write

$$\Delta = \inf_{\mathbf{x}\in S} \|F(\mathbf{x})\|, \quad M_0 = \sup_{\mathbf{x}\in U} \|F(\mathbf{x})\|, \quad M_1 = \sup_{\mathbf{x}\in U} \|DF(\mathbf{x})\|, \tag{61}$$

$$\omega(\varepsilon) = \sup\{\|DF(\mathbf{x}) - DF(\mathbf{y})\| : \mathbf{x} \in S,\ \mathbf{y} \in U,\ \|\mathbf{x} - \mathbf{y}\| \le \varepsilon\}$$

and

$$M^0 = \sup_{\mathbf{x}\in S} \|\mathcal{P}^0(\mathbf{x})\|, \; M^s = \sup_{\mathbf{x}\in S} \|\mathcal{P}^s(\mathbf{x})\|, \; M^u = \sup_{\mathbf{x}\in S} \|\mathcal{P}^u(\mathbf{x})\|.$$

In order to prove the proposition, we also need the following lemma.

Lemma 7.9. *Let S be a compact hyperbolic set for* Eq.(1) *(with U convex) as in Definition* 7.1. *Let $G(\mathbf{x})$ be a C^1 vectorfield defined on U and σ a nonnegative* *number such that for* $\mathbf{x}$ *in* U

$$\|G(\mathbf{x}) - F(\mathbf{x})\| + \|DG(\mathbf{x}) - DF(\mathbf{x})\| \le \sigma.$$

Suppose $\mathbf{x}(t)$ *is a solution of* Eq.(1) *in* S, $\mathbf{y}(t)$ *a solution of the system*

$$\dot{\mathbf{y}} = G(\mathbf{y})$$

and $\alpha(t)$ *a continuous real valued function such that for* $a \le t \le b$

$$\langle \mathbf{y}(\alpha(t)) - \mathbf{x}(t),\ F(\mathbf{x}(t)) \rangle = 0 \tag{62}$$

and for some positive number ε

$$\|\mathbf{y}(\alpha(t)) - \mathbf{x}(t)\| \le \varepsilon.$$

Then if

$$\sigma + M_1\varepsilon < \Delta,$$

$\alpha(t)$ *is continuously differentiable and* $\alpha'(t)$ *is given by*

$$1 - \frac{\langle \mathbf{y}(\alpha(t)) - \mathbf{x}(t), DF(\mathbf{x}(t))F(\mathbf{x}(t)) \rangle + \langle G(\mathbf{y}(\alpha(t))) - F(\mathbf{x}(t)), F(\mathbf{x}(t)) \rangle}{\langle G(\mathbf{y}(\alpha(t))), F(\mathbf{x}(t)) \rangle}.$$

Proof. Consider the continuously differentiable real function

$$g(\alpha, t) = \langle \mathbf{y}(\alpha) - \mathbf{x}(t), F(\mathbf{x}(t)) \rangle.$$

We estimate

$$\begin{aligned} \frac{\partial g}{\partial \alpha}(\alpha(t), t) &= \langle G(\mathbf{y}(\alpha(t))),\ F(\mathbf{x}(t)) \rangle \\ &\ge \|F(\mathbf{x}(t))\|^2 - \|G(\mathbf{y}(\alpha(t))) - F(\mathbf{x}(t))\|\,\|F(\mathbf{x}(t))\| \\ &\ge \|F(\mathbf{x}(t))\|\,[\Delta - (\sigma + M_1\varepsilon)] \\ &> 0. \end{aligned}$$

Then it follows from the continuity of $\alpha(t)$ and the implicit function theorem that $\alpha(t)$ is continuously differentiable. Moreover, if we differentiate Eq.(62) with respect to t, we obtain the equation

$$\langle G(\mathbf{y}(\alpha(t))), F(\mathbf{x}(t)) \rangle \alpha'(t) - \|F(\mathbf{x}(t))\|^2 + \langle \mathbf{y}(\alpha(t)) - \mathbf{x}(t), DF(\mathbf{x}(t))F(\mathbf{x}(t)) \rangle = 0$$

and thus the required formula for $\alpha'(t)$.

Now we are ready to state and prove the proposition. Actually, for later use, we prove a result slightly more general than what is needed here.

Proposition 7.10. *Let S be a compact hyperbolic set for Eq.(1) as in Definition 7.1 (with U convex). Choose the numbers β_1, β_2 so that*

$$0 < \beta_1 < \alpha_1, \ \ 0 < \beta_2 < \alpha_2.$$

Let $G(\mathbf{x})$ be a C^1 vectorfield defined on U and σ a nonnegative number such that for $\mathbf{x}$ in U

$$\|G(\mathbf{x}) - F(\mathbf{x})\| + \|DG(\mathbf{x}) - DF(\mathbf{x})\| \le \sigma.$$

Let $\mathbf{x}(t)$ be a solution of Eq.(1) *in S, $\mathbf{y}(t)$ and $\mathbf{z}(t)$ solutions of the equation*

$$\dot{\mathbf{y}} = G(\mathbf{y}),$$

and let $\alpha(t)$ and $\beta(t)$ be continuous real-valued functions such that for $a \le t \le b$

$$\langle \mathbf{y}(\alpha(t)) - \mathbf{x}(t), F(\mathbf{x}(t)) \rangle = \langle \mathbf{z}(\beta(t)) - \mathbf{x}(t), F(\mathbf{x}(t)) \rangle = 0 \tag{63}$$

and for some positive number ε

$$\|\mathbf{y}(\alpha(t)) - \mathbf{x}(t)\| \le \varepsilon, \ \ \|\mathbf{z}(\beta(t)) - \mathbf{x}(t)\| \le \varepsilon. \tag{64}$$

Then if σ and ε are sufficiently small depending only on K_1, K_2, M^s, M^u, α_1, α_2, β_1, β_2, M_1, Δ, $\omega(\cdot)$, there exist constants L_1, L_2 depending only on K_1, K_2, M^s, M^u, α_1, α_2 such that the inequality

$$\begin{aligned} \|\mathbf{y}(\alpha(t)) - \mathbf{z}(\beta(t))\| \ &\le L_1 e^{-\beta_1(t-a)} \|\mathbf{y}(\alpha(a)) - \mathbf{z}(\beta(a))\| \\ &\quad + L_2 e^{-\beta_2(b-t)} \|\mathbf{y}(\alpha(b)) - \mathbf{z}(\beta(b))\| \end{aligned} \tag{65}$$

holds for $a \le t \le b$. Also if Eqs.(63) *and* (64) *hold for $t \ge a$, then*

$$\|\mathbf{y}(\alpha(t)) - \mathbf{z}(\beta(t))\| \le L_1 e^{-\beta_1(t-a)} \|\mathbf{y}(\alpha(a)) - \mathbf{z}(\beta(a))\| \tag{66}$$

holds for $t \ge a$.

Proof. Provided $\sigma + M_1\varepsilon < \Delta$, it follows from Lemma 7.9 that $\alpha(t)$, $\beta(t)$ are continuously differentiable and satisfy

$$\alpha'(t) = 1 -$$
$$\frac{\langle \mathbf{y}(\alpha(t)) - \mathbf{x}(t), DF(\mathbf{x}(t))F(\mathbf{x}(t)) \rangle + \langle G(\mathbf{y}(\alpha(t))) - F(\mathbf{x}(t)), F(\mathbf{x}(t)) \rangle}{\langle G(\mathbf{y}(\alpha(t))), F(\mathbf{x}(t)) \rangle},$$

$$\beta'(t) = 1 - \frac{\langle \mathbf{z}(\beta(t)) - \mathbf{x}(t), DF(\mathbf{x}(t))F(\mathbf{x}(t)) \rangle + \langle G(\mathbf{z}(\beta(t))) - F(\mathbf{x}(t)), F(\mathbf{x}(t)) \rangle}{\langle G(\mathbf{z}(\beta(t))), F(\mathbf{x}(t)) \rangle}.$$

Let $S(t)$ be as in the considerations after Proposition 7.3 (with $\mathbf{x}(0) = \mathbf{x}$). Since $\mathbf{y}(\alpha(t)) - \mathbf{z}(\beta(t))$ is orthogonal to $F(\mathbf{x}(t)) = F(\phi^t(\mathbf{x}))$, we may write

$$\mathbf{y}(\alpha(t)) - \mathbf{z}(\beta(t)) = S(t)\mathbf{w}(t),$$

where $\mathbf{w}(t)$ is in $I\!R^{n-1}$. Since $S^*(t)S(t) = I$,

$$\mathbf{w}(t) = S^*(t)[\mathbf{y}(\alpha(t)) - \mathbf{z}(\beta(t))]$$

and so

$$\begin{aligned}
\dot{\mathbf{w}}(t) = \;& S^*(t)[G(\mathbf{y}(\alpha(t)))\alpha'(t) - G(\mathbf{z}(\beta(t)))\beta'(t)] + \dot{S}^*(t)S(t)\mathbf{w}(t) \\
= \;& S^*(t)\,[G(\mathbf{y}(\alpha(t)))\alpha'(t) - G(\mathbf{z}(\beta(t)))\beta'(t)] - S^*(t)\dot{S}(t)\mathbf{w}(t) \\
= \;& A(t)\mathbf{w}(t) + S^*(t)\,\{DF(\mathbf{x}(t))(\alpha'(t) - 1)S(t)\mathbf{w}(t) \qquad (67)\\
& + [G(\mathbf{y}(\alpha(t))) - G(\mathbf{z}(\beta(t))) - DF(\mathbf{x}(t))S(t)\mathbf{w}(t)]\,\alpha'(t) \\
& + [G(\mathbf{z}(\beta(t))) - F(\mathbf{x}(t))](\alpha'(t) - \beta'(t))\},
\end{aligned}$$

where

$$A(t) = S^*(t)\left[DF(\mathbf{x}(t)S(t) - \dot{S}(t)\right]$$

and we have used the fact that $S^*(t)F(\mathbf{x}(t)) = 0$.

We estimate the terms on the right side of Eq.(67). First note that

$$\begin{aligned}
|\langle G(\mathbf{y}(\alpha(t))), F(\mathbf{x}(t)) \rangle| \;&\geq\; \|F(\mathbf{x}(t))\|^2 - (\sigma + M_1\varepsilon)\|F(\mathbf{x}(t))\| \\
&\geq\; \|F(\mathbf{x}(t))\|[\Delta - (\sigma + M_1\varepsilon)] \\
&\geq\; \Delta\|F(\mathbf{x}(t))\|/2,
\end{aligned}$$

provided

$$2(\sigma + M_1\varepsilon) \leq \Delta.$$

So

$$|\alpha'(t) - 1| \leq \frac{(M_1\varepsilon + \sigma + M_1\varepsilon)\|F(\mathbf{x}(t)\|}{\Delta\|F(\mathbf{x}(t))\|/2} = 2\Delta^{-1}(\sigma + 2M_1\varepsilon). \qquad (68)$$

Also

$$\begin{aligned}
& G(\mathbf{y}(\alpha(t))) - G(\mathbf{z}(\beta(t))) - DF(\mathbf{x}(t))S(t)\mathbf{w}(t) \qquad (69)\\
& = \int_0^1 [DG(\theta\mathbf{y}(\alpha(t)) + (1-\theta)\mathbf{z}(\beta(t))) - DF(\mathbf{x}(t))]d\theta\, S(t)\mathbf{w}(t),
\end{aligned}$$

where

$$\left\| \int_0^1 [DG(\theta \mathbf{y}(\alpha(t)) + (1-\theta)\mathbf{z}(\beta(t))) - DF(\mathbf{x}(t))] d\theta \right\| \leq \sigma + \omega(\varepsilon). \qquad (70)$$

Next we observe that

$$\alpha'(t) - \beta'(t) = -I_1 - I_2 + I_3,$$

where

$$I_1 = \frac{\langle \mathbf{y}(\alpha(t)) - \mathbf{z}(\beta(t)), DF(\mathbf{x}(t))F(\mathbf{x}(t)) \rangle}{\langle G(\mathbf{y}(\alpha(t))), F(\mathbf{x}(t)) \rangle},$$

$$I_2 = \frac{\langle G(\mathbf{y}(\alpha(t))) - G(\mathbf{z}(\beta(t))), F(\mathbf{x}(t)) \rangle}{\langle G(\mathbf{y}(\alpha(t))), F(\mathbf{x}(t)) \rangle}$$

and

$$I_3 = \langle G(\mathbf{y}(\alpha(t))) - G(\mathbf{z}(\beta(t))), F(\mathbf{x}(t)) \rangle \times$$

$$\frac{\langle \mathbf{z}(\beta(t)) - \mathbf{x}(t), DF(\mathbf{x}(t))F(\mathbf{x}(t)) \rangle + \langle G(\mathbf{z}(\beta(t))) - F(\mathbf{x}(t)), F(\mathbf{x}(t)) \rangle}{\langle G(\mathbf{y}(\alpha(t))), F(\mathbf{x}(t)) \rangle \langle G(\mathbf{z}(\beta(t))), F(\mathbf{x}(t)) \rangle}.$$

So noting that

$$\mathbf{y}(\alpha(t)) - \mathbf{z}(\beta(t)) = S(t)\mathbf{w}(t)$$

and

$$G(\mathbf{y}(\alpha(t))) - G(\mathbf{z}(\beta(t))) = \int_0^1 [DG(\theta \mathbf{y}(\alpha(t)) + (1-\theta)\mathbf{z}(\beta(t)))] d\theta \, S(t)\mathbf{w}(t),$$

we find that

$$\alpha'(t) - \beta'(t) = \gamma(t)\mathbf{w}(t), \qquad (71)$$

where

$$\|\gamma(t)\| \leq 2\Delta^{-1}[M_1 + (M_1 + \sigma) + 2\Delta^{-1}(2M_1 + \sigma)(M_1 + \sigma)]. \qquad (72)$$

Then it follows from Eqs.(67), (68), (69), (70), (71) and (72) that

$$\dot{\mathbf{w}}(t) = [A(t) + B(t)]\, \mathbf{w}(t), \qquad (73)$$

where

$$\begin{aligned} \|B(t)\| \quad &\leq M_1 \cdot 2\Delta^{-1}(\sigma + 2M_1\varepsilon) + (\sigma + \omega(\varepsilon))[1 + 2\Delta^{-1}(\sigma + 2M_1\varepsilon)] \\ &+ (\sigma + M_1\varepsilon)2\Delta^{-1}[2M_1 + \sigma + 2\Delta^{-1}(2M_1 + \sigma)(M_1 + \sigma)]. \end{aligned}$$

From the considerations after Proposition 7.3, we know that the equation $\dot{\mathbf{w}} = A(t)\mathbf{w}$ has an exponential dichotomy on $[a, b]$ with constants $K_1 M^s$, $K_2 M^u$ and exponents α_1, α_2. Then it follows from Lemma 7.4 that provided σ and ε are sufficiently small depending on K_1, K_2, M^s, M^u, α_1, α_2, β_1, β_2, M_1, Δ and $\omega(\cdot)$, system (73) has an exponential dichotomy on $[a, b]$ with exponents β_1, β_2 and with constants L_1, L_2 and bounds N^s, N^u on the projections $Q(t)$, $I - Q(t)$ depending only on K_1, K_2, M^s, M^u, α_1, α_2. Then for $a \le t \le b$

$$\|\mathbf{w}(t)\| \le \|W(t)W^{-1}(a)Q(a)\mathbf{w}(a)\| + \|W(t)W^{-1}(b)(I - Q(b))\mathbf{w}(b)\|,$$

where $W(t)$ is a fundamental matrix for Eq.(73). Hence, for $a \le t \le b$,

$$\|\mathbf{w}(t)\| \le L_1 e^{-\beta_1(t-a)}\|\mathbf{w}(a)\| + L_2 e^{-\beta_2(b-t)}\|\mathbf{w}(b)\|.$$

Thus (65) holds.

If Eqs.(63) and (64) hold for $t \ge a$, then Eq.(65) holds for $a \le t \le b$ for all $b \ge a$ and we may let $b \to \infty$ to get Eq.(66). Thus the proof of Proposition 7.10 is complete.

Now we are ready to prove the expansiveness, which we encapsulate in the following theorem. Note that the quantities M_0, M_1, M^s, M^u, Δ and $\omega(\cdot)$ are defined at the beginning of this section.

Theorem 7.11. *Let S be a compact hyperbolic set for Eq.(1) as in Definition 7.1, where $U \subset \mathbb{R}^n$ is open and convex, and let $F : U \to \mathbb{R}^n$ be a C^1 vector field. Then S is expansive. In fact, if $\mathbf{x}$ and $\mathbf{y}$ are points in U with $\mathbf{x} \in S$ such that*

$$\|\phi^{\alpha(t)}(\mathbf{y}) - \phi^t(\mathbf{x})\| \le d \quad \text{for all } t,$$

where $\alpha(t)$ is a continuous real valued function and d is a positive number sufficiently small depending on K_1, K_2, M^s, M^u, α_1, α_2, M_0, M_1, Δ and $\omega(\cdot)$, there exists a number α_0 with

$$|\alpha_0 - \alpha(0)| \le 4\Delta^{-1}d$$

such that

$$\mathbf{x} = \phi^{\alpha_0}(\mathbf{y}).$$

Also if, in addition,

$$\langle \phi^{\alpha(t)}(\mathbf{y}) - \phi^t(\mathbf{x}), F(\phi^t(\mathbf{x})) \rangle = 0 \tag{74}$$

for some t, we have

$$\alpha(t) = \alpha_0 + t. \tag{75}$$

Proof. For convenience, let us write

$$\mathbf{x}(t) = \phi^t(\mathbf{x}), \quad \mathbf{y}(t) = \phi^t(\mathbf{y}).$$

Assuming

$$2M_1 d \le \Delta \quad \text{and} \quad 8M_0 M_1 d < \Delta^2,$$

we can apply Lemma 6.2 with $\mathbf{y}(\alpha(t))$ as $\mathbf{x}$, $\mathbf{x}(t)$ as $\mathbf{y}$, $F(\mathbf{x}(t))$ as $\mathbf{v}$ and $\alpha = \Delta/2M_0M_1$ to deduce the existence of $\tau = \tau(t)$ such that

$$|\tau| \le 4|\langle \mathbf{y}(\alpha(t)) - \mathbf{x}(t), F(\mathbf{x}(t)) \rangle| \, / \, \|F(\mathbf{x}(t))\|^2 \le 4\Delta^{-1}d$$

and

$$\langle \mathbf{y}(\alpha(t)+\tau) - \mathbf{x}(t), F(\mathbf{x}(t)) \rangle = 0.$$

Also this τ is the unique τ satisfying the last equation in $|\tau| \le \alpha$. Next note that

$$\begin{aligned} \|\mathbf{y}(\alpha(t)+\tau(t)) - \mathbf{x}(t)\| &\le \|\mathbf{y}(\alpha(t)+\tau(t)) - \mathbf{y}(\alpha(t))\| + \|\mathbf{y}(\alpha(t)) - \mathbf{x}(t)\| \\ &\le M_0|\tau(t)| + d \\ &\le M_0 \cdot 4\Delta^{-1}d + d \\ &= (1 + 4M_0\Delta^{-1})d. \end{aligned}$$

Next we show that $\tau(t)$ is continuous. To this end, set

$$g(t,\tau) = \langle \mathbf{y}(\alpha(t)+\tau) - \mathbf{x}(t), F(\mathbf{x}(t)) \rangle.$$

Then $g(t,\tau)$ is a continuous real-valued function with continuous derivative given by

$$\begin{aligned} \frac{\partial g}{\partial \tau}(t,\tau) &= \langle \dot{\mathbf{y}}(\alpha(t)+\tau), F(\mathbf{x}(t)) \rangle \\ &= \langle F(\mathbf{y}(\alpha(t)+\tau)), F(\mathbf{x}(t)) \rangle \\ &= \|F(\mathbf{x}(t))\|^2 + \langle F(\mathbf{y}(\alpha(t)+\tau)) - F(\mathbf{x}(t)), F(\mathbf{x}(t)) \rangle. \end{aligned}$$

So

$$\begin{aligned} \frac{\partial g}{\partial \tau}(t,\tau(t)) &\ge \|F(\mathbf{x}(t)\|^2 - M_1 \, \| \, \mathbf{y}(\alpha(t)+\tau(t)) - \mathbf{x}(t) \, \| \, \|F(\mathbf{x}(t))\| \\ &\ge \|F(\mathbf{x}(t))\| [\Delta - M_1(1 + 4M_0\Delta^{-1})d] \\ &> 0, \end{aligned}$$

by the conditions imposed on d. So since $\tau = \tau(t)$ is the unique solution of $g(t,\tau) = 0$ in $|\tau| \le \alpha$ and $|\tau(t)| < \alpha$, it follows from the implicit function theorem and the uniqueness of τ that $\tau(t)$ is continuous.

Now set

$$\overline{\alpha}(t) = \alpha(t) + \tau(t).$$

Then we have proved that for all t

$$\|\mathbf{y}(\overline{\alpha}(t)) - \mathbf{x}(t)\| \leq (1 + 4M_0\Delta^{-1})d \quad \text{and} \quad \langle \mathbf{y}(\overline{\alpha}(t)) - \mathbf{x}(t), F(\mathbf{x}(t)) \rangle = 0$$

and that $\overline{\alpha}(t)$ is continuous. Then it follows from Proposition 7.10, provided d is sufficiently small depending only on K_1, K_2, M^s, M^u, α_1, α_2, M_1, Δ, $\omega(\cdot)$, that there exist constants L_1, L_2 depending only on K_1, K_2, M^s, M^u, α_1, α_2 such that for any real a and b with $a \leq b$, the inequality

$$\|\mathbf{y}(\overline{\alpha}(t)) - \mathbf{x}(t)\| \leq L_1 e^{-\alpha_1(t-a)/2}\|\mathbf{y}(\overline{\alpha}(a)) - \mathbf{x}(a)\| + L_2 e^{-\alpha_2(b-t)/2}\|\mathbf{y}(\overline{\alpha}(b)) - \mathbf{x}(b)\|$$

holds for $a \leq t \leq b$. Letting $a \to -\infty$ and $b \to \infty$, we conclude that

$$\mathbf{y}(\overline{\alpha}(t)) = \mathbf{x}(t) \tag{76}$$

for all t. Hence, in particular, $\mathbf{x} = \phi^{\overline{\alpha}(0)}(\mathbf{y})$ where

$$|\overline{\alpha}(0) - \alpha(0)| = |\tau(0)| \leq 4\Delta^{-1}d.$$

Thus the first part of the theorem follows with $\alpha_0 = \overline{\alpha}(0)$.

Now it follows from Eq.(76) that

$$\phi^t(\mathbf{x}) = \phi^{\overline{\alpha}(t)-\overline{\alpha}(0)}(\mathbf{y}(\overline{\alpha}(0))) = \phi^{\overline{\alpha}(t)-\alpha_0}(\mathbf{x}).$$

So if $\phi^t(\mathbf{x})$ is not periodic,

$$\overline{\alpha}(t) = \alpha_0 + t \tag{77}$$

for all t. Suppose $\phi^t(\mathbf{x})$ has minimal period T. Then

$$\overline{\alpha}(t) - \alpha_0 - t = kT$$

where, by continuity, the integer k does not depend on t. Taking $t = 0$, we see that $k = 0$. So Eq.(77) holds in any case. Then if Eq.(74) holds for some t, it follows by uniqueness of $\tau(t)$ that $\overline{\alpha}(t) = \alpha(t)$ and so Eq.(75) holds. This completes the proof of the theorem.

Note that it is clear from the conclusions of the theorem that the flow ϕ is also expansive on S in the Bowen-Walters [1972] sense, as described after Definition 7.8, with $\delta = \Delta d/4$.

Next we show that a flow is expansive on a compact invariant set without equilibria if and only if it has the following *discrete expansivity* property.

Definition 7.12. Let $F : U \to \mathbb{R}^n$ be a C^1 vector field, let ϕ be the flow associated with Eq.(1) and let S be an invariant set. Then Eq.(1) has the *discrete expansivity* property on S if there is a positive number d such that if $\{t_k\}_{k=-\infty}^{\infty}$ and $\{\tau_k\}_{k=-\infty}^{\infty}$ are sequences of real numbers such that

$$t_k \to \pm\infty \quad \text{as} \quad k \to \pm\infty,$$

$$0 < t_{k+1} - t_k \leq d, \quad |\tau_{k+1} - \tau_k| \leq d \quad \text{for all} \quad k,$$

and if $\mathbf{x}$, $\mathbf{y}$ in S satisfy

$$\|\phi^{t_k}(\mathbf{x}) - \phi^{\tau_k}(\mathbf{y})\| \leq d \quad \text{for all} \quad k,$$

then $\mathbf{y} = \phi^{\tau}(\mathbf{x})$ for some τ.

Proposition 7.13. *Let S be a compact invariant set for Eq.(1), where $F : U \to \mathbb{R}^n$ is a C^1 vector field with U open and convex, such that $F(\mathbf{x})$ does not vanish on S. Then Eq.(1) is expansive on S if and only if it has the discrete expansivity property on S.*

Proof. Suppose, first, that Eq.(1) has the discrete expansivity property on S. Let $\mathbf{x}$, $\mathbf{y}$ be points in S and $\alpha(t)$ a continuous real-valued function such that

$$\|\phi^{\alpha(t)}(\mathbf{y}) - \phi^{t}(\mathbf{x})\| \leq d$$

for all t, where ϕ is the flow corresponding to Eq.(1) and d the constant in Definition 7.12. It is clear that we can find a sequence $\{t_k\}_{k=-\infty}^{\infty}$ satisfying

$$0 < t_{k+1} - t_k \leq d$$

for all k and such that

$$|\alpha(t_{k+1}) - \alpha(t_k)| \leq d$$

for all k. In fact, we take $t_0 = 0$ and for $k \geq 0$ (for $k \leq 0$, we go in the opposite direction) if t_k has been determined, we take

$$t_{k+1} = \sup\{t : 0 < t - t_k \leq d, \ |\alpha(t) - \alpha(t_k)| \leq d\}.$$

Suppose the increasing sequence t_k has a finite limit B. Then choose δ, $0 < \delta < d$, such that

$$|\alpha(t) - \alpha(B)| < d/2 \quad \text{for} \quad |t - B| \leq \delta.$$

Next choose t_k such that $B - t_k < \delta$. Then $B - t_{k+1} < \delta$ holds also and so

$$|\alpha(t_{k+1}) - \alpha(t_k)| \leq |\alpha(t_{k+1}) - \alpha(B)| + |\alpha(t_k) - \alpha(B)| < d.$$

However, we also have

$$t_{k+1} - t_k < \delta < d.$$

This contradicts the definition of t_{k+1}. Hence $t_k \to \infty$ as $k \to \infty$. Similarly, $t_k \to -\infty$ as $k \to -\infty$. Then, from Definition 7.12 with $\tau_k = \alpha(t_k)$, we conclude that $\mathbf{y} = \phi^\tau(\mathbf{x})$. So the flow is expansive on S.

Suppose, conversely, that the flow is expansive on S as in Definition 7.8. Let $\{t_k\}_{k=-\infty}^{\infty}$, $\{\tau_k\}_{k=-\infty}^{\infty}$ be sequences and $\mathbf{x}$, $\mathbf{y}$ points in S as in Definition 7.12, where d is to be determined in the course of the proof. We have to show that $\mathbf{y} = \phi^\tau(\mathbf{x})$ for some τ.

First we apply Lemma 6.2 to Eq.(1) with $\phi^{t_k}(\mathbf{x})$ as $\mathbf{x}$, $\phi^{\tau_k}(\mathbf{y})$ as $\mathbf{y}$, $F(\phi^{t_k}(\mathbf{x}))$ as $\mathbf{v}$ and $\alpha = \Delta/2M_0M_1$, where M_0, M_1 and Δ are as in Eq.(61). Then if

$$d < \min\{\Delta/2M_1,\ \Delta^2/8M_0M_1\},$$

there exists s satisfying

$$|s| \leq 4\Delta^{-1}d$$

such that

$$\langle\, \phi^{s+\tau_k}(\mathbf{y}) - \phi^{t_k}(\mathbf{x}),\ F(\phi^{t_k}(\mathbf{x}))\,\rangle = 0.$$

Moreover, this is the only s in $|s| \leq \alpha$ satisfying the last equation. If we set

$$\tau_k' = \tau_k + s,$$

then for all k,

$$\langle\, \phi^{\tau_k'}(\mathbf{y}) - \phi^{t_k}(\mathbf{x}),\ F(\phi^{t_k}(\mathbf{x}))\,\rangle = 0,$$

$$|\tau_{k+1}' - \tau_k'| \leq (8\Delta^{-1} + 1)d$$

and

$$\|\phi^{\tau_k'}(\mathbf{y}) - \phi^{t_k}(\mathbf{x})\| \leq M_0|\tau_k' - \tau_k| + d \leq (4M_0\Delta^{-1} + 1)d.$$

Next we apply Lemma 6.2 again for $t_k \leq t \leq t_{k+1}$ with $\phi^t(\mathbf{x})$ as $\mathbf{x}$, $\phi^{t-t_k+\tau_k'}(\mathbf{y})$ as $\mathbf{y}$, $F(\phi^t(\mathbf{x}))$ as $\mathbf{v}$ and $\alpha = \Delta/2M_0M_1$. Note that if $t_k \leq t \leq t_{k+1}$, it follows from Gronwall's lemma that

$$\|\phi^{t-t_k+\tau_k'}(\mathbf{y}) - \phi^t(\mathbf{x})\| \leq e^{M_1 d}\|\phi^{\tau_k'}(\mathbf{y}) - \phi^{t_k}(\mathbf{x})\| \leq e^{M_1 d}(4M_0\Delta^{-1} + 1)d.$$

Then if

$$e^{M_1 d}(4M_0\Delta^{-1} + 1)d < \min\{\Delta/2M_1, \Delta^2/8M_0M_1\},$$

there exists s satisfying

$$|s| \leq 4\Delta^{-1}e^{M_1 d}(4M_0\Delta^{-1} + 1)d$$

such that

$$\langle\, \phi^{s+t-t_k+\tau_k'}(\mathbf{y}) - \phi^t(\mathbf{x}),\ F(\phi^t(\mathbf{x}))\,\rangle = 0.$$

Moreover, this is the only s in $|s| \leq \alpha$ satisfying the last equation.

We write

$$s = s_k(t).$$

To show that $s_k(t)$ is continuous in $t_k \leq t \leq t_{k+1}$, we define the function

$$g(t, s) = \langle \phi^{s+t-t_k+\tau'_k}(\mathbf{y}) - \phi^t(\mathbf{x}), F(\phi^t(\mathbf{x})) \rangle.$$

Then for $s = s_k(t)$,

$$\begin{aligned}
&\frac{\partial g}{\partial s}(t, s) \\
&= \langle F(\phi^{s+t-t_k+\tau'_k}(\mathbf{y})), F(\phi^t(\mathbf{x})) \rangle \\
&\geq \|F(\phi^t(\mathbf{x}))\|[\|F(\phi^t(\mathbf{x}))\| - \|F(\phi^{s+t-t_k+\tau'_k}(\mathbf{y})) - F(\phi^t(\mathbf{x}))\|] \\
&\geq \|F(\phi^t(\mathbf{x}))\|[\Delta - M_1(M_0|s| + \|\phi^{t-t_k+\tau'_k}(\mathbf{y}) - \phi^t(\mathbf{x})\|)] \\
&\geq \|F(\phi^t(\mathbf{x}))\|[\Delta - M_1(4M_0\Delta^{-1} + 1)e^{M_1 d}(4M_0\Delta^{-1} + 1)d] \\
&> 0.
\end{aligned}$$

It follows from the implicit function theorem and the uniqueness of $s_k(t)$ that $s_k(t)$ is, indeed, continuous.

Next note it follows from the uniqueness also that

$$s_k(t_k) + \tau'_k = \tau'_k, \; s_k(t_{k+1}) + t_{k+1} - t_k + \tau'_k = \tau'_{k+1}, \tag{78}$$

provided that

$$(8\Delta^{-1} + 2)d \leq \Delta/2M_0M_1.$$

Now we define the function $\alpha : \mathbb{R} \to \mathbb{R}$ by

$$\alpha(t) = s_k(t) + t - t_k + \tau'_k \quad \text{if} \quad t_k \leq t \leq t_{k+1}.$$

It follows from Eq.(78) that $\alpha(t)$ is well-defined and continuous. Also for all k, if $t_k \leq t \leq t_{k+1}$,

$$\begin{aligned}
\|\phi^{\alpha(t)}(\mathbf{y}) - \phi^t(\mathbf{x})\| &= \|\phi^{s_k(t)+t-t_k+\tau'_k}(\mathbf{y}) - \phi^{t-t_k}(\phi^{t_k}(\mathbf{x}))\| \\
&\leq M_0|s_k(t)| + \|\phi^{t-t_k+\tau'_k}(\mathbf{y}) - \phi^{t-t_k}(\phi^{t_k}(\mathbf{x}))\| \\
&\leq (4M_0\Delta^{-1} + 1)e^{M_1 d}(4M_0\Delta^{-1} + 1)d.
\end{aligned}$$

So if we also ask that $(4M_0\Delta^{-1} + 1)^2 e^{M_1 d} d$ not exceed an expansivity constant for S, it follows that $\mathbf{y} = \phi^\tau(\mathbf{x})$. This completes the proof of the proposition.

Let us remark here that Bowen and Walters [1972] also define a discrete expansivity property (see Theorem 3(iv) in their paper). It is the same as Definition 7.12 except that they add the extra conditions that $t_0 = \tau_0 = 0$ and that for any $\varepsilon > 0$, the positive number d can be chosen so that $|\tau| < \varepsilon$. It is easy to see that the proof of Proposition 7.13 can be modified to show that this discrete expansivity property is equivalent to their definition of expansiveness as described following Definition 7.8.

7.6 ROUGHNESS OF HYPERBOLIC SETS

Let $F : U \to I\!R^n$ be a C^1 vector field with U open and convex, and suppose S is a compact hyperbolic set for Eq.(1) as in Definition 7.1. In general, S may not be isolated but if we consider a sufficiently "tight" closed neighborhood of S we shall show that the maximal invariant set inside the neighborhood is still hyperbolic. Moreover, we shall show this holds for a C^1 perturbation of the vector field F.

So let O be a bounded open neighborhood of S with $\overline{O} \subset U$ and let $G : U \to I\!R^n$ be another C^1 vector field such that

$$\sup_{\mathbf{x}\in U} \|G(\mathbf{x}) - F(\mathbf{x})\| + \sup_{\mathbf{x}\in U} \|DG(\mathbf{x}) - DF(\mathbf{x})\| \leq \sigma, \tag{79}$$

where σ is a positive number the size of which is to be determined. We define

$$S_O = \{\mathbf{x} \in \overline{O} : \psi^t(\mathbf{x}) \in \overline{O} \text{ for all real } t\}, \tag{80}$$

where ψ^t is the flow corresponding to the system

$$\dot{\mathbf{x}} = G(\mathbf{x}). \tag{81}$$

Clearly, S_O is a compact invariant set for Eq.(81) and S_O is the maximal invariant set in $\overline{O}$.

We show the following theorem.

Theorem 7.14. *Let S be a compact hyperbolic set for Eq.*(1) *as in Definition* 7.1. *Choose numbers β_1, β_2 such that*

$$0 < \beta_1 < \alpha_1 \quad \text{and} \quad 0 < \beta_2 < \alpha_2.$$

Then there exist positive numbers σ_0 and d_0 depending only on F, S, β_1 and β_2 such that if O is an open neighborhood of S satisfying

$$d = \max_{\mathbf{x}\in\overline{O}} \operatorname{dist}(\mathbf{x}, S) \leq d_0$$

and $G : U \to I\!R^n$ is a C^1 vector field satisfying Eq.(79) with $\sigma \le \sigma_0$, the set S_O defined in Eq.(80) is a compact hyperbolic set for Eq.(81) *with exponents β_1, β_2, and with constants and bounds on the projections depending only on F, S, β_1 and β_2.*

Proof. First we make a preliminary estimate. Let $\mathbf{x}$ and $\mathbf{y}$ be points in U such that the solutions $\phi^t(\mathbf{x})$ and $\psi^t(\mathbf{y})$ of (1) and (81) respectively are defined in an interval J containing 0. We derive an inequality for $\|\phi^t(\mathbf{x}) - \psi^t(\mathbf{y})\|$. Note that if $t \in J$

$$\begin{aligned}\|\phi^t(\mathbf{x}) - \psi^t(\mathbf{y})\| &= \left\|\mathbf{x} - \mathbf{y} + \int_0^t [F(\phi^s(\mathbf{x})) - G(\psi^s(\mathbf{y}))]ds\right\| \\ &\le \|\mathbf{x} - \mathbf{y}\| + \left|\int_0^t \|F(\phi^s(\mathbf{x})) - F(\psi^s(\mathbf{y}))\|ds\right| \\ &\quad + \left|\int_0^t \|F(\psi^s(\mathbf{y})) - G(\psi^s(\mathbf{y}))\|ds\right| \\ &\le \|\mathbf{x} - \mathbf{y}\| + \sigma|t| + M_1 \left|\int_0^t \|\phi^s(\mathbf{x}) - \psi^s(\mathbf{y})\|ds\right|,\end{aligned}$$

where

$$M_1 = \sup_{\mathbf{x} \in U} \|DF(\mathbf{x})\|.$$

So, by Gronwall's lemma (conf. Coppel [1965, p.19]),

$$\|\phi^t(\mathbf{x}) - \psi^t(\mathbf{y})\| \le \frac{\sigma}{M_1}(e^{M_1|t|} - 1) + \|\mathbf{x} - \mathbf{y}\|e^{M_1|t|} \quad \text{for } t \in J. \tag{82}$$

Having finished with this preliminary estimate, now let $\mathbf{x} \in S_O$ and write

$$\mathbf{x}_k = \psi^k(\mathbf{x}) \quad \text{for } k \in \mathbf{Z}.$$

Since $\operatorname{dist}(\mathbf{x}_k, S) \le d$, there exists $\mathbf{y}_k$ in S such that

$$\|\mathbf{x}_k - \mathbf{y}_k\| \le d.$$

Then

$$\begin{aligned}\|\mathbf{y}_{k+1} - \phi^1(\mathbf{y}_k)\| &\le \|\mathbf{y}_{k+1} - \mathbf{x}_{k+1}\| + \|\psi^1(\mathbf{x}_k) - \phi^1(\mathbf{y}_k)\| \\ &\le d + (\sigma + \|\mathbf{x}_k - \mathbf{y}_k\|)e^{M_1} \quad \text{by Eq.(82)} \\ &\le (1 + e^{M_1})d + \sigma e^{M_1}.\end{aligned}$$

Hence

$$\|\mathbf{y}_{k+1} - \phi^1(\mathbf{y}_k)\| \le \delta$$

for all integers k with

$$\delta = (1 + e^{M_1})d + \sigma e^{M_1}. \tag{83}$$

This means that the sequence $\{\mathbf{y}_k\}_{k=-\infty}^{\infty}$ is a *discrete δ pseudo orbit* in the sense of the following definition.

Definition 7.15. If δ is a positive number, a sequence of points $\{\mathbf{y}_k\}_{k=-\infty}^{\infty}$ in U is said to be a *discrete δ pseudo orbit* for Eq.(1) if there is a sequence $\{h_k\}_{k=-\infty}^{\infty}$ of positive times with $\sup h_k < \infty$, $\inf h_k > 0$ such that

$$\|\mathbf{y}_{k+1} - \phi^{h_k}(\mathbf{y}_k)\| \le \delta \quad \text{for} \quad k \in \mathbf{Z}.$$

We now prove a lemma, which reflects the second way of using the hyperbolicity to get exponential dichotomies. An essential tool here is Lemma 2.17. As usual, we denote by $\mathcal{P}^0(\mathbf{x})$, $\mathcal{P}^s(\mathbf{x})$ and $\mathcal{P}^u(\mathbf{x})$ the projections associated with the splitting (5).

Lemma 7.16. *Let S be a compact hyperbolic set for the differential equation* (1) *with U open and convex, as in Definition* 7.1, *and let $\{\mathbf{y}_k\}_{k=-\infty}^{\infty}$ be a discrete δ pseudo orbit in S with associated times $\{h_k\}_{k=-\infty}^{\infty}$ satisfying*

$$0 < h_{min} \le h_k \le h_{max}.$$

We define a sequence $\{a_k\}_{k=-\infty}^{\infty}$ of real numbers by the recurrence relation

$$a_{k+1} = a_k + h_k$$

with $a_0 = 0$ and then we define the piecewise continuously differentiable function

$$\mathbf{y}(t) = \phi^{t-a_k}(\mathbf{y}_k)$$

for $a_k \le t < a_{k+1}$, $k \in \mathbf{Z}$. Suppose $\tilde{\beta}_1$, $\tilde{\beta}_2$ and λ are positive numbers satisfying

$$\tilde{\beta}_1 < \alpha_1,\ \tilde{\beta}_2 < \alpha_2,\ \lambda < \min\{\tilde{\beta}_1, \tilde{\beta}_2\}.$$

Then if δ is sufficiently small depending on F, S, h_{min}, h_{max}, $\tilde{\beta}_1$, $\tilde{\beta}_2$ and λ, there are real-valued functions $\omega_+(\cdot)$ and $\omega_-(\cdot)$ depending on F, S, $\tilde{\beta}_1$, $\tilde{\beta}_2$ and λ with $lim_{\delta \to 0+} \omega_\pm(\delta) = 0$ such that the equation

$$\dot{\mathbf{x}} = [DF(\mathbf{y}(t)) + \lambda]\mathbf{x} \tag{84}$$

has an exponential dichotomy on $(-\infty, \infty)$ with exponents $\tilde{\beta}_1 - \lambda$, $\lambda/4$ and with projection $\overline{\mathrm{P}}(t)$ satisfying $\operatorname{rank} \overline{\mathrm{P}}(t) = \operatorname{rank} \mathcal{P}^s$ *and*

$$\|\overline{\mathrm{P}}(t) - \mathcal{P}^s(\mathbf{y}(t))\| \le \omega_+(\delta),$$

and the equation

$$\dot{\mathbf{x}} = [DF(\mathbf{y}(t)) - \lambda]\mathbf{x} \tag{85}$$

has an exponential dichotomy on $(-\infty, \infty)$ *with exponents* $\lambda/4$, $\tilde{\beta}_2 - \lambda$, *and with projection* $\overline{Q}(t)$ *satisfying* rank $\overline{Q}(t) = n -$ rank $\mathcal{P}^u$ *and*

$$\|\overline{Q}(t) - [\mathcal{P}^0(\mathbf{y}(t)) + \mathcal{P}^s(\mathbf{y}(t))]\| \leq \omega_-(\delta).$$

In both cases the associated constants depend only on F, S, $\tilde{\beta}_1$, $\tilde{\beta}_2$ *and* λ. *Also we have*

$$\overline{P}(t)\overline{Q}(t) = \overline{Q}(t)\overline{P}(t) = \overline{P}(t) \tag{86}$$

and $\overline{Q}(t) - \overline{P}(t)$ *is a projection of rank* 1.

Proof. First we observe that for $k \geq m$

$$\|\mathbf{y}_k - \phi^{a_k - a_m}(\mathbf{y}_m)\| \leq (1 + \tilde{M}_1 + \cdots \tilde{M}_1^{k-m-1})\delta = (\tilde{M}_1 - 1)^{-1}(\tilde{M}_1^{k-m} - 1)\delta,$$

where both sides are interpreted as zero when $k = m$ and

$$\tilde{M}_1 = e^{M_1 h_{max}}$$

with

$$M_1 = \sup_{\mathbf{x} \in U} \|DF(\mathbf{x})\|.$$

This inequality follows by induction on k using the estimate

$$\begin{aligned}
&\|\mathbf{y}_{k+1} - \phi^{a_{k+1} - a_m}(\mathbf{y}_m)\| \\
&\quad \leq \|\mathbf{y}_{k+1} - \phi^{h_k}(\mathbf{y}_k)\| + \|\phi^{h_k}(\mathbf{y}_k) - \phi^{h_k}(\phi^{a_k - a_m}(\mathbf{y}_m))\| \\
&\quad \leq \delta + \tilde{M}_1 \|\mathbf{y}_k - \phi^{a_k - a_m}(\mathbf{y}_m)\|,
\end{aligned}$$

where we used Gronwall's lemma to get the bound

$$\|D\phi^{h_k}(\mathbf{y})\| \leq e^{M_1 h_k}.$$

Suppose now $t \geq a$. Then there are integers $m \leq k$ such that

$$a_m \leq a < a_{m+1}, \; a_k \leq t < a_{k+1}.$$

So, by Gronwall's lemma again,

$$\begin{aligned}
\|\mathbf{y}(t) - \phi^{t-a}(\mathbf{y}(a))\| &= \|\phi^{t-a_k}(\mathbf{y}_k) - \phi^{t-a_m}(\mathbf{y}_m)\| \\
&= \|\phi^{t-a_k}(\mathbf{y}_k) - \phi^{t-a_k}(\phi^{a_k - a_m}(\mathbf{y}_m))\| \\
&\leq e^{M_1 h_{max}} \|\mathbf{y}_k - \phi^{a_k - a_m}(\mathbf{y}_m))\| \\
&\leq \tilde{M}_1(\tilde{M}_1 - 1)^{-1}(\tilde{M}_1^{k-m} - 1)\delta.
\end{aligned}$$

Hence if $t \geq a$,

$$\|\mathbf{y}(t) - \phi^{t-a}(\mathbf{y}(a))\| \leq \tilde{M}_1(\tilde{M}_1 - 1)^{-1}(\tilde{M}_1^{\frac{t-a}{h_{min}}+1} - 1)\delta. \tag{87}$$

Next note it follows from the remarks before Lemma 7.4 that for any a the equation

$$\dot{\mathbf{x}} = [DF(\phi^{t-a}(\mathbf{y}(a))) - \lambda]\mathbf{x} \tag{88}$$

has an exponential dichotomy on $(-\infty, \infty)$ with projections

$$\mathcal{P}^0(\phi^{t-a}(\mathbf{y}(a))) + \mathcal{P}^s(\phi^{t-a}(\mathbf{y}(a))),$$

exponents λ, $\alpha_2 - \lambda$ and constants L_1^+, L_2^+ depending only on F and S.

With a view to eventually applying Lemma 2.17, we choose n_0 as the least positive integer such that

$$2L_1^+(2L_1^+ + L_2^+) < e^{n_0\lambda/4},\ 2L_2^+(2L_2^+ + L_1^+) < e^{n_0(\alpha_2 - \tilde{\beta}_2)/2} \tag{89}$$

and then set

$$\delta_1 = \tilde{M}_1(\tilde{M}_1 - 1)^{-1}(\tilde{M}_1^{\frac{n_0}{h_{min}}+1} - 1)\delta.$$

Note, using Eq.(87), if $(i-1)n_o \leq t \leq in_0$,

$$\|DF(\mathbf{y}(t)) - DF(\phi^{t-(i-1)n_0}(\mathbf{y}((i-1)n_0)))\| \leq \omega(\delta_1),$$

where

$$\omega(\delta_1) = \sup\{\|DF(\mathbf{y}) - DF(\mathbf{x})\| : \mathbf{x}, \mathbf{y} \in S,\ \|\mathbf{y} - \mathbf{x}\| \leq \delta_1\}.$$

Then, provided δ is sufficiently small depending on F, S, h_{min}, h_{max}, $\tilde{\beta}_2$ and λ, it follows from Lemma 7.4 and the remark about Eq.(88) with $a = (n-1)n_0$, that Eq.(85) has an exponential dichotomy on $[(i-1)n_0, in_0]$ with exponents $\lambda/2$, $(\alpha_2 + \tilde{\beta}_2)/2 - \lambda$ and projection $Q^{(i)}(t)$ satisfying

$$\|Q^{(i)}(t)\,\mathcal{P}^0(\phi^{t-(i-1)n_0}(\mathbf{y}((i-1)n_0))) - \mathcal{P}^s(\phi^{t-(i-1)n_0}(\mathbf{y}((i-1)n_0)))\|$$
$$\leq N\omega(\delta_1), \tag{90}$$

where the constants in the dichotomy are $2L_1^+(2L_1^+ + L_2^+)$, $2L_2^+(2L_2^+ + L_1^+)$ and N depends only on F, S and λ. Next note it follows from Eqs.(87) and (90) that

$$\|Q^{(i)}(t) - \mathcal{P}^0(\mathbf{y}(t)) - \mathcal{P}^s(\mathbf{y}(t))\| \leq N\omega(\delta_1) + \omega^0(\delta_1) + \omega^s(\delta_1), \tag{91}$$

where $\omega^0(\cdot)$ and $\omega^s(\cdot)$ are the moduli of continuity of $\mathcal{P}^0(\cdot)$ and $\mathcal{P}^s(\cdot)$ on S.

Now denote by $Y(t)$ a fundamental matrix for the equation

$$\dot{\mathbf{x}} = DF(\mathbf{y}(t))\mathbf{x}.$$

Then, in virtue of Proposition 7.3, our conclusions about Eq.(85) imply that the difference equation

$$\mathbf{u}_{k+1} = Y(k+1)Y^{-1}(k)e^{-\lambda}\mathbf{u}_k \tag{92}$$

has for all $i \in \boldsymbol{Z}$ an exponential dichotomy on $[(i-1)n_0, in_0]$ with exponents $e^{-\lambda/2}$, $e^{-(\frac{\alpha_2+\tilde{\beta}_2}{2}-\lambda)}$, projection $Q_k^{(i)} = Q^{(i)}(k)$ satisfying

$$\|Q_k^{(i)} - \mathcal{P}^0(\mathbf{y}(k)) - \mathcal{P}^s(\mathbf{y}(k))\| \leq N\omega(\delta_1) + \omega^0(\delta_1) + \omega^s(\delta_1)$$

and constants $2L_1^+(2L_1^+ + L_2^+)$ and $2L_2^+(2L_2^+ + L_1^+)$. Now, in view of Eq.(89), we may apply Lemma 2.17 to deduce that when δ is sufficiently small depending on F, S, h_{min}, h_{max}, $\tilde{\beta}_2$ and λ, Eq.(92) has an exponential dichotomy on $(-\infty, \infty)$ with exponents $e^{-\lambda/4}$, $e^{-(\tilde{\beta}_2-\lambda)}$, projection Q_k satisfying

$$\|Q_k - Q_k^{(i)}\| \leq \tilde{N}[N\omega(\delta_1) + \omega^0(\delta_1) + \omega^s(\delta_1)] \tag{93}$$

for $(i-1)n_0 \leq k \leq in_0$, $i \in \boldsymbol{Z}$, where $\tilde{N}$ and the constants in the dichotomy depend only on F, S, λ, and $\tilde{\beta}_2$. Then if we define

$$\overline{Q}(t) = Y(t)Y^{-1}(0)Q_0Y(0)Y^{-1}(t),$$

it follows from Proposition 7.3 that Eq.(85) has an exponential dichotomy on $(-\infty, \infty)$ with projection $\overline{Q}(t)$, exponents $\lambda/4$, $\tilde{\beta}_2 - \lambda$ and constants depending only on F, S, $\tilde{\beta}_2$ and λ. Now given t there exist integers k and i such that

$$(i-1)n_0 \leq k \leq t \leq k+1 \leq in_0.$$

Then, using the invariance of the projections, Gronwall's lemma and Eqs.(91) and (93), we estimate

$$\begin{aligned}
&\|\overline{Q}(t) - \mathcal{P}^0(\mathbf{y}(t)) - \mathcal{P}^s(\mathbf{y}(t))\| \\
&\leq \|\overline{Q}(t) - Q^{(i)}(t)\| + \|Q^{(i)}(t) - \mathcal{P}^0(\mathbf{y}(t)) - \mathcal{P}^s(\mathbf{y}(t))\| \\
&= \|Y(t)Y^{-1}(k)[Q_k - Q_k^{(i)}]Y(k)Y^{-1}(t)\| + \|Q^{(i)}(t) - \mathcal{P}^0(\mathbf{y}(t)) - \mathcal{P}^s(\mathbf{y}(t))\| \\
&\leq e^{2M_1}\|Q_k - Q_k^{(i)}\| + N\omega(\delta_1) + \omega^0(\delta_1) + \omega^s(\delta_1) \\
&\leq e^{2M_1}\tilde{N}[N\omega(\delta_1) + \omega^0(\delta_1) + \omega^s(\delta_1)] + N\omega(\delta_1) + \omega^0(\delta_1) + \omega^s(\delta_1) \\
&= \omega_-(\delta).
\end{aligned}$$

Similarly we show when δ is sufficiently small depending on F, S, h_{min}, h_{max}, $\tilde{\beta}_1$ and λ that Eq.(84) has an exponential dichotomy on $(-\infty, \infty)$ with exponents $\tilde{\beta}_1 - \lambda$, $\lambda/4$ and projection $\overline{P}(t)$ satisfying

$$\|\overline{P}(t) - \mathcal{P}^s(\mathbf{y}(t))\| \leq \omega_+(\delta),$$

where $\omega_+(\cdot)$ and the constants in the dichotomy depend only on F, S, $\tilde{\beta}_1$ and λ.

Finally if $\xi \in \mathcal{R}(\overline{P}(0))$ and $t \geq 0$,

$$\begin{aligned}\|Y(t)Y^{-1}(0)e^{-\lambda t}\xi\| &= e^{-2\lambda t}\|Y(t)Y^{-1}(0)e^{\lambda t}\xi\| \\ &\leq e^{-2\lambda t}\text{constant} \cdot \|\xi\|e^{-(\tilde{\beta}_1-\lambda)t} \\ &\to 0 \text{ as } t \to \infty,\end{aligned}$$

where we have used the dichotomy property of Eq.(84). By Proposition 7.3, this means that $\xi \in \mathcal{R}(\overline{Q}(0))$ and so

$$\mathcal{R}(\overline{P}(0)) \subset \mathcal{R}(\overline{Q}(0)).$$

Similarly, taking $\xi \in \mathcal{N}(\overline{Q}(0))$ and considering $\|Y(t)Y^{-1}(0)e^{\lambda t}\xi\|$ for $t \leq 0$, we find that

$$\mathcal{N}(\overline{Q}(0)) \subset \mathcal{N}(\overline{P}(0)).$$

Then, using the invariance of the projections, the relations (86) easily follow. These relations imply that $\overline{Q}(t) - \overline{P}(t)$ is a projection and that $I\!R^n$ is the direct sum of $\mathcal{R}(\overline{P}(t))$, $\mathcal{N}(\overline{Q}(t))$ and $\mathcal{R}(\overline{Q}(t) - \overline{P}(t))$. So the range of $\overline{Q}(t) - \overline{P}(t)$ has dimension 1. Thus the proof of Lemma 7.16 is complete.

We apply this lemma to the sequence $\{\mathbf{y}_k\}_{k=-\infty}^{\infty}$ defined before Definition 7.15. For this sequence $h_k = 1$ and $a_k = k$. We choose

$$\tilde{\beta}_i = \frac{\alpha_i + \beta_i}{2} \quad \text{and} \quad \lambda = \frac{1}{2}\min\{\tilde{\beta}_1, \tilde{\beta}_2\}.$$

Provided δ in Eq.(83) is sufficiently small depending on F, S, β_1 and β_2, we deduce that Eqs.(84) and (85) have exponential dichotomies as described in Lemma 7.16. Note if $k \leq t < k+1$ and $k \in \boldsymbol{Z}$,

$$\begin{aligned}\|DG(\psi^t(\mathbf{x})) - DF(\mathbf{y}(t))\| &= \|DG(\psi^{t-k}(\mathbf{x}_k)) - DF(\phi^{t-k}(\mathbf{y}_k))\| \\ &\leq \|DG(\psi^{t-k}(\mathbf{x}_k)) - DF(\psi^{t-k}(\mathbf{x}_k))\| \\ &\quad + \|DF(\psi^{t-k}(\mathbf{x}_k)) - DF(\phi^{t-k}(\mathbf{y}_k))\| \\ &\leq \sigma + \omega(\|\psi^{t-k}(\mathbf{x}_k) - \phi^{t-k}(\mathbf{y}_k)\|) \\ &\leq \sigma + \omega([\sigma + \|\mathbf{x}_k - \mathbf{y}_k\|]e^{M_1}) \quad \text{by Eq.(82)} \\ &\leq \sigma + \omega((\sigma + d)e^{M_1}),\end{aligned}$$

where now

$$\omega(\varepsilon) = \sup\{\|DF(\mathbf{x}) - DF(\mathbf{y})\| : \mathbf{x} \in S, \|\mathbf{x} - \mathbf{y}\| \leq \varepsilon\}.$$

By Lemma 7.4 again, if σ and d are sufficiently small depending on F, S, β_1 and β_2 , the equation

$$\dot{\mathbf{y}} = [DG(\psi^t(\mathbf{x})) + \lambda]\mathbf{y}, \tag{94}$$

has an exponential dichotomy on $(-\infty, \infty)$ with exponents $\beta_1 - \lambda$ and $\lambda/8$, constants $\tilde{L}_1$, $\tilde{L}_2$ depending only on F, S, β_1 and β_2, and projections $P_{\mathbf{x}}(t)$ having the same rank as $\mathcal{P}^s$. Similarly this holds for the equation

$$\dot{\mathbf{y}} = [DG(\psi^t(\mathbf{x})) - \lambda]\mathbf{y} \tag{95}$$

and then the exponents are $\lambda/8$ and $\beta_2 - \lambda$, the projections $Q_{\mathbf{x}}(t)$ have the same rank as $\mathcal{P}^0 + \mathcal{P}^s$ and we may take the same constants $\tilde{L}_1$ and $\tilde{L}_2$.

We now use these dichotomy properties to determine the stable and unstable bundles. First let ξ be in $\mathcal{R}(P_{\mathbf{x}}(0))$. Then $\mathbf{y}(t) = e^{\lambda t} D\psi^t(\mathbf{x})\xi$ is a solution of Eq.(94) with $\mathbf{y}(0)$ in $\mathcal{R}(P_{\mathbf{x}}(0))$ and so

$$\|\mathbf{y}(t)\| \leq \tilde{L}_1 e^{-(\beta_1 - \lambda)t)} \|\mathbf{y}(0)\|$$

for $t \geq 0$. So if $\xi \in \mathcal{R}(P_{\mathbf{x}}(0))$,

$$\|D\psi^t(\mathbf{x})\xi\| \leq \tilde{L}_1 e^{-\beta_1 t} \|\xi\| \tag{96}$$

for $t \geq 0$. Similarly, using Eq.(95), we find that if $\xi \in \mathcal{N}(Q_{\mathbf{x}}(0))$,

$$\|D\psi^{-t}(\mathbf{x})\xi\| \leq \tilde{L}_2 e^{-\beta_2 t} \|\xi\| \tag{97}$$

for $t \geq 0$.

These relations hold for all $\mathbf{x} \in S_O$. For fixed s, if we replace $\mathbf{x}$ in Eq.(94) by $\psi^s(\mathbf{x})$, we obtain the same equation with t replaced by $t + s$. This equation has an exponential dichotomy on $(-\infty, \infty)$ with projections $P_{\mathbf{x}}(t + s)$. By the uniqueness of the projection (conf. Proposition 7.3 (ii)), this means that for all t and s

$$P_{\mathbf{x}}(t + s) = P_{\psi^s(\mathbf{x})}(t).$$

Thus

$$P_{\mathbf{x}}(t) = P_{\psi^t(\mathbf{x})}(0).$$

However, by the invariance of the projections,

$$D\psi^t(\mathbf{x})e^{\lambda t} P_{\mathbf{x}}(0) = P_{\mathbf{x}}(t) D\psi^t(\mathbf{x})e^{\lambda t}.$$

That is, for $\mathbf{x} \in S_O$ and all t,

$$D\psi^t(\mathbf{x}) P_{\mathbf{x}}(0) = P_{\psi^t(\mathbf{x})}(0) D\psi^t(\mathbf{x}). \tag{98}$$

Similarly,

$$D\psi^t(\mathbf{x})Q_{\mathbf{x}}(0) = Q_{\psi^t(\mathbf{x})}(0)D\psi^t(\mathbf{x}). \tag{99}$$

Next we note that if $\xi \in \mathcal{R}(P_{\mathbf{x}}(0))$, then

$$e^{-\lambda t}D\psi^t(\mathbf{x})\xi = e^{-2\lambda t}\cdot e^{\lambda t}D\psi^t(\mathbf{x})\xi$$

is a solution of Eq.(95) bounded on $t \geq 0$. So by Proposition 7.3

$$\mathcal{R}(P_{\mathbf{x}}(0)) \subset \mathcal{R}(Q_{\mathbf{x}}(0)). \tag{100}$$

Similarly,

$$\mathcal{N}(Q_{\mathbf{x}}(0)) \subset \mathcal{N}(P_{\mathbf{x}}(0)). \tag{101}$$

Next since

$$\|G(\psi^t(\mathbf{x}))\| \leq \|F(\psi^t(\mathbf{x}))\| + \|G(\psi^t(\mathbf{x})) - F(\psi^t(\mathbf{x}))\| \leq M_0 + \sigma,$$

where

$$M_0 = \sup_{\mathbf{x}\in U} \|F(\mathbf{x})\|,$$

it follows that $e^{-\lambda t}G(\psi^t(\mathbf{x}))$ is a solution of Eq.(95) bounded in $t \geq 0$. So by Proposition 7.3,

$$G(\mathbf{x}) \in \mathcal{R}(Q_{\mathbf{x}}(0)). \tag{102}$$

On the other hand, $e^{\lambda t}G(\psi^t(\mathbf{x}))$ is a solution of Eq.(94) bounded in $t \leq 0$. So by Proposition 7.3 again,

$$G(\mathbf{x}) \in \mathcal{N}(P_{\mathbf{x}}(0)). \tag{103}$$

We also assume

$$M_1 d + \sigma < \Delta$$

so that

$$\|G(\mathbf{x})\| \geq \|F(\mathbf{x})\| - \|G(\mathbf{x}) - F(\mathbf{x})\| \geq \Delta - M_1 d - \sigma > 0. \tag{104}$$

Now it follows from Eqs.(100), (101), (102), (103) and (104), and the dimensions of the subspaces that

$$I\!R^n = \text{span}\{G(\mathbf{x})\} \oplus \mathcal{R}(P_{\mathbf{x}}(0)) \oplus \mathcal{N}(Q_{\mathbf{x}}(0)) \tag{105}$$

with $Q_{\mathbf{x}}(0) - P_{\mathbf{x}}(0)$, $P_{\mathbf{x}}(0)$ and $I - Q_{\mathbf{x}}(0)$ as the corresponding projections.

Hence for $\mathbf{x} \in S_O$, we take

$$E^s(\mathbf{x}) = \mathcal{R}(P_{\mathbf{x}}(0)), E^u(\mathbf{x}) = \mathcal{N}(Q_{\mathbf{x}}(0)).$$

Then it follows from Eq.(105) that

$$I\!R^n = \text{span}\{G(\mathbf{x})\} \oplus E^s(\mathbf{x}) \oplus E^u(\mathbf{x}).$$

The invariance follows from Eqs.(98) and (99) and Eqs.(96), (97) imply that for $t \geq 0$

$$\|D\psi^t(\mathbf{x})\xi\| \leq \tilde{L}_1 e^{-\beta_1 t}\|\xi\| \quad \text{for} \quad \xi \in E^s(\mathbf{x})$$

and

$$\|D\psi^{-t}(\mathbf{x})\xi\| \leq \tilde{L}_2 e^{-\beta_2 t}\|\xi\| \quad \text{for} \quad \xi \in E^u(\mathbf{x}).$$

As follows from Eq.(105), the projections onto the stable and unstable subspaces are $P_{\mathbf{x}}(0)$ and $I - Q_{\mathbf{x}}(0)$ respectively. These have bounds $\tilde{L}_1$ and $\tilde{L}_2$ respectively, which depend only on F, S, β_1 and β_2. Thus the proof of the theorem is complete.

8. TRANSVERSAL HOMOCLINIC ORBITS AND HYPERBOLIC SETS IN DIFFERENTIAL EQUATIONS

In this chapter, we show how to construct a hyperbolic set somewhat more complicated than a single hyperbolic periodic orbit. In Chapter 10 we shall use shadowing to show that the dynamics in the neighbourhood of this hyperbolic set is chaotic.

8.1 THE HYPERBOLIC SET ASSOCIATED WITH A TRANSVERSAL HOMOCLINIC ORBIT

Let $F : U \to \mathbb{R}^n$ be a C^1 vector field defined in a convex open set U in $\mathbb{R}^n$ and denote by ϕ the flow associated with the system

$$\dot{\mathbf{x}} = F(\mathbf{x}). \tag{1}$$

Now let $\mathbf{u}(t)$ be a hyperbolic periodic solution of Eq.(1) with minimal period T. The stable and unstable manifolds $W^s(\mathbf{u})$ and $W^u(\mathbf{u})$ were defined and studied in Chapter 6.

Definition 8.1. Let $\mathbf{u}(t)$ be a hyperbolic periodic orbit of Eq.(1). A point $\mathbf{p}_0 \in U$ is said to be a *homoclinic point* with respect to $\mathbf{u}$ if $\mathbf{p}_0 \neq \mathbf{u}(t)$ for all t and $\mathbf{p}_0$ is in the intersection $W^s(\mathbf{u}) \cap W^u(\mathbf{u})$ of the stable and unstable manifolds of $\mathbf{u}(t)$. $\mathbf{p}_0$ is said to be a *transversal homoclinic point* if, in addition,

$$T_{\mathbf{p}_0} W^s(\mathbf{u}) \cap T_{\mathbf{p}_0} W^u(\mathbf{u}) = \operatorname{span}\{F(\mathbf{p}_0)\}. \tag{2}$$

Let $\gamma^+(\mathbf{p}_0)$ and $\gamma^-(\mathbf{p}_0)$ be the forward and backward asymptotic phases of $\mathbf{p}_0$. Then it follows from the discussion in Chapter 6 that

$$T_{\mathbf{p}_0} W^s(\mathbf{u}) = \operatorname{span}\{F(\mathbf{p}_0)\} \oplus T_{\mathbf{p}_0} W^{s,\gamma^+(\mathbf{p}_0)}(\mathbf{u})$$

and

$$T_{\mathbf{p}_0} W^u(\mathbf{u}) = \operatorname{span}\{F(\mathbf{p}_0)\} \oplus T_{\mathbf{p}_0} W^{s,\gamma^-(\mathbf{p}_0)}(\mathbf{u}).$$

So the transversality condition (2) is equivalent to

$$\mathbb{R}^n = \operatorname{span}\{F(\mathbf{p}_0)\} \oplus T_{\mathbf{p}_0} W^{s,\gamma^+(\mathbf{p}_0)}(\mathbf{u}) \oplus T_{\mathbf{p}_0} W^{s,\gamma^-(\mathbf{p}_0)}(\mathbf{u}). \tag{3}$$

Now we state and prove the main theorem of this section.

Theorem 8.2. *Let $\mathbf{p}_0$ be a transversal homoclinic point with respect to the hyperbolic periodic orbit $\mathbf{u}(t)$ of the autonomous system* (1). *Then the set*

$$S = \{\mathbf{u}(t) : -\infty < t < \infty\} \cup \{\phi^t(\mathbf{p}_0) : -\infty < t < \infty\}$$

is a compact hyperbolic set.

Proof. Clearly S is compact since

$$\text{dist}(\phi^t(\mathbf{p}_0), \mathbf{u}) = \min_{0 \le s \le T} \|\phi^t(\mathbf{p}_0) - \mathbf{u}(s)\| \to 0 \quad \text{as} \quad |t| \to \infty.$$

Also $\phi^t(S) = S$ for all t. As shown in Chapter 7 the set $\{\mathbf{u}(t) : -\infty < t < \infty\}$ is compact hyperbolic with splitting

$$I\!R^n = \text{span}\{F(\mathbf{u}(t))\} \oplus D\phi^t(\mathbf{u}(0))(E^s) \oplus D\phi^t(\mathbf{u}(0))(E^u), \tag{4}$$

where E^s and E^u are the generalized eigenspaces of $D\phi^T(\mathbf{u}(0))$ corresponding to the eigenvalues inside and outside the unit circle. Now, by the invariance properties,

$$D\phi^t(\mathbf{p}_0))F(\mathbf{p}_0) = F(\mathbf{p}(t)),$$

$$D\phi^t(\mathbf{p}_0)(T_{\mathbf{p}_0} W^{s,\gamma^+(\mathbf{p}_0)}(\mathbf{u})) = T_{\mathbf{p}(t)} W^{s,\gamma^+(\mathbf{p}(t))}(\mathbf{u}),$$

$$D\phi^t(\mathbf{p}_0)(T_{\mathbf{p}_0} W^{u,\gamma^-(\mathbf{p}_0)}(\mathbf{u})) = T_{\mathbf{p}(t)} W^{s,\gamma^-(\mathbf{p}(t))}(\mathbf{u}),$$

where $\mathbf{p}(t) = \phi^t(\mathbf{p}_0)$, and so it follows from Eq.(3) that for all t

$$I\!R^n = \text{span}\{F(\mathbf{p}(t))\} \oplus E^s(\mathbf{p}(t)) \oplus E^u(\mathbf{p}(t)), \tag{5}$$

where

$$E^s(\mathbf{p}(t)) = T_{\mathbf{p}(t)} W^{s,\gamma^+(\mathbf{p}(t))}(\mathbf{u}), \; E^u(\mathbf{p}(t)) = T_{\mathbf{p}(t)} W^{u,\gamma^-(\mathbf{p}(t))}(\mathbf{u}). \tag{6}$$

The subspaces in Eq.(5) have the same dimension as those in Eq.(4) and the splitting is invariant. So to complete the proof that S is hyperbolic, all we need do is verify the exponential estimates for the splitting (5) along $\mathbf{p}(t)$.

To this end, we prove the following proposition.

Proposition 8.3. *Let $\mathbf{u}(t)$ be a hyperbolic periodic solution of Eq.*(1), *let $\mathbf{p}_0 \in W^s(\mathbf{u})$ and let the subspace V_0 satisfy*

$$I\!R^n = \text{span}\{F(\mathbf{p}_0)\} \oplus T_{\mathbf{p}_0} W^{s,\gamma^+(\mathbf{p}_0)}(\mathbf{u}) \oplus V_0.$$

Then there are positive numbers L_1, L_2, α_1, α_2 and a number τ^+ such that for $s \ge \tau^+$,

$$\|D\phi^t(\phi^s(\mathbf{p}_0))\xi\| \le L_1 e^{-\alpha_1 t}\|\xi\| \;\text{ for } \xi \in T_{\phi^s(\mathbf{p}_0)} W^{s,\gamma^+(\phi^s(\mathbf{p}_0))}(\mathbf{u}),\; t \ge 0 \tag{7}$$

and

$$\|D\phi^{-t}(\phi^s(\mathbf{p}_0))\xi\| \le L_2 e^{-\alpha_2 t}\|\xi\| \text{ for } \xi \in D\phi^s(\mathbf{p}_0)(V_0),\ 0 \le t \le s - \tau^+. \quad (8)$$

Proof. For given $\varepsilon > 0$, it follows from Proposition 6.5 that there exists a real number τ such that $\mathbf{y} = \phi^\tau(\mathbf{p}_0)$ is in $W^{s,\varepsilon}(\mathbf{x}_0)$, where $\mathbf{x}_0 = \mathbf{u}(0)$ and $W^{s,\varepsilon}(\mathbf{x}_0)$ is the local stable manifold of $\mathbf{x}_0$ considered as a fixed point of the Poincaré map P. Let ε be chosen as in the proof of Theorem 6.7. Then, as in that proof, there is a sequence of real numbers $\{t_k\}_{k=0}^\infty$ with $\lim t_k = \infty$ and $t_0 = 0$ such that

$$\mathbf{y}_k = P^k(\mathbf{y}) = \phi^{t_k}(\mathbf{y}),\ \ t_{k+1} - t_k = \tau(\mathbf{y}_k) \ \text{ for } k \ge 0.$$

Recall from the end of Chapter 6 that

$$T_\mathbf{y} W^{s,\gamma^+(\mathbf{y})}(\mathbf{u}) = \{-(\gamma^{+'}(\mathbf{y})\mathbf{h})F(\mathbf{y}) + \mathbf{h} : \mathbf{h} \in T_\mathbf{y} W^{s,\varepsilon}(\mathbf{x}_0)\}.$$

Next let

$$\mathbf{v}_0 = -(\gamma^{+'}(\mathbf{y})\mathbf{h})F(\mathbf{y}) + \mathbf{h}$$

be a vector in $T_\mathbf{y} W^{s,\gamma^+(\mathbf{y})}(\mathbf{u})$. Using the facts that

$$\gamma^{+'}(\mathbf{y})\mathbf{h} = \lim_{k\to\infty} - s_k'(\mathbf{y})\mathbf{h} = -\sum_{j=0}^{\infty} \tau'(\mathbf{y}_j) DP^j(\mathbf{y})\mathbf{h}$$

(conf. Eq.(14) in Chapter 6) and

$$D\varphi^{\tau(\mathbf{y}_k)}(\mathbf{y}_k)F(\mathbf{y}_k) = F(\mathbf{y}_{k+1}),$$

and Eq.(6) in Chapter 6 with $\mathbf{x} = \mathbf{y}_k$, it is easy to show by induction on k that

$$\mathbf{v}_k := D\phi^{t_k}(\mathbf{y})\mathbf{v}_0 = \alpha_k F(\mathbf{y}_k) + \mathbf{w}_k,$$

where

$$\alpha_k = \sum_{j=k}^{\infty} \tau'(\mathbf{y}_j)\mathbf{w}_j, \quad \mathbf{w}_k = DP^k(\mathbf{y})\mathbf{h}. \quad (9)$$

Observe that

$$\begin{aligned}
\|\mathbf{w}_k\| &= \|\mathbf{v}_k - \alpha_k F(\mathbf{y}_k)\| \\
&= \left\| \mathbf{v}_k - \frac{\langle F(\mathbf{x}_0), \mathbf{v}_k \rangle}{\langle F(\mathbf{x}_0), F(\mathbf{y}_k) \rangle} F(\mathbf{y}_k) \right\| \\
&\le \left(1 + \frac{\|F(\mathbf{x}_0)\|\|F(\mathbf{y}_k)\|}{|\langle F(\mathbf{x}_0), F(\mathbf{y}_k)\rangle|}\right) \|\mathbf{v}_k)\| \\
&\le \left(1 + \frac{\|F(\mathbf{x}_0)\| \cdot \frac{3}{2}\|F(\mathbf{x}_0)\|}{\frac{1}{2}\|F(\mathbf{x}_0)\|^2}\right) \|\mathbf{v}_k)\|,
\end{aligned}$$

provided we assume ε has been chosen so small that

$$||F(\mathbf{x}) - F(\mathbf{x}_0)|| \le \frac{1}{2}||F(\mathbf{x}_0)|| \tag{10}$$

when $||\mathbf{x}-\mathbf{x}_0|| \le \varepsilon$ (note that since $\mathbf{y} \in W^{s,\varepsilon}(\mathbf{x}_0)$, $||\mathbf{y}_k - \mathbf{x}_0|| = ||P^k(\mathbf{y}) - \mathbf{x}_0|| \le \varepsilon$ for $k \ge 0$). Thus for $k \ge 0$,

$$||\mathbf{w}_k|| \le 4||\mathbf{v}_k||. \tag{11}$$

Next, if ε is sufficiently small, it follows from Eq.(2) in Proposition 3.3 that there are positive constants K_1, β_1 with $\beta_1 < 1$ such that

$$||\mathbf{w}_k|| \le K_1 \beta_1^{k-m} ||\mathbf{w}_m|| \quad \text{for } 0 \le m \le k. \tag{12}$$

Now it follows from Eq.(9) that

$$|\alpha_k| \le N \sum_{j=k}^{\infty} ||\mathbf{w}_j||,$$

where

$$N = \sup_{k=0}^{\infty} |\tau'(\mathbf{y}_k)|.$$

So, using Eq.(12),

$$|\alpha_k| \le N \sum_{j=k}^{\infty} K_1 \beta_1^{j-k} ||\mathbf{w}_k|| = N K_1 (1 - \beta_1)^{-1} ||\mathbf{w}_k||.$$

Then

$$||\mathbf{v}_k|| = ||\alpha_k F(\mathbf{y}_k) + \mathbf{w}_k|| \le [M_0 N K_1 (1 - \beta_1)^{-1} + 1] ||\mathbf{w}_k||,$$

where

$$M_0 = \sup_{k=0}^{\infty} ||F(\mathbf{y}_k)||.$$

Continuing, using Eq.(11) and Eq.(12) again, we find that for $0 \le m \le k$

$$\begin{aligned} ||\mathbf{v}_k|| &\le K_1 [M_0 N K_1 (1 - \beta_1)^{-1} + 1] \beta_1^{k-m} ||\mathbf{w}_m|| \\ &\le 4 K_1 [M_0 N K_1 (1 - \beta_1)^{-1} + 1] \beta_1^{k-m} ||\mathbf{v}_m||. \end{aligned}$$

Thus we have found that

$$||\mathbf{v}_k|| \le K_2 \beta_1^{k-m} ||\mathbf{v}_m|| \quad \text{for } 0 \le m \le k, \tag{13}$$

where

$$K_2 = 4K_1[M_0NK_1(1-\beta_1)^{-1}+1].$$

Now if we write

$$\mathbf{v}(t) = D\phi^t(\mathbf{y})\mathbf{v}_0,$$

then what we have shown in Eq.(13) is that

$$\|\mathbf{v}(t_k)\| \le K_2\beta_1^{k-m}\|\mathbf{v}(t_m)\| \quad \text{for } 0 \le m \le k.$$

Recall from the proof of Theorem 6.7 that $t_k - kT \to -\gamma^+(\mathbf{y})$ as $k \to \infty$ and so

$$\frac{t_k}{T} - k \to 0 \quad \text{as} \quad k \to \infty.$$

This implies that there exists an integer k_0 such that if $k \ge k_0$,

$$-\frac{1}{2} < \frac{t_k}{T} - k < \frac{1}{2}$$

and so

$$\frac{t_k}{T} - \frac{1}{2} < k < \frac{t_k}{T} + \frac{1}{2}. \tag{14}$$

Hence for $k_0 \le m \le k$,

$$\|\mathbf{v}(t_k)\| \le K_2\beta_1^{-1}e^{-\alpha_1(t_k-t_m)}\|\mathbf{v}(t_m)\|, \tag{15}$$

where

$$\alpha_1 = -T^{-1}\log\beta_1.$$

Next if $t_{k_0} \le s \le t$, there exist integers k, m such that $k_0 \le m \le k$ and $t_m \le s \le t_{m+1}$, $t_k \le t \le t_{k+1}$. Then, by Gronwall's lemma and the fact that $t_{k+1} - t_k \le T$ if $k \ge k_0$,

$$\|\mathbf{v}(t)\| \le e^{M_1(t-t_k)}\|\mathbf{v}(t_k)\| \le e^{M_1T}\|\mathbf{v}(t_k)\|,$$

where

$$M_1 = \sup_{-\infty<t<\infty} \|DF(\phi^t(\mathbf{p}_0))\|.$$

Using Eq.(15), it follows that

$$\begin{aligned}
\|\mathbf{v}(t)\| &\le e^{M_1T}K_2\beta^{-1}e^{-\alpha_1(t_k-t_m)}\|\mathbf{v}(t_m)\| \\
&\le e^{M_1T}K_2\beta^{-1}e^{-\alpha_1(t_k-t_m)}e^{M_1(s-t_m)}\|\mathbf{v}(s)\| \\
&= e^{M_1T}K_2\beta^{-1}e^{-\alpha_1(t_k-t)}e^{-\alpha_1(s-t_m)}e^{M_1(s-t_m)}e^{-\alpha_1(t-s)}\|\mathbf{v}(s)\|.
\end{aligned}$$

Hence

$$\|\mathbf{v}(t)\| \leq K_2\beta_1^{-1}e^{(2M_1+\alpha_1)T}e^{-\alpha_1(t-s)}\|\mathbf{v}(s)\| \quad \text{for } t_{k_0} \leq s \leq t.$$

If we put

$$L_1 = K_2\beta_1^{-1}e^{(2M_1+\alpha_1)T}, \quad \tau^+ = t_{k_0} + \tau,$$

this means that

$$\|D\phi^t(\phi^s(\mathbf{p}_0))\xi\| \leq L_1 e^{-\alpha_1 t}\|\xi\| \tag{16}$$

for ξ in $T_{\phi^s(\mathbf{p}_0)}W^{s,\gamma^+(\phi^s(\mathbf{p}_0))}(\mathbf{u})$, $t \geq 0$, $s \geq \tau^+$ and so the estimates in Eq.(7) are proved.

Now we prove the exponential estimates in Eq.(8). By the assumption on V_0 and Eq.(17) in Chapter 6,

$$\begin{aligned} I\!R^n &= \operatorname{span}\{F(\mathbf{y})\} \oplus T_\mathbf{y}W^{s,\gamma^+(\mathbf{y})}(\mathbf{u}) \oplus \overline{V}_0 \\ &= \operatorname{span}\{F(\mathbf{y})\} \oplus T_\mathbf{y}W^{s,\varepsilon}(\mathbf{x}_0) \oplus \overline{V}_0, \end{aligned}$$

where, as above, $\mathbf{y} = \phi^\tau(\mathbf{p}_0) \in W^{s,\varepsilon}(\mathbf{x}_0)$ and

$$\overline{V}_0 = D\phi^\tau(\mathbf{p}_0)(V_0).$$

With a view to using Proposition 3.3, we note that if $\mathbf{v} \in I\!R^n$, it can be written uniquely in the form

$$\mathbf{v} = \alpha F(\mathbf{y}) + \mathbf{w},$$

where $\mathbf{w} \in$ span $\{F(\mathbf{x}_0)\}^\perp$ and

$$\alpha = \frac{\langle F(\mathbf{x}_0), \mathbf{v}\rangle}{\langle F(\mathbf{x}_0), F(\mathbf{y})\rangle},$$

where $\langle F(\mathbf{x}_0), F(\mathbf{y})\rangle \neq 0$, provided we assume ε is so small that Eq.(10) holds for $\|\mathbf{x} - \mathbf{x}_0\| \leq \varepsilon$. Then if Q is the projection of $I\!R^n$ onto span $\{F(\mathbf{x}_0)\}^\perp$ along span $\{F(\mathbf{y})\}$, we have

$$Q\mathbf{v} = \mathbf{w}.$$

If we define the subspace of span$\{F(\mathbf{x}_0)\}^\perp$

$$\tilde{V}_0 = Q(\overline{V}_0),$$

it is clear that

$$T_\mathbf{y}W^{s,\varepsilon}(\mathbf{x}_0) \oplus \tilde{V}_0 = \operatorname{span}\{F(\mathbf{x}_0)\}^\perp.$$

Now if ε is sufficiently small it follows from Proposition 3.3 that there exist positive constants K_2 and β_2 with $\beta_2 < 1$ such that if we put

$$\mathbf{w}_k = DP^k(\mathbf{y})\mathbf{w} \quad \text{for } k \geq 0$$

when $\mathbf{w} \in \tilde{V}_0$, then

$$\|\mathbf{w}_k\| \leq K_2 \beta_2^{m-k} \|\mathbf{w}_m\| \quad \text{for } 0 \leq k \leq m. \tag{17}$$

If $\mathbf{v}_0 \in \overline{V}_0$, it can be written in the form

$$\mathbf{v}_0 = \alpha F(\mathbf{y}) + \mathbf{w},$$

where α is a scalar and $\mathbf{w} \in \tilde{V}_0$. As in the first part of the proof, $P^k(\mathbf{y}) = \mathbf{y}_k = D\phi^{t_k}(\mathbf{y})$. Then it is easy to show by induction using Eq.(6) in Chapter 6 that

$$\mathbf{v}_k := D\phi^{t_k}(\mathbf{y})\mathbf{v}_0 = \alpha_k F(\mathbf{y}_k) + \mathbf{w}_k,$$

where

$$\alpha_k = \alpha - \sum_{j=0}^{k-1} \tau'(\mathbf{y}_j)\mathbf{w}_j, \quad \mathbf{w}_k = DP^k(\mathbf{y})\mathbf{w}.$$

Note also, using Eq.(10) and by similar reasoning to that before Eq.(10), that

$$|\alpha| = \left| \frac{\langle F(\mathbf{x}_0), \mathbf{v}_0 \rangle}{\langle F(\mathbf{x}_0), F(\mathbf{y}) \rangle} \right| \leq 2 \frac{\|\mathbf{v}_0\|}{\|F(\mathbf{x}_0)\|} = 2 \frac{\|Q^{-1}\mathbf{w}\|}{\|F(\mathbf{x}_0)\|},$$

where Q^{-1} is the inverse of $Q : \overline{V}_0 \to \tilde{V}_0$, and that

$$\|\mathbf{w}_k\| \leq 4\|\mathbf{v}_k\|.$$

We define M_0 and N as in the first part of the proof. Then if $\mathbf{v}_0 \in \overline{V}_0$ and $0 \leq k \leq m$,

$$
\begin{aligned}
\|\mathbf{v}_k\| &\le M_0|\alpha_k| + \|\mathbf{w}_k\| \\
&= M_0 \left| \alpha - \sum_{j=0}^{k-1} \tau'(\mathbf{y}_j)\mathbf{w}_j \right| + \|\mathbf{w}_k\| \\
&\le M_0 \left[\frac{2\|Q^{-1}\|\|\mathbf{w}\|}{\|F(\mathbf{x}_0)\|} + N \sum_{j=0}^{k-1} \|\mathbf{w}_j\| \right] + \|\mathbf{w}_k\| \\
&\le M_0 \left[\frac{2\|Q^{-1}\|}{\|F(\mathbf{x}_0)\|} K_2 \beta_2^m \|\mathbf{w}_m\| + N \sum_{j=0}^{k-1} K_2 \beta_2^{m-j} \|\mathbf{w}_m\| \right] + K_2 \beta_2^{m-k} \|\mathbf{w}_m\| \\
&\qquad \text{using Eq.(17)} \\
&\le K_2 \left[M_0 \left(\frac{2\|Q^{-1}\|}{\|F(\mathbf{x}_0)\|} + N\beta_2(1-\beta_2)^{-1} \right) + 1 \right] \beta_2^{m-k} \|\mathbf{w}_m\| \\
&\le 4K_2 \left[M_0 \left(\frac{2\|Q^{-1}\|}{\|F(\mathbf{x}_0)\|} + N\beta_2(1-\beta_2)^{-1} \right) + 1 \right] \beta_2^{m-k} \|\mathbf{v}_m\|.
\end{aligned}
$$

Hence if $\mathbf{v}_0 \in \overline{V}_0$ and $\mathbf{v}_k = D\phi^{t_k}(\mathbf{y})\mathbf{v}_0$,

$$\|\mathbf{v}_k\| \le K_3 \beta_2^{m-k} \|\mathbf{v}_m\| \quad \text{for } 0 \le k \le m,$$

where

$$K_3 = 4K_2 \left[M_0 \left(\frac{2\|Q^{-1}\|}{\|F(\mathbf{x}_0)\|} + N\beta_2(1-\beta_2)^{-1} \right) + 1 \right].$$

Now write

$$\mathbf{v}(t) = D\phi^t(\mathbf{y})\mathbf{v}_0.$$

What we have just shown is that if $0 \le k \le m$

$$\|\mathbf{v}(t_k)\| \le K_3 \beta_2^{m-k} \|\mathbf{v}(t_m)\|.$$

Then, using Eq.(14), if $k_0 \le k \le m$,

$$\|\mathbf{v}(t_k)\| \le K_3 \beta_2^{\frac{t_m - t_k}{T} - 1} \|\mathbf{v}(t_m)\| = K_3 \beta_2^{-1} e^{-\alpha_2(t_m - t_k)} \|\mathbf{v}(t_m)\|,$$

where

$$\alpha_2 = -T^{-1} \log \beta_2.$$

Now suppose $t_{k_0} \leq t \leq s$. Then there exist integers k and m such that $k_0 \leq k \leq m$ and $t_k \leq t \leq t_{k+1}$, $t_m \leq s \leq t_{m+1}$. So, by Gronwall's lemma with M_1 as defined previously in the proof,

$$\begin{aligned}
\|\mathbf{v}(t)\| &\leq e^{M_1(t-t_k)} \|\mathbf{v}(t_k)\| \\
&\leq e^{M_1 T} K_3 \beta_2^{-1} e^{-\alpha_2(t_m - t_k)} \|\mathbf{v}(t_m)\| \\
&\leq e^{M_1 T} K_3 \beta_2^{-1} e^{\alpha_2(s-t_m)} e^{\alpha_2(t_k - t)} e^{-\alpha_2(s-t)} e^{M_1(s-t_m)} \|\mathbf{v}(s)\| \\
&\leq K_3 \beta_2^{-1} e^{(2M_1+\alpha_2)T} e^{-\alpha_2(s-t)} \|\mathbf{v}(s)\|.
\end{aligned}$$

If we put

$$L_2 = K_3 \beta_2^{-1} e^{(2M_1+\alpha_2)T},$$

this means that inequality (8) is established with $\tau^+ = t_{k_0} + \tau$ and the proof of the proposition is complete.

Let us remark here that it follows from the proposition that if $\mathbf{p}_0 \in W^s(\mathbf{u})$

$$T_{\mathbf{p}_0} W^{s,\gamma^+(\mathbf{p}_0)}(\mathbf{u}) = \{\xi : D\phi^t(\mathbf{p}_0)\xi \to \mathbf{0} \;\text{ as } t \to \infty\}.$$

Also if $\mathbf{p}_0 \in W^u(\mathbf{u})$, it can be similarly shown that

$$T_{\mathbf{p}_0} W^{u,\gamma^-(\mathbf{p}_0)}(\mathbf{u}) = \{\xi : D\phi^t(\mathbf{p}_0)\xi \to \mathbf{0} \;\text{ as } t \to -\infty\}.$$

To complete the proof of Theorem 8.2, we observe that in Proposition 8.3 we may take $V_0 = T_{\mathbf{p}_0} W^{u,\gamma^-(\mathbf{p}_0)}(\mathbf{u})$ and so, in terms of the notation just given before the proposition, we have

$$\|D\phi^t(\mathbf{p}(s))\xi\| \leq L_1 e^{-\alpha_1 t} \|\xi\| \quad \text{for } \xi \in E^s(\mathbf{p}(s)),\; s \geq \tau^+,\; t \geq 0 \tag{18}$$

and

$$\|D\phi^{-t}(\mathbf{p}(s))\xi\| \leq L_2 e^{-\alpha_2 t} \|\xi\| \quad \text{for } \xi \in E^u(\mathbf{p}(s)),\; s \geq \tau^+,\; 0 \leq t \leq s - \tau^+.$$

Similarly, we can show there exists a real number τ^- and positive constants $L_1, \alpha_1, L_2, \alpha_2$ (assumed without loss of generality to be the same as those in the last two equations) such that

$$\|D\phi^t(\mathbf{p}(s))\xi\| \le L_1 e^{-\alpha_1 t}\|\xi\| \quad \text{for } \xi \in E^s(\mathbf{p}(s)),\ s \le \tau^-,\ 0 \le t \le \tau^- - s \tag{19}$$

and

$$\|D\phi^{-t}(\mathbf{p}(s))\xi\| \le L_2 e^{-\alpha_2 t}\|\xi\| \quad \text{for } \xi \in E^u(\mathbf{p}(s)),\ s \le \tau^-,\ t \ge 0.$$

Now, if $\tau^- \ge \tau^+$, the inequalities for $E^s(\mathbf{p}(s))$ and $E^u(\mathbf{p}(s))$ hold for all s and $t \ge 0$ with L_1 and L_2 replaced by their squares and the proof of the theorem is finished. So suppose $\tau^- < \tau^+$. Let us write

$$\mathbf{v}(t) = D\phi^t(\mathbf{p}(0))\xi$$

for $\xi \in E^s(\mathbf{p}(0))$. Then inequalities (18) and (19) mean that

$$\|\mathbf{v}(t)\| \le L_1 e^{-\alpha_1(t-s)}\|\mathbf{v}(s)\|$$

for $s \le t \le \tau^-$ and $\tau^+ \le s \le t$. Now if $s \le t$ there are four additional cases: $s \le \tau^- < t < \tau^+$, $s \le \tau^- < \tau^+ \le t$, $\tau^- < s \le t < \tau^+$ and $\tau^- < s < \tau^+ \le t$. We just consider the second case as the other cases can be treated similarly. If $s \le \tau^- < \tau^+ \le t$,

$$\begin{aligned}
\|\mathbf{v}(t)\| &\le L_1 e^{-\alpha_1(t-\tau^+)}\|\mathbf{v}(\tau^+)\| \\
&\le L_1 e^{-\alpha_1(t-\tau^+)} e^{M_1(\tau^+-\tau^-)}\|\mathbf{v}(\tau^-)\| \\
&\le L_1 e^{-\alpha_1(t-\tau^+)} e^{M_1(\tau^+-\tau^-)} L_1 e^{-\alpha_1(\tau^- - s)}\|\mathbf{v}(s)\| \\
&= L_1^2 e^{(M_1+\alpha_1)(\tau^+-\tau^-)} e^{-\alpha_1(t-s)}\|\mathbf{v}(s)\|.
\end{aligned}$$

So we find that (in this and the other cases)

$$\|\mathbf{v}(t)\| \le L_1^2 e^{(M_1+\alpha_1)(\tau^+-\tau^-)} e^{-\alpha_1(t-s)}\|\mathbf{v}(s)\| \quad \text{for } s \le t.$$

This means that for all s and $t \ge 0$,

$$\|D\phi^t(\mathbf{p}(s))\xi\| \le L_1^2 e^{(M_1+\alpha_1)(\tau^+-\tau^-)} e^{-\alpha_1 t}\|\xi\| \quad \text{for } \xi \in E^s(\mathbf{p}(s)).$$

Similarly, we can show that for all s and $t \ge 0$,

$$\|D\phi^{-t}(\mathbf{p}(s))\xi\| \le L_2^2 e^{(M_1+\alpha_2)(\tau^+-\tau^-)} e^{-\alpha_2 t}\|\xi\| \quad \text{for } \xi \in E^u(\mathbf{p}(s)).$$

So the proof of the theorem is completed.

8.2 THE CONSTRUCTION OF A DIFFERENTIAL EQUATION WITH A TRANSVERSAL HOMOCLINIC ORBIT

We consider a coupled system

$$\begin{aligned} \dot{\mathbf{x}} &= F(\mathbf{x}) \\ \dot{\mathbf{y}} &= g(\mathbf{y}) + \varepsilon \mathbf{x}, \end{aligned} \tag{20}$$

where ε is a small parameter, $F : U \to \mathbb{R}^2$ is a C^2 vector field such that Eq.(1) has a hyperbolic periodic solution $\mathbf{u}(t)$ with minimal period T, $g : \mathbb{R}^2 \to \mathbb{R}^2$ is a C^2 vector field such that $\mathbf{0}$ is a saddle point of

$$\dot{\mathbf{y}} = g(\mathbf{y})$$

and $\zeta(t)$ is an associated homoclinic orbit (as in Section 3.3).

Now consider the equation

$$\dot{\mathbf{y}} = g(\mathbf{y}) + \varepsilon \mathbf{u}(t), \tag{21}$$

and denote by $\phi(t, \xi, \varepsilon)$ the solution of Eq.(21) satisfying $\mathbf{y}(0) = \xi$. Then, for ε sufficiently small, the mapping

$$f_\varepsilon(\mathbf{x}) = \phi(T, \mathbf{x}, \varepsilon)$$

is a C^2 diffeomorphism of an open subset of $\mathbb{R}^2$ onto another open set. It is shown as in Section 3.2 that when ε is sufficiently small, then f_ε has a hyperbolic fixed point $\xi(\varepsilon)$ near $\mathbf{0}$. Moreover $\xi(\varepsilon)$ is a C^2 function with $\xi(0) = \mathbf{0}$. It follows that

$$\mathbf{v}(t, \varepsilon) = \phi(t, \xi(\varepsilon), \varepsilon)$$

is a T-periodic solution of Eq.(21) with

$$||\mathbf{v}(t, \varepsilon)|| \to 0 \text{ as } \varepsilon \to 0$$

uniformly in t. Also, as noted in Example 1.5, the Floquet multipliers of the variational equation

$$\dot{\mathbf{y}} = Dg(\mathbf{v}(t, \varepsilon))\mathbf{y} \tag{22}$$

lie off the unit circle since $\xi(\varepsilon)$ is hyperbolic; furthermore, since $\mathbf{u}(t)$ is hyperbolic, all but one of the Floquet multipliers of

$$\dot{\mathbf{x}} = DF(\mathbf{u}(t))\mathbf{x} \tag{23}$$

lie off the unit circle. However the Floquet multipliers of the system

$$\begin{aligned} \dot{\mathbf{x}} &= DF(\mathbf{u}(t))\mathbf{x} \\ \dot{\mathbf{y}} &= \varepsilon \mathbf{x} + Dg(\mathbf{v}(t, \varepsilon))\mathbf{y} \end{aligned}$$

are just the Floquet multipliers of Eqs.(22) and (23) combined together. It follows that $(\mathbf{u}(t),\ \mathbf{v}(t,\varepsilon))$ is a hyperbolic periodic orbit of Eq.(20).

Next consider the Melnikov function associated with Eq.(21)

$$\Delta(\alpha) = \int_{-\infty}^{\infty} e^{-\int_0^t \mathrm{Tr}\, Dg(\zeta(s+\alpha))ds} g(\zeta(t+\alpha)) \wedge \mathbf{u}(t)\, dt$$

and suppose there exists α_0 such that

$$\Delta(\alpha_0) = 0, \ \Delta'(\alpha_0) \neq 0.$$

Then, by Theorem 3.7, if ε is sufficiently small and nonzero, the mapping f_ε has a transversal homoclinic point $\mathbf{y}(\varepsilon)$ near $\zeta(\alpha_0)$ associated with the hyperbolic fixed point $\xi(\varepsilon)$. That is,

$$\|f_\varepsilon^k(\mathbf{y}(\varepsilon)) - \xi(\varepsilon)\| \to 0 \text{ as } |k| \to \infty \tag{24}$$

and

$$T_{\mathbf{y}(\varepsilon)} W^s(\xi(\varepsilon)) \cap T_{\mathbf{y}(\varepsilon)} W^u(\xi(\varepsilon)) = \{\mathbf{0}\}.$$

Now let

$$\mathbf{p}(t,\varepsilon) = \phi(t, \mathbf{y}(\varepsilon), \epsilon).$$

Then, using a Gronwall lemma argument, it follows from Eq.(24) that

$$\|\mathbf{p}(t,\epsilon) - \mathbf{v}(t,\varepsilon)\| \to 0 \text{ as } |t| \to \infty.$$

This means that the solution $(\mathbf{u}(t),\ \mathbf{p}(t,\varepsilon))$ of Eq.(20) satisfies

$$\|(\mathbf{u}(t), \mathbf{p}(t,\varepsilon)) - (\mathbf{u}(t), \mathbf{v}(t,\varepsilon))\| \to 0 \text{ as } |t| \to \infty.$$

So $(\mathbf{u}(0), \mathbf{p}(0,\varepsilon))$ is a homoclinic point with respect to the hyperbolic periodic orbit $(\mathbf{u}(t),\ \mathbf{v}(t,\varepsilon))$.

To show the transversality, consider the variational equation along the solution $(\mathbf{u}(t),\ \mathbf{p}(t,\varepsilon))$, that is,

$$\begin{aligned} \dot{\mathbf{x}} &= DF(\mathbf{u}(t))\mathbf{x} \\ \dot{\mathbf{y}} &= \varepsilon\mathbf{x} + Dg(\mathbf{p}(t,\varepsilon))\mathbf{y}. \end{aligned}$$

In view of the remark after the proof of Proposition 8.3, to prove the transversality we need only show that the trivial solution is the only solution of this equation which tends to zero as $|t| \to \infty$. Since $\mathbf{u}(t)$ is hyperbolic, multiples of $\dot{\mathbf{u}}(t)$ are the only bounded solutions of

$$\dot{\mathbf{x}} = DF(\mathbf{u}(t))\mathbf{x}.$$

So the trivial solution is the only solution of this equation which tends to zero as $|t| \to \infty$.

Next, since $\mathbf{y}(\varepsilon)$ is a transversal homoclinic point for f_ε, we know from the remark at the end of Section 3.1 that $\xi = \mathbf{0}$ is the only vector such that $Df_\varepsilon^k(\mathbf{y}(\varepsilon))\xi \to \mathbf{0}$ as $|k| \to \infty$, that is,

$$\frac{\partial \phi}{\partial \xi}(k, \mathbf{y}(\varepsilon), \varepsilon)\xi \to \mathbf{0}$$

as $|k| \to \infty$. By a simple argument using Gronwall's lemma, this is equivalent to

$$\frac{\partial \phi}{\partial \xi}(t, \mathbf{y}(\varepsilon), \varepsilon)\xi \to \mathbf{0}$$

as $|t| \to \infty$. This means that the trivial solution is the only solution of

$$\dot{\mathbf{y}} = Dg(\mathbf{p}(t, \varepsilon))\mathbf{y}$$

which tends to zero as $|t| \to \infty$. Thus the transversality is established and so we have proved that when the Melnikov function $\Delta(\alpha)$ has a simple zero, Eq.(20) has a hyperbolic periodic orbit with associated transversal homoclinic point when $\varepsilon \neq 0$ is sufficiently small.

Now we give an explicit example of this situation.

Example. The planar autonomous system

$$\dot{\mathbf{x}} = F(\mathbf{x})$$

given by

$$\begin{aligned} \dot{x}_1 &= -x_1 + x_2 + \frac{x_1}{\sqrt{x_1^2 + x_2^2}} \\ \dot{x}_2 &= -x_1 - x_2 + \frac{x_2}{\sqrt{x_1^2 + x_2^2}} \end{aligned}$$

has the periodic solution $\mathbf{u}(t) = (\cos t, -\sin t)$. One calculates the trace $\operatorname{Tr} DF(\mathbf{u}(t)) = -1$. Then, by Jacobi's formula, the product of the Floquet multipliers is

$$e^{\int_0^{2\pi} \operatorname{Tr} DF(\mathbf{u}(t))dt} = e^{-2\pi}.$$

Since one Floquet multiplier is 1, the other must be $e^{-2\pi}$. So $\mathbf{u}(t)$ is hyperbolic.

Next, as we saw in the example at the end of Chapter 3, $(0, 0)$ is a saddle for the system

$$\dot{\mathbf{y}} = g(\mathbf{y})$$

given by

$$\begin{aligned}\dot{y}_1 &= y_2\\ \dot{y}_2 &= y_1 - 2y_1^3\end{aligned}$$

with associated saddle connexion $\zeta(t) = (\xi(t), \dot{\xi}(t))$ where $\xi(t) = \operatorname{sech} t$. In this case the Melnikov function for Eq.(21) is

$$\begin{aligned}\Delta(\alpha) &= -\int_{-\infty}^{\infty} \dot{\xi}(t+\alpha)\sin t + \ddot{\xi}(t+\alpha)\cos t\,dt\\ &= -\int_{-\infty}^{\infty} \dot{\xi}(t)\sin(t-\alpha) + \ddot{\xi}(t)\cos(t-\alpha)dt\\ &= -\int_{-\infty}^{\infty} \dot{\xi}(t)(\sin t\cos\alpha - \cos t\sin\alpha) + \ddot{\xi}(t)(\cos t\cos\alpha + \sin t\sin\alpha)dt\\ &= I_1\cos\alpha + I_2\sin\alpha,\end{aligned}$$

where

$$I_1 = -\int_{-\infty}^{\infty} \dot{\xi}(t)\sin t + \ddot{\xi}(t)\cos t\,dt \quad\text{and}\quad I_2 = \int_{-\infty}^{\infty} \dot{\xi}(t)\cos t - \ddot{\xi}(t)\sin t\,dt.$$

Note that $I_2 = 0$ since its integrand is odd. We calculate I_1 as follows:

$$\begin{aligned}I_1 &= \int_{-\infty}^{\infty} \frac{d}{dt}(\dot{\xi}(t)\cos t) - 2\ddot{\xi}(t)\cos t\,dt\\ &= -2\int_{-\infty}^{\infty} \ddot{\xi}(t)\cos t\,dt\\ &= -2\left(\dot{\xi}(t)\cos t\,\Big|_{-\infty}^{\infty} - \int_{-\infty}^{\infty} \dot{\xi}(t)(-\sin t)\,dt\right)\\ &= -2\int_{-\infty}^{\infty} \dot{\xi}(t)\sin t\,dt\\ &= -2\left(\xi(t)\sin t\,\Big|_{-\infty}^{\infty} - \int_{-\infty}^{\infty} \xi(t)\cos t\,dt\right)\\ &= 2\int_{-\infty}^{\infty} \xi(t)\cos t\,dt\end{aligned}$$

and so

$$I_1 = 2\int_{-\infty}^{\infty} \operatorname{sech} t\cos t\,dt \neq 0.$$

Hence we have shown that

$$\Delta(\alpha) = I_1 \cos\alpha,$$

where $I_1 \neq 0$. It is clear that this function has simple zeros and hence we may conclude that, for ε nonzero and sufficiently small, the system

$$\begin{aligned}
\dot{x}_1 &= -x_1 + x_2 + \frac{x_1}{\sqrt{x_1^2 + x_2^2}} \\
\dot{x}_2 &= -x_1 - x_2 + \frac{x_2}{\sqrt{x_1^2 + x_2^2}} \\
\dot{y}_1 &= y_2 + \varepsilon x_1 \\
\dot{y}_2 &= y_1 - 2y_1^3 + \varepsilon x_2
\end{aligned}$$

has a hyperbolic periodic orbit with associated transversal homoclinic point.

Note that Szmolyan [1991] gives conditions ensuring the existence of a transversal homoclinic orbit to a hyperbolic periodic orbit in a singularly perturbed system. Also Deng and Sakamoto [1995] show the existence of orbits homoclinic to a hyperbolic periodic orbit.

9. SHADOWING THEOREMS FOR HYPERBOLIC SETS OF DIFFERENTIAL EQUATIONS

For differential equations, it is not immediately clear how the shadowing theorem should be formulated and, in particular, how pseudo orbits should be defined. We consider two possibilities: first a discrete pseudo orbit such as would be obtained by numerically computing the solutions of a differential equation and then a continuous pseudo orbit which is needed for theoretical purposes. Shadowing theorems are proved for both kinds of pseudo orbits. We also study both discrete and continuous versions of expansivity. Then shadowing is used to show that there is a topological conjugacy between the flow on an isolated hyperbolic set and the flow on the nearby hyperbolic set for a perturbed flow. Also we use shadowing to show that isolated hyperbolic sets have the asymptotic phase property.

9.1 THE DISCRETE SHADOWING THEOREM

Let U be a convex open set in $\mathbb{R}^n$, let $F : U \to \mathbb{R}^n$ be a C^1 vector field and let ϕ be the flow associated with the equation

$$\dot{\mathbf{x}} = F(\mathbf{x}). \tag{1}$$

Definition 9.1. If δ is a positive number, a sequence $\{\mathbf{y}_k\}_{k=-\infty}^{\infty}$ of points in U is said to be a *discrete δ pseudo orbit* for Eq.(1) if there is a sequence $\{h_k\}_{k=-\infty}^{\infty}$ of positive times with $\sup h_k < \infty$, $\inf h_k > 0$ such that

$$\|\mathbf{y}_{k+1} - \phi^{h_k}(\mathbf{y}_k)\| \le \delta \ \text{ for } \ k \in \mathbf{Z}.$$

We ask the question: are there points from a true solution of Eq.(1) near a discrete δ pseudo orbit in the sense of the following definition?

Definition 9.2. A discrete δ pseudo orbit $\{\mathbf{y}_k\}_{k=-\infty}^{\infty}$ of Eq.(1) with associated times $\{h_k\}_{k=-\infty}^{\infty}$ is said to be *ε-shadowed* by a true orbit of Eq.(1) if there are sequences $\{\mathbf{x}_k\}_{k-\infty}^{\infty}$ and $\{t_k\}_{k=-\infty}^{\infty}$ such that $\mathbf{x}_{k+1} = \phi^{t_k}(\mathbf{x}_k)$ and

$$\|\mathbf{x}_k - \mathbf{y}_k\| \le \varepsilon \ \text{ and } \ |t_k - h_k| \le \varepsilon$$

for all k.

Our aim in this section is to prove the following theorem which gives sufficient conditions under which discrete δ pseudo orbits of Eq.(1) are shadowed by true orbits. Actually we prove the more general result that discrete δ pseudo orbits of Eq.(1) are shadowed by true orbits of perturbed equations.

Theorem 9.3. *Let $U \subset \mathbb{R}^n$ be an open convex set and let $F : U \to \mathbb{R}^n$ be a C^1 vector field. Suppose $S \subset U$ is a compact hyperbolic set for Eq.(1). Let $\{\mathbf{y}_k\}_{k=-\infty}^{\infty}$ be a discrete δ pseudo orbit of Eq.(1) in S with associated times $\{h_k\}_{k=-\infty}^{\infty}$ satisfying*

$$0 < h_{min} \leq h_k \leq h_{max}.$$

Then there exist positive constants δ_0, σ_0 and M depending only on F, S, h_{min} and h_{max} such that if $G : U \to \mathbb{R}^n$ is a C^1 vector field satisfying

$$\|F(\mathbf{x}) - G(\mathbf{x})\| + \|DF(\mathbf{x}) - DG(\mathbf{x})\| \leq \sigma \text{ for } \mathbf{x} \in U$$

with $\sigma \leq \sigma_0$, and if $\delta \leq \delta_0$, the discrete δ pseudo orbit $\{\mathbf{y}_k\}_{k=-\infty}^{\infty}$ is ε-shadowed by a unique true orbit $\{\mathbf{x}_k\}_{k=-\infty}^{\infty}$ of the equation

$$\dot{\mathbf{x}} = G(\mathbf{x}) \tag{2}$$

with $\varepsilon = M(\delta + \sigma)$ and such that

$$\langle \mathbf{x}_k - \mathbf{y}_k, F(\mathbf{y}_k) \rangle = 0 \text{ for all } k.$$

Proof. First we impose a condition on σ. We choose ε_0 so that $0 < \varepsilon_0 \leq \frac{1}{2}\text{dist}(S, \partial U)$, $\varepsilon_0 < h_{min}$ and so that when $\text{dist}(\mathbf{x}, S) \leq \varepsilon_0$, the solution $\phi^t(\mathbf{x})$ is defined for $|t| \leq h_{max} + \varepsilon_0$ and satisfies $\text{dist}(\phi^t(\mathbf{x}), S) \leq \frac{1}{2}\text{dist}(S, \partial U)$. Then we suppose σ satisfies

$$M_2\sigma < \frac{1}{2}\text{dist}(S, \partial U),$$

where

$$M_2 = \frac{1}{M_1}(e^{M_1(h_{max}+\varepsilon_0)} - 1)$$

with

$$M_1 = \sup_{\mathbf{x}\in U} \|DF(\mathbf{x})\|.$$

Denote by $\psi^t(\mathbf{x})$ the flow corresponding to Eq.(2). We assert that if

$$\text{dist}(\mathbf{x}, S) \leq \varepsilon_0,$$

the solution $\psi^t(\mathbf{x})$ is defined for $|t| \leq h_{max} + \varepsilon_0$ and satisfies

$$\|\psi^t(\mathbf{x}) - \phi^t(\mathbf{x})\| \leq M_2\sigma.$$

To prove the assertion, suppose $\psi^t(\mathbf{x})$ is defined on an interval containing 0 and contained in $[-h_{\max} - \varepsilon_0, h_{\max} + \varepsilon_0]$ and such that

$$\|\psi^t(\mathbf{x}) - \phi^t(\mathbf{x})\| \leq \frac{1}{2}\text{dist}(S, \partial U)$$

in the interval. Then if t is in the interval, it follows from Eq.(82) in Chapter 7 that

$$\|\psi^t(\mathbf{x}) - \phi^t(\mathbf{x})\| \leq \frac{\sigma}{M_1}(e^{M_1(h_{max}+\varepsilon_0)} - 1) = M_2\sigma. \quad (3)$$

Then, since $M_2\sigma < \frac{1}{2}\mathrm{dist}(S, \partial U)$, the assertion follows by a standard argument.

Now let $\{\mathbf{y}_k\}_{k=-\infty}^{\infty}$ be a discrete δ pseudo orbit of Eq.(1) in the compact hyperbolic set S with associated times $\{h_k\}_{k=-\infty}^{\infty}$ satisfying $h_{min} \leq h_k \leq h_{max}$. We seek a solution $\{\mathbf{x}_k\}_{k=-\infty}^{\infty}$, $\{t_k\}_{k=-\infty}^{\infty}$ of the equation

$$\mathbf{x}_{k+1} = \psi^{t_k}(\mathbf{x}_k) \ \text{ for } \ k \in \boldsymbol{Z}$$

such that for $k \in \boldsymbol{Z}$,

$$\|\mathbf{x}_k - \mathbf{y}_k\| \leq \varepsilon, \ \ |t_k - h_k| \leq \varepsilon$$

and

$$\langle \mathbf{x}_k - \mathbf{y}_k, F(\mathbf{y}_k) \rangle = 0.$$

Denote by X the Banach space of pairs of bounded sequences $(\mathbf{x}, \mathbf{t}) = (\{\mathbf{x}_k\}_{k=-\infty}^{\infty}, \{t_k\}_{k=-\infty}^{\infty})$ with $\mathbf{x}_k \in \mathbb{R}^n$, $t_k \in \mathbb{R}$ and with norm

$$\|(\mathbf{x}, \mathbf{t})\| = \max\{\|\mathbf{x}\|, \|\mathbf{t}\|\},$$

where

$$\|\mathbf{x}\| = \sup_{k=-\infty}^{\infty} \|\mathbf{x}_k\|, \ \ \|\mathbf{t}\| = \sup_{k=-\infty}^{\infty} |t_k|.$$

Let O be the open subset of X consisting of those pairs $(\mathbf{x}, \mathbf{t})$ such that $\|\mathbf{x}-\mathbf{y}\| < \varepsilon_0$ and $\|\mathbf{t}-\mathbf{h}\| < \varepsilon_0$, where $\mathbf{y} = \{\mathbf{y}_k\}_{k=-\infty}^{\infty}$ and $\mathbf{h} = \{h_k\}_{k=-\infty}^{\infty}$. Then we define a C^1 mapping $\mathcal{G} : O \to X$ as follows. If $(\mathbf{x}, \mathbf{t}) = (\{\mathbf{x}_k\}_{k=-\infty}^{\infty}, \{t_k\}_{k=-\infty}^{\infty}) \in O$, then

$$\mathcal{G}(\mathbf{x}, \mathbf{t}) = (\{\mathbf{x}_{k+1} - \psi^{t_k}(\mathbf{x}_k)\}_{k=-\infty}^{\infty}, \ \{\langle \mathbf{x}_k - \mathbf{y}_k, F(\mathbf{y}_k) \rangle\}_{k=-\infty}^{\infty}). \quad (4)$$

Note $\mathcal{G}$ is well-defined because of our choice of ε_0 and the condition we imposed on σ. The theorem will be proved if we can show there exists a unique solution $(\mathbf{x}, \mathbf{t})$ of the equation

$$\mathcal{G}(\mathbf{x}, \mathbf{t}) = 0$$

with $\|(\mathbf{x}, \mathbf{t}) - (\mathbf{y}, \mathbf{h})\| \leq \varepsilon$.

To do this we apply Lemma 4.4. It is easy to see that $D\mathcal{G}(\mathbf{y}, \mathbf{h}) = L$, where the linear operator $L : X \to X$ is defined as follows: if $(\mathbf{z}, \mathbf{s}) \in X$, where $\mathbf{z} = \{\mathbf{z}_k\}_{k=-\infty}^{\infty}$ and $\mathbf{s} = \{s_k\}_{k=-\infty}^{\infty}$, then

$$L(\mathbf{z}, \mathbf{s})$$

$$= (\{\mathbf{z}_{k+1} - D\psi^{h_k}(\mathbf{y}_k)\mathbf{z}_k - G(\psi^{h_k}(\mathbf{y}_k))s_k\}_{k=-\infty}^{\infty}, \ \{\langle \mathbf{z}_k, F(\mathbf{y}_k) \rangle\}_{k=-\infty}^{\infty}).$$

To apply Lemma 4.4, we must show that under suitable conditions the operator L is invertible and we must find a bound for $\|L^{-1}\|$.

Invertibility of L and estimation of $\|L^{-1}\|$: We define the numerical sequence $\{a_k\}_{k=-\infty}^{\infty}$ by the recurrence relation

$$a_{k+1} = a_k + h_k$$

with $a_0 = 0$, and then define the piecewise continuously differentiable function

$$\mathbf{y}(t) = \phi^{t-a_k}(\mathbf{y}_k)$$

for $a_k \le t < a_{k+1}$, $k \in \boldsymbol{Z}$. We take

$$\beta_1 = \alpha_1/2,\ \beta_2 = \alpha_2/2,\ \lambda = \min\{\alpha_1, \alpha_2\}/4.$$

Then we apply Lemma 7.16 with $\tilde{\beta}_1 = \beta_1 + \lambda$ and $\tilde{\beta}_2 = \beta_2 + \lambda$. So, provided δ is sufficiently small depending on F, S, h_{min} and h_{max}, it follows from Lemma 7.16 that there are constants L_1 and L_2 depending on F and S and projections $\overline{P}(t)$ and $\overline{Q}(t)$ such that if $Y(t)$ is a fundamental matrix for the equation

$$\dot{\mathbf{x}} = DF(\mathbf{y}(t))\mathbf{x}, \tag{5}$$

then for all s and t

$$Y(t)Y^{-1}(s)\overline{P}(s) = \overline{P}(t)Y(t)Y^{-1}(s),\ Y(t)Y^{-1}(s)\overline{Q}(s) = \overline{Q}(t)Y(t)Y^{-1}(s)$$

and

$$\|Y(t)Y^{-1}(s)\overline{P}(s)\| \le L_1 e^{-\beta_1(t-s)} \quad \text{for} \quad s \le t, \tag{6}$$

$$\|Y(t)Y^{-1}(s)(I - \overline{Q}(s))\| \le L_2 e^{-\beta_2(s-t)} \quad \text{for} \quad s \ge t, \tag{7}$$

Note also that $\overline{Q}(t) - \overline{P}(t)$ is also a projection of rank 1 and that $\overline{P}(t)$, $I - \overline{Q}(t)$ and $\overline{Q}(t) - \overline{P}(t)$ are mutually orthogonal projections.

Now we define

$$P_k = \overline{P}(a_k), Q_k = \overline{Q}(a_k)$$

and suppose

$$\mathcal{R}(Q_k - P_k) = \text{span}\{\mathbf{v}_k\},$$

where $\mathbf{v}_k$ is a unit vector. Note $D\phi^{h_k}(\mathbf{y}_k)\mathbf{v}_k$ is a multiple of $\mathbf{v}_{k+1}$ by invariance of $\overline{Q}(t) - \overline{P}(t)$ under Eq.(5). Then we define S_k to be an $n \times (n-1)$ matrix such that the $n \times n$ matrix $[\mathbf{v}_k \quad S_k]$ is orthogonal. We shall prove that the difference equation

$$\mathbf{w}_{k+1} = S_{k+1}^* D\phi^{h_k}(\mathbf{y}_k)S_k\mathbf{w}_k, \tag{8}$$

for which the transition matrix is

$$\Phi(k,m) = S_k^* D\phi^{a_k - a_m}(\mathbf{y}_m)S_m = S_k^* Y(a_k)Y^{-1}(a_m)S_m,$$

has an exponential dichotomy on $(-\infty, \infty)$ with projection

$$R_k = S_k^* P_k S_k = S_k^* Q_k S_k.$$

The last two expressions are equal because $\mathcal{R}(Q_k - P_k) = \text{span}\{\mathbf{v}_k\}$ and $S_k^* \mathbf{v}_k = 0$. Also R_k is a projection since

$$R_k^2 = S_k^* P_k S_k S_k^* P_k S_k = S_k^* P_k (I - \mathbf{v}_k \mathbf{v}_k^*) P_k S_k = S_k^* P_k S_k = R_k,$$

using the fact that $P_k \mathbf{v}_k = \mathbf{0}$. Moreover, R_k is invariant with respect to Eq.(8) since

$$\begin{aligned}
S_{k+1}^* D\phi^{h_k}(\mathbf{y}_k) S_k R_k &= S_{k+1}^* D\phi^{h_k}(\mathbf{y}_k) S_k S_k^* P_k S_k \\
&= S_{k+1}^* D\phi^{h_k}(\mathbf{y}_k)(I - \mathbf{v}_k \mathbf{v}_k^*) P_k S_k \\
&= S_{k+1}^* D\phi^{h_k}(\mathbf{y}_k) P_k S_k \\
&\qquad \text{since } D\phi^{h_k}(\mathbf{y}_k)\mathbf{v}_k \text{ is a multiple of } \mathbf{v}_{k+1} \\
&= S_{k+1}^* Y(a_{k+1}) Y^{-1}(a_k) P_k S_k \\
&= S_{k+1}^* P_{k+1} Y(a_{k+1}) Y^{-1}(a_k) S_k \\
&\qquad \text{by the invariance of } \overline{P}(t) \\
&= S_{k+1}^* P_{k+1} S_{k+1} S_{k+1}^* Y(a_{k+1}) Y^{-1}(a_k) S_k \\
&= R_{k+1} S_{k+1}^* D\phi^{h_k}(\mathbf{y}_k) S_k.
\end{aligned}$$

Next it follows from inequalities (6) and (7) that if $\Phi(k, m)$ is the transition matrix for Eq.(8) then for $m \leq k$

$$\begin{aligned}
\|\Phi(k,m) R_m\| &= \|S_k^* Y(a_k) Y^{-1}(a_m) S_m S_m^* P_m S_m\| \\
&= \|S_k^* Y(a_k) Y^{-1}(a_m) P_m S_m\| \quad \text{since} \quad S_k^* \mathbf{v}_k = \mathbf{0} \\
&\leq \|Y(a_k) Y^{-1}(a_m) P_m\| \\
&\leq L_1 e^{-\beta_1 (a_k - a_m)} \\
&\leq L_1 e^{-\frac{\beta_1}{h_{min}}(k-m)}
\end{aligned}$$

and, similarly, for $m \geq k$,

$$\|\Phi(k,m)(I - R_m)\| \leq \|Y(a_k) Y^{-1}(a_m)(I - Q_m)\| \leq L_2 e^{-\frac{\beta_2}{h_{min}}(m-k)}.$$

Hence we have proved our assertion that Eq.(8) has an exponential dichotomy on $(-\infty, \infty)$ with projections R_k.

Note we also have from Lemma 7.16 that

$$\|P_k - \mathcal{P}^s(\mathbf{y}_k)\| \le \omega_+(\delta), \|I - Q_k - \mathcal{P}^u(\mathbf{y}_k)\| \le \omega_-(\delta),$$

where $\omega_+(\cdot)$ and $\omega_-(\cdot)$ depend only on F, S, h_{min}, h_{max} and and have the property that $\lim_{\delta \to 0^+} \omega_\pm(\delta) = 0$. We use this to derive an inequality for

$$\left\| \mathbf{v}_k - \frac{F(\mathbf{y}_k)}{\|F(\mathbf{y}_k)\|} \right\|.$$

Note we can find a scalar α, assumed without loss of generality to be nonnegative, such that

$$\mathbf{v}_k = \alpha \frac{F(\mathbf{y}_k)}{\|F(\mathbf{y}_k)\|} + [\mathcal{P}^s(\mathbf{y}_k) + \mathcal{P}^u(\mathbf{y}_k)]\mathbf{v}_k. \tag{9}$$

Then

$$\begin{aligned} |1-\alpha| &= \left| \|\mathbf{v}_k\| - \left\| \alpha \frac{F(\mathbf{y}_k)}{\|F(\mathbf{y}_k)\|} \right\| \right| \\ &\le \left\| \mathbf{v}_k - \alpha \frac{F(\mathbf{y}_k)}{\|F(\mathbf{y}_k)\|} \right\| \\ &= \|[\mathcal{P}^s(\mathbf{y}_k) + \mathcal{P}^u(\mathbf{y}_k)]\mathbf{v}_k\| \\ &= \|[\mathcal{P}^s(\mathbf{y}_k) + \mathcal{P}^u(\mathbf{y}_k) - (P_k + (I - Q_k))]\mathbf{v}_k\| \\ &\le \omega_+(\delta) + \omega_-(\delta). \end{aligned} \tag{10}$$

So if we assume

$$\omega_+(\delta) + \omega_-(\delta) < 1,$$

it follows that $\alpha > 0$. Then it follows from Eq.(9) and the reasoning in Eq.(10) that

$$\left\| \mathbf{v}_k - \frac{F(\mathbf{y}_k)}{\|F(\mathbf{y}_k)\|} \right\| \le \left\| \mathbf{v}_k - \alpha \frac{F(\mathbf{y}_k)}{\|F(\mathbf{y}_k)\|} \right\| + |\alpha - 1| \le 2[\omega_+(\delta) + \omega_-(\delta)]. \tag{11}$$

Now we approximate L by the operator T given by

$$T(\mathbf{z}, \mathbf{s}) =$$

$$(\{\mathbf{z}_{k+1} - D\phi^{h_k}(\mathbf{y}_k)\mathbf{z}_k - s_k\|F(\mathbf{y}_{k+1})\|\mathbf{v}_{k+1}\}_{k=-\infty}^{\infty}, \{\langle \mathbf{z}_k, \|F(\mathbf{y}_k)\|\mathbf{v}_k \rangle\}_{k=-\infty}^{\infty}).$$

In order to estimate $\|L - T\|$, we first estimate

$$\begin{aligned}
&\| \, \|F(\mathbf{y}_{k+1})\|\mathbf{v}_{k+1} - G(\psi^{h_k}(\mathbf{y}_k))\| \\
&\leq \| \, \|F(\mathbf{y}_{k+1})\|\mathbf{v}_{k+1} - F(\mathbf{y}_{k+1})\| + \|F(\mathbf{y}_{k+1}) - F(\psi^{h_k}(\mathbf{y}_k))\| \\
&\qquad + \|F(\psi^{h_k}(\mathbf{y}_k)) - G(\psi^{h_k}(\mathbf{y}_k))\| \\
&\leq 2M_0[\omega_+(\delta) + \omega_-(\delta)] + M_1\|\mathbf{y}_{k+1} - \psi^{h_k}(\mathbf{y}_k)\| + \sigma \\
&\qquad \text{using Eq.(11) with } M_0 = \sup_{\mathbf{x}\in S}\|F(\mathbf{x})\| \\
&\leq 2M_0[\omega_+(\delta) + \omega_-(\delta)] + M_1\|\mathbf{y}_{k+1} - \phi^{h_k}(\mathbf{y}_k)\| \\
&\qquad + M_1\|\phi^{h_k}(\mathbf{y}_k) - \psi^{h_k}(\mathbf{y}_k)\| + \sigma \\
&\leq 2M_0[\omega_+(\delta) + \omega_-(\delta)] + M_1\left[\delta + \frac{\sigma}{M_1}(e^{M_1 h_{max}} - 1)\right] + \sigma,
\end{aligned}$$

using Eq.(82) in Chapter 7. Hence we have the estimate

$$\begin{aligned}
&\| \, \|F(\mathbf{y}_{k+1})\|\mathbf{v}_{k+1} - G(\psi^{h_k}(\mathbf{y}_k))\| \\
&\qquad\qquad \leq 2M_0[\omega_+(\delta) + \omega_-(\delta)] + M_1\delta + e^{M_1 h_{max}}\sigma.
\end{aligned}$$

We next estimate the norm of $D\phi^t(\mathbf{y}) - D\psi^t(\mathbf{y}) = W(t)$ for $\mathbf{y}$ in S and $0 \leq t \leq h_{max}$. Note that

$$\dot{W}(t) = DF(\phi^t(\mathbf{y}))D\phi^t(\mathbf{y}) - DG(\psi^t(\mathbf{y}))D\psi^t(\mathbf{y}).$$

So $W(t)$ is given by the formula

$$\int_0^t DF(\phi^s(\mathbf{y}))[D\phi^s(\mathbf{y}) - D\psi^s(\mathbf{y})]ds + \int_0^t [DF(\phi^s(\mathbf{y})) - DG(\psi^s(\mathbf{y}))]D\psi^s(\mathbf{y})ds$$

and thus for $0 \leq t \leq h_{max}$, using Gronwall's lemma to estimate $\|D\psi^s(\mathbf{y})\|$ and then using Eq.(82) in Chapter 7,

$$\begin{aligned}
\|W(t)\| &\leq \int_0^t M_1\|W(s)\|ds + \int_0^t [\sigma + \omega(\|\phi^s(\mathbf{y}) - \psi^s(\mathbf{y})\|)]\, e^{(M_1+\sigma)s}ds, \\
&\leq \int_0^t [\sigma + \omega(\sigma s e^{M_1 s})]\, e^{(M_1+\sigma)s}ds + M_1\int_0^t \|W(s)\|ds \\
&\leq h_{max}[\sigma + \omega\left(\sigma h_{max} e^{M_1 h_{max}}\right)]\, e^{(M_1+\sigma)h_{max}} + M_1\int_0^t \|W(s)\|ds,
\end{aligned}$$

where
$$\omega(\delta) = \sup\{\|DF(\mathbf{x}) - DF(\mathbf{y})\| : \mathbf{x}, \mathbf{y} \in U, \|\mathbf{x} - \mathbf{y}\| \le \delta\}.$$
Hence, by Gronwall's lemma,
$$\|W(t)\| \le h_{max}\left[\sigma + \omega(\sigma h_{max} e^{M_1 h_{max}})\right] e^{(M_1+\sigma)h_{max}+M_1 t}.$$
That is,
$$\|D\psi^t(\mathbf{y}) - D\phi^t(\mathbf{y})\| \le h_{max}\left[\sigma + \omega(\sigma h_{max} e^{M_1 h_{max}})\right] e^{(2M_1+\sigma)h_{max}}$$
for $\mathbf{y}$ in S and $0 \le t \le h_{max}$. In particular,
$$\|D\psi^{h_k}(\mathbf{y}_k) - D\phi^{h_k}(\mathbf{y}_k)\| \le h_{max}\left[\sigma + \omega(\sigma h_{max} e^{M_1 h_{max}})\right] e^{(2M_1+\sigma)h_{max}}.$$

Next note that it follows from Eq.(11) that
$$\|\, \|F(\mathbf{y}_k)\|\mathbf{v}_k - F(\mathbf{y}_k)\| \le 2M_0[\omega_+(\delta) + \omega_-(\delta)].$$

Putting these estimates together, we find that
$$\begin{aligned} \|L - T\| \; &\le h_{max}\left[\sigma + \omega(\sigma h_{max} e^{M_1 h_{max}})\right] e^{(2M_1+\sigma)h_{max}} \\ &+ 2M_0[\omega_+(\delta) + \omega_-(\delta)] + M_1\delta + e^{M_1 h_{max}}\sigma. \end{aligned} \tag{12}$$

To verify that T is invertible, given $(\mathbf{g}, \ell) = (\{\mathbf{g}_k\}_{k=-\infty}^{\infty}, \{\ell_k\}_{k=-\infty}^{\infty})$ in X, we have to solve the equations
$$\mathbf{z}_{k+1} = D\phi^{h_k}(\mathbf{y}_k)\mathbf{z}_k + s_k\|F(\mathbf{y}_{k+1})\|\mathbf{v}_{k+1} + \mathbf{g}_k \tag{13}$$
and
$$\langle \mathbf{z}_k, \|F(\mathbf{y}_k)\|\mathbf{v}_k \rangle = \ell_k \tag{14}$$
for bounded sequences $\mathbf{z}_k$ and s_k. If we set
$$\mathbf{z}_k = \lambda_k \mathbf{v}_k + S_k \mathbf{w}_k, \tag{15}$$
then we see that
$$\langle \mathbf{z}_k, \mathbf{v}_k \rangle = \lambda_k$$
and so from Eq.(14)
$$\lambda_k = \frac{\ell_k}{\|F(\mathbf{y}_k)\|}. \tag{16}$$
Multiplying Eq.(13) by $\mathbf{v}_{k+1}^*$, we get
$$s_k = \frac{1}{\|F(\mathbf{y}_{k+1})\|}\{\lambda_{k+1} - \mathbf{v}_{k+1}^* D\phi^{h_k}(\mathbf{y}_k)(\lambda_k \mathbf{v}_k + S_k \mathbf{w}_k) - \mathbf{v}_{k+1}^* \mathbf{g}_k\}. \tag{17}$$

Multiplying Eq.(13) by S_{k+1}^* and using the fact that $D\phi^{h_k}(\mathbf{y}_k)\mathbf{v}_k$ is a multiple of $\mathbf{v}_{k+1}$, we get

$$\mathbf{w}_{k+1} = A_k\mathbf{w}_k + S_{k+1}^*\mathbf{g}_k, \tag{18}$$

where

$$A_k = S_{k+1}^* D\phi^{h_k}(\mathbf{y}_k)S_k.$$

Then, using Lemma 4.5, it follows from the fact that Eq.(8) has an exponential dichotomy on $(-\infty, \infty)$ with constants L_1, L_2 and exponents β_1/h_{min}, β_2/h_{min} that Eq.(18) has a unique bounded solution $\mathbf{w}_k$ with

$$||\mathbf{w}|| \leq C_1||\mathbf{g}||,$$

where

$$C_1 = L_1\left(1 - e^{-\frac{\beta_1}{h_{min}}}\right)^{-1} + L_2\frac{\beta_2}{h_{min}}\left(1 - e^{-\frac{\beta_2}{h_{min}}}\right)^{-1}.$$

Hence $T(\mathbf{z}, \mathbf{s}) = (\mathbf{g}, \ell)$ has the unique solution $(\mathbf{z}, \mathbf{s})$ in X where $\mathbf{s}$ is as given in Eq.(17) and $\mathbf{z}$ is given in Eqs.(15), (16) so that

$$||\mathbf{z}|| \leq \Delta^{-1}||\ell|| + C_1||\mathbf{g}||$$

and

$$||\mathbf{s}|| \leq \Delta^{-1}[\Delta^{-1}||\ell|| + e^{M_1 h_{max}}\Delta^{-1}||\ell|| + e^{M_1 h_{max}}C_1||\mathbf{g}|| + ||\mathbf{g}||],$$

where

$$\Delta = \inf_{\mathbf{x}\in S}||F(\mathbf{x})||.$$

So T is invertible and

$$||T^{-1}|| \leq C_2,$$

C_2 being a quantity depending on Δ, C_1, M_1 and h_{max}.

Next we suppose δ and σ are so small that the right side of Eq.(12) does not exceed $1/2C_2$. Then it follows that L is invertible and

$$||L^{-1}|| \leq C/2,$$

where $C = 4C_2$.

Completion of the proof of the theorem: We apply Lemma 4.4 to the mapping $\mathcal{G}$ defined in Eq.(4). As just shown, $L^{-1} = D\mathcal{G}(\mathbf{y}, \mathbf{h})^{-1}$ exists and

$$||L^{-1}|| \leq C/2,$$

provided δ and σ are sufficiently small depending on F, S, h_{max} and h_{min}.

Also, using Eq.(82) in Chapter 7,

$$\|\mathcal{G}(\mathbf{y},\mathbf{h})\| \le \delta + e^{M_1 h_{\max}} h_{\max}\sigma \le \delta_1 := \max(e^{M_1 h_{\max}} h_{\max}, 1)(\delta + \sigma).$$

Next note that if $(\mathbf{x},\mathbf{t}) \in O$ and $(\mathbf{z},\mathbf{s}) \in X$,

$$D\mathcal{G}(\mathbf{x},\mathbf{t})(\mathbf{z},\mathbf{s})$$
$$= \left(\{\mathbf{z}_{k+1} - D\psi^{t_k}(\mathbf{x}_k)\mathbf{z}_k - G(\psi^{t_k}(\mathbf{x}_k))s_k\}_{k=-\infty}^{\infty}, \{\langle \mathbf{z}_k, F(\mathbf{y}_k)\rangle\}_{k=-\infty}^{\infty}\right).$$

So, if $\|(\mathbf{x},\mathbf{t}) - (\mathbf{y},\mathbf{h})\| \le C\delta_1$ (which we assume is less than ε_0),

$$\|[D\mathcal{G}(\mathbf{x},\mathbf{t}) - D\mathcal{G}(\mathbf{y},\mathbf{h})](\mathbf{z},\mathbf{s})\|$$
$$= \sup_{k\in\mathbf{Z}} \|[D\psi^{t_k}(\mathbf{x}_k) - D\psi^{h_k}(\mathbf{y}_k)]\mathbf{z}_k + [G(\psi^{t_k}(\mathbf{x}_k)) - G(\psi^{h_k}(\mathbf{y}_k))]s_k\|.$$

Hence

$$\|[D\mathcal{G}(\mathbf{x},\mathbf{t}) - D\mathcal{G}(\mathbf{y},\mathbf{h})](\mathbf{z},\mathbf{s})\|$$
$$\le \sup_{k\in\mathbf{Z}} \left[\|D\psi^{t_k}(\mathbf{x}_k) - D\psi^{h_k}(\mathbf{y}_k)\| + (M_1+\sigma)\|\psi^{t_k}(\mathbf{x}_k) - \psi^{h_k}(\mathbf{y}_k)\|\right]. \quad (19)$$

With a view to estimating the terms on the right side, we note that if $\|\mathbf{x} - \mathbf{y}_k\| \le C\delta_1$ and $0 \le t \le h_{\max} + \varepsilon_0$,

$$\begin{aligned}\|\psi^t(\mathbf{x}) - \psi^t(\mathbf{y}_k)\| &= \|\mathbf{x} - \mathbf{y}_k + \int_0^t [G(\psi^s(\mathbf{x})) - G(\psi^s(\mathbf{y}_k))]ds\| \\ &\le \|\mathbf{x} - \mathbf{y}_k\| + (M_1+\sigma)\int_0^t \|\psi^s(\mathbf{x}) - \psi^s(\mathbf{y}_k)\|ds\end{aligned}$$

and so, by Gronwall's lemma,

$$\|\psi^t(\mathbf{x}) - \psi^t(\mathbf{y}_k)\| \le \|\mathbf{x} - \mathbf{y}_k\| e^{(M_1+\sigma)t}. \quad (20)$$

Then

$$\begin{aligned}\|\psi^t(\mathbf{x}) - \psi^{h_k}(\mathbf{y}_k)\| &\le \|\psi^t(\mathbf{x}) - \psi^t(\mathbf{y}_k)\| + \|\psi^t(\mathbf{y}_k) - \psi^{h_k}(\mathbf{y}_k)\| \\ &\le \|\mathbf{x} - \mathbf{y}_k\| e^{(M_1+\sigma)t} + \left\|\int_{h_k}^t G(\psi^s(\mathbf{y}_k))ds\right\| \\ &\le Ce^{(M_1+\sigma)(h_{max}+\varepsilon_0)}\delta_1 + (M_0+\sigma)|t - h_k|.\end{aligned}$$

Hence if $\|\mathbf{x} - \mathbf{y}_k\| \le C\delta_1$ and $|t - h_k| \le C\delta_1$,

$$\|\psi^t(\mathbf{x}) - \psi^{h_k}(\mathbf{y}_k)\| \le \delta_2 := C[e^{(M_1+\sigma)(h_{max}+\varepsilon_0)} + M_0 + \sigma]\delta_1. \quad (21)$$

Next, using Eq.(20), we note that for the same $\mathbf{x}$ and $\mathbf{t}$,

$$\begin{aligned}
&\|D\psi^t(\mathbf{x}) - D\psi^t(\mathbf{y}_k)\| \\
&\quad = \left\| \int_0^t [DG(\psi^s(\mathbf{x}))D\psi^s(\mathbf{x}) - DG(\psi^s(\mathbf{y}_k))D\psi^s(\mathbf{y}_k)]ds \right\| \\
&\quad \le \int_0^t \|DG(\psi^s(\mathbf{x}))\| \, \|D\psi^s(\mathbf{x}) - D\psi^s(\mathbf{y}_k)\| ds \\
&\qquad\quad + \int_0^t \|DG(\psi^s(\mathbf{x})) - DG(\psi^s(\mathbf{y}_k))\| \, \|D\psi^s(\mathbf{y}_k)\| ds \\
&\quad \le (M_1 + \sigma) \int_0^t \|D\psi^s(\mathbf{x}) - D\psi^s(\mathbf{y}_k)\| ds \\
&\qquad\quad + \int_0^t [2\sigma + \omega(\|\psi^s(\mathbf{x}) - \psi^s(\mathbf{y}_k)\|)] e^{(M_1+\sigma)s} ds \\
&\quad \le (M_1 + \sigma) \int_0^t \|D\psi^s(\mathbf{x}) - D\psi^s(\mathbf{y}_k)\| ds \\
&\qquad\quad + [2\sigma + \omega(\delta_2)] \, e^{(M_1+\sigma)(h_{max}+\varepsilon_0)}(h_{max} + \varepsilon_0).
\end{aligned}$$

So for $\|\mathbf{x} - \mathbf{y}_k\| \le C\delta_1$ and $|t - h_k| \le C\delta_1$,

$$\|D\psi^t(\mathbf{x}) - D\psi^t(\mathbf{y}_k)\| \le [2\sigma + \omega(\delta_2)] e^{2(M_1+\sigma)(h_{max}+\varepsilon_0)}(h_{max} + \varepsilon_0).$$

Then, for the same $\mathbf{x}$ and t again,

$$\begin{aligned}
&\|D\psi^t(\mathbf{x}) - D\psi^{h_k}(\mathbf{y}_k)\| \\
&\quad \le \|D\psi^t(\mathbf{x}) - D\psi^t(\mathbf{y}_k)\| + \|D\psi^t(\mathbf{y}_k) - D\psi^{h_k}(\mathbf{y}_k)\| \\
&\quad \le \|D\psi^t(\mathbf{x}) - D\psi^t(\mathbf{y}_k)\| + \left\| \int_{h_k}^t DG(\psi^s(\mathbf{y}_k)) D\psi^s(\mathbf{y}_k) ds \right\| \\
&\quad \le \|D\psi^t(\mathbf{x}) - D\psi^t(\mathbf{y}_k)\| + \left| \int_{h_k}^t (M_1 + \sigma) e^{(M_1+\sigma)s} ds \right| \\
&\quad \le \|D\psi^t(\mathbf{x}) - D\psi^t(\mathbf{y}_k)\| + (M_1 + \sigma) e^{(M_1+\sigma)(h_{max}+\varepsilon_0)} |t - h_k|.
\end{aligned}$$

So if $\|\mathbf{x} - \mathbf{y}_k\| \le C\delta_1$ and $|t - h_k| \le C\delta_1$,

$$\|D\psi^t(\mathbf{x}) - D\psi^{h_k}(\mathbf{y}_k)\| \le \delta_3, \tag{22}$$

where

$$\delta_3 = [2\sigma + \omega(\delta_2)]\, e^{2(M_1+\sigma)(h_{max}+\varepsilon_0)}(h_{max} + \varepsilon_0)$$
$$+ C(M_1 + \sigma)e^{(M_1+\sigma)(h_{max}+\varepsilon_0)}\delta_1.$$

Then it follows from Eqs.(19), (21) and (22) that when $\|(\mathbf{x}, \mathbf{t}) - (\mathbf{y}, \mathbf{h})\| \leq C\delta_1$,

$$\|D\mathcal{G}(\mathbf{x}, \mathbf{t}) - D\mathcal{G}(\mathbf{y}, \mathbf{h})\| \leq \delta_4 := (M_1 + \sigma)\delta_2 + \delta_3.$$

So, if in addition to the conditions imposed on σ and δ before, we also ask that

$$C\delta_4 \leq 1,$$

Lemma 4.4 can be applied and the theorem follows with

$$M = C \max\{e^{M_1 h_{max}} h_{max}, 1\}.$$

9.2 THE CONTINUOUS SHADOWING THEOREM

We can also prove a continuous version of the shadowing theorem, which is useful in certain contexts. In this theorem the pseudo orbit is allowed to stray outside S but not too far.

Theorem 9.4. *Let $U \in \mathbb{R}^n$ be a convex open set and let $F : U \to \mathbb{R}^n$ be a continuously differentiable vector field. Suppose $S \subset U$ is a compact hyperbolic set for Eq.(1) as in Definition 7.1. Let $\mathbf{y}(t)$ be a continuously differentiable vector valued function with $\mathbf{y}(t) \in U$ such that for some nonnegative numbers d and δ*

$$\text{dist}(\mathbf{y}(t), S) \leq d, \quad \|\dot{\mathbf{y}}(t) - F(\mathbf{y}(t))\| \leq \delta$$

for all t. Then there exist positive numbers δ_0, σ_0, d_0 and M, depending only on F and S, such that if $G : U \to \mathbb{R}^n$ is a continuously differentiable vector field satisfying

$$\|F(\mathbf{x}) - G(\mathbf{x})\| + \|DF(\mathbf{x}) - DG(\mathbf{x})\| \leq \sigma \ \text{for} \ \mathbf{x} \in U$$

with $\sigma \leq \sigma_0$ and if $\delta \leq \delta_0$, $d \leq d_0$, there exists a unique solution $\mathbf{x}(t)$ of Eq.(2) and a unique continuously differentiable real-valued function $\alpha(t)$ with $\alpha(0) = 0$ such that for all t

$$\|\mathbf{x}(\alpha(t)) - \mathbf{y}(t)\| \leq M(\delta + \sigma + d)$$

and

$$\langle \mathbf{x}(\alpha(t)) - \mathbf{y}(t), F(\mathbf{y}(t)) \rangle = 0.$$

Moreover,

$$|\alpha'(t) - 1| \leq M(\delta + \sigma + d).$$

Proof. The basic idea of the proof is to use the discrete shadowing theorem and then define $\alpha(t)$ appropriately. For $k \in \boldsymbol{Z}$ choose $\mathbf{y}_k \in S$ such that

$$\|\mathbf{y}(k) - \mathbf{y}_k\| \leq d.$$

Then, using Gronwall's lemma,

$$\begin{aligned}
\|\mathbf{y}_{k+1} - \phi^1(\mathbf{y}_k)\| &\leq \|\mathbf{y}_{k+1} - \mathbf{y}(k+1)\| + \|\mathbf{y}(k+1) - \phi^1(\mathbf{y}(k))\| \\
&\quad + \|\phi^1(\mathbf{y}(k)) - \phi^1(\mathbf{y}_k)\| \\
&\leq d + e^{M_1}\delta + e^{M_1} d \\
&= e^{M_1}\delta + (1 + e^{M_1})d,
\end{aligned}$$

where

$$M_1 = \sup_{\mathbf{x} \in U} \|DF(\mathbf{x})\|.$$

Let $\bar{\delta}_0$, $\bar{\sigma}_0$, $\overline{M}$ correspond to F, S, $h_{min} = h_{max} = 1$ in Theorem 9.3. Then if

$$\sigma \leq \bar{\sigma}_0 \quad \text{and} \quad e^{M_1}\delta + (1 + e^{M_1})d \leq \bar{\delta}_0,$$

the discrete pseudo orbit $\{\mathbf{y}_k\}_{k=-\infty}^{\infty}$ is ε-shadowed by points $\{\mathbf{x}_k\}_{k=-\infty}^{\infty}$ on a true orbit of Eq.(2) with $\mathbf{x}_{k+1} = \psi^{t_k}(\mathbf{x}_k)$ and

$$\varepsilon = \overline{M}[e^{M_1}\delta + (1 + e^{M_1})d + \sigma].$$

First we modify the $\mathbf{x}_k$'s. For $k \in \boldsymbol{Z}$ we seek a number τ_k such that

$$\langle\, \psi^{\tau_k}(\mathbf{x}_k) - \mathbf{y}(k), F(\mathbf{y}(k))\,\rangle = 0.$$

Assuming

$$M_1 d + \sigma \leq \Delta/4, \tag{23}$$

where

$$\Delta = \inf_{\mathbf{x} \in S} \|F(\mathbf{x})\|,$$

we apply Lemma 6.2 to Eq.(2) with $\mathbf{y}(k)$ as $\mathbf{y}$, $\mathbf{x}_k$ as $\mathbf{x}$, $F(\mathbf{y}(k))$ as $\mathbf{v}$ and $\alpha = \Delta/4(M_0 + \sigma)(M_1 + \sigma)$, where

$$M_0 = \sup_{\mathbf{x} \in U} \|F(\mathbf{x})\|.$$

Note that

$$\|\mathbf{x}_k - \mathbf{y}(k)\| \leq \varepsilon + d. \tag{24}$$

So if

$$\varepsilon + d < \min\{\Delta/4(M_1+\sigma), \Delta^2/32(M_0+\sigma)(M_1+\sigma)\},$$

it follows from Lemma 6.2 that there exists $\tau = \tau_k$ satisfying

$$|\tau_k| \le 8\Delta^{-1}\|\mathbf{x}_k - \mathbf{y}(k)\| \quad (<\alpha) \tag{25}$$

and

$$\langle \psi^{\tau}(\mathbf{x}_k) - \mathbf{y}(k), F(\mathbf{y}(k)) \rangle = 0. \tag{26}$$

Moreover τ_k is the unique number τ satisfying Eq.(26) in $|\tau| \le \alpha$. Then we define

$$\overline{\mathbf{x}}_k = \psi^{\tau_k}(\mathbf{x}_k)$$

so that

$$\langle \overline{\mathbf{x}}_k - \mathbf{y}(k), F(\mathbf{y}(k)) \rangle = 0.$$

It follows from Eqs.(24) and (25) that

$$|\tau_k| \le 8\Delta^{-1}(\varepsilon + d)$$

and hence

$$\begin{aligned} \|\overline{\mathbf{x}}_k - \mathbf{y}(k)\| &\le \|\psi^{\tau_k}(\mathbf{x}_k) - \mathbf{x}_k\| + \|\mathbf{x}_k - \mathbf{y}(k)\| \\ &\le (M_0+\sigma)|\tau_k| + \|\mathbf{x}_k - \mathbf{y}(k)\| \\ &\le \left[8(M_0+\sigma)\Delta^{-1} + 1\right](\varepsilon + d)\,. \end{aligned} \tag{27}$$

Also note that

$$\overline{\mathbf{x}}_{k+1} = \psi^{s_k}(\overline{\mathbf{x}}_k),$$

where

$$s_k = -\tau_k + t_k + \tau_{k+1},$$

and we estimate

$$|s_k - 1| \le |t_k - 1| + |\tau_k| + |\tau_{k+1}| \le \varepsilon + 16\Delta^{-1}(\varepsilon + d).$$

Now for each $k \in \boldsymbol{Z}$, we seek a function $\alpha_k : [0,1] \to I\!R$ such that

$$\langle \psi^{\alpha_k(t)}(\overline{\mathbf{x}}_k) - \mathbf{y}(t+k), F(\mathbf{y}(t+k)) \rangle = 0$$

for $0 \le t \le 1$. Again we apply Lemma 6.2 to Eq.(2) with $\mathbf{y}(t+k)$ as $\mathbf{y}$, $\psi^t(\overline{\mathbf{x}}_k)$ as $\mathbf{x}$, $F(\mathbf{y}(t+k))$ as $\mathbf{v}$ and $\alpha = \Delta/4(M_0+\sigma)(M_1+\sigma)$. Note that, by Gronwall's lemma,

$$\|\psi^t(\overline{\mathbf{x}}_k) - \mathbf{y}(t+k)\| \le e^{M_1}(\|\overline{\mathbf{x}}_k - \mathbf{y}(k)\| + \sigma + \delta) \tag{28}$$

if $0 \le t \le 1$. Hence, using Eqs.(23), (27) and (28), if

$$e^{M_1}[(8(M_0+\sigma)\Delta^{-1}+1)(\varepsilon+d)+\sigma+\delta] < \min\{\Delta/4(M_1+\sigma), \Delta^2/32(M_0+\sigma)(M_1+\sigma)\}, \tag{29}$$

it follows from Lemma 6.2 that there exists $\tau = \tau_k(t)$ satisfying

$$|\tau_k(t)| \le 8\Delta^{-1}\|\psi^t(\overline{\mathbf{x}}_k) - \mathbf{y}(t+k)\| \ (< \alpha) \tag{30}$$

and

$$\langle\, \psi^{t+\tau}(\overline{\mathbf{x}}_k) - \mathbf{y}(t+k), F(\mathbf{y}(t+k))\,\rangle = 0. \tag{31}$$

Moreover $\tau_k(t)$ is the unique number τ satisfying Eq.(31) in $|\tau| \le \alpha$. In particular, by uniqueness, it follows that

$$\tau_k(0) = 0, \ \tau_k(1) = s_k - 1 \tag{32}$$

provided that

$$\varepsilon + 16\Delta^{-1}(\varepsilon + 2d) \le \Delta/4(M_0+\sigma)(M_1+\sigma).$$

Next we prove that $\tau_k(t)$ is continuously differentiable in $[0,1]$. If we define

$$g(t,\tau) = \langle\, \psi^{t+\tau}(\overline{\mathbf{x}}_k) - \mathbf{y}(t+k), F(\mathbf{y}(t+k))\,\rangle,$$

we see that

$$\begin{aligned} \frac{\partial g}{\partial \tau}(t,\tau) &= \langle\, G(\psi^{t+\tau}(\overline{\mathbf{x}}_k)), F(\mathbf{y}(t+k))\,\rangle \\ &\ge [\|F(\mathbf{y}(t+k))\| - \|G(\psi^{t+\tau}(\overline{\mathbf{x}}_k)) - F(\mathbf{y}(t+k))\|]\ \|F(\mathbf{y}(t+k))\|. \end{aligned}$$

Now, using Eqs.(27), (28), (29) and (30), we find that for $\tau = \tau_k(t)$,

$$\begin{aligned} &\|G(\psi^{t+\tau}(\overline{\mathbf{x}}_k)) - F(\mathbf{y}(t+k))\| \\ &\quad \le \|G(\psi^{t+\tau}(\overline{\mathbf{x}}_k)) - F(\psi^{t+\tau}(\overline{\mathbf{x}}_k))\| + \|F(\psi^{t+\tau}(\overline{\mathbf{x}}_k)) - F(\psi^{t}(\overline{\mathbf{x}}_k))\| \\ &\qquad + \|F(\psi^{t}(\overline{\mathbf{x}}_k)) - F(\mathbf{y}(t+k))\| \\ &\quad \le \sigma + M_1[8(M_0+\sigma)\Delta^{-1}+1]\|\psi^t(\overline{\mathbf{x}}_k) - \mathbf{y}(t+k)\| \\ &\quad \le \sigma + \Delta/2. \end{aligned}$$

Hence, using Eq.(23),

$$\begin{aligned} \frac{\partial g}{\partial \tau}(t,\tau_k(t)) &\ge (\Delta - M_1 d - \sigma - \Delta/2)\|F(\mathbf{y}(t+k))\| \\ &\ge \frac{\Delta}{4}\|F(\mathbf{y}(t+k))\|. \end{aligned} \tag{33}$$

So

$$\frac{\partial g}{\partial \tau}(t, \tau_k(t)) > 0$$

and it follows from the implicit function theorem and the uniqueness of $\tau_k(t)$ that $\tau_k(t)$ is continuously differentiable on $[0, 1]$.

Then if we define

$$\alpha_k(t) = t + \tau_k(t) \text{ for } 0 \le t \le 1,$$

we see that $\alpha_k(t)$ is continuously differentiable and that for $0 \le t \le 1$,

$$\langle \psi^{\alpha_k(t)}(\overline{\mathbf{x}}_k) - \mathbf{y}(t+k), F(\mathbf{y}(t+k)) \rangle = 0 \tag{34}$$

and, using Eq.(30),

$$\|\psi^{\alpha_k(t)}(\overline{\mathbf{x}}_k) - \mathbf{y}(t+k)\| \le [8(M_0+\sigma)\Delta^{-1} + 1]\,\|\psi^t(\overline{\mathbf{x}}_k) - \mathbf{y}(t+k)\|. \tag{35}$$

Also it follows from Eq.(32) that

$$\alpha_k(0) = 0, \;\; \alpha_k(1) = s_k.$$

To obtain a formula for $\alpha_k'(t)$, we differentiate Eq.(34) with respect to t and get

$$\begin{aligned} &\langle G(\psi^{\alpha_k(t)}(\overline{\mathbf{x}}_k))\alpha_k'(t) - \dot{\mathbf{y}}(t+k), F(\mathbf{y}(t+k)) \rangle \\ &\quad + \langle \psi^{\alpha_k(t)}(\overline{\mathbf{x}}_k) - \mathbf{y}(t+k), DF(\mathbf{y}(t+k))\dot{\mathbf{y}}(t+k) \rangle = 0. \end{aligned} \tag{36}$$

Now it follows from Eq.(33) that

$$\langle G(\psi^{\alpha_k(t)}(\overline{\mathbf{x}}_k)), F(\mathbf{y}(t+k)) \rangle \ge \frac{\Delta}{4}\|F(\mathbf{y}(t+k))\| > 0. \tag{37}$$

So we may solve Eq.(36) for $\alpha_k'(t)$ to get

$$\alpha_k'(t) = \tag{38}$$

$$\frac{\langle \dot{\mathbf{y}}(t+k), F(\mathbf{y}(t+k)) \rangle - \langle \psi^{\alpha_k(t)}(\overline{\mathbf{x}}_k) - \mathbf{y}(t+k), DF(\mathbf{y}(t+k))\dot{\mathbf{y}}(t+k) \rangle}{\langle G(\psi^{\alpha_k(t)}(\overline{\mathbf{x}}_k)), F(\mathbf{y}(t+k)) \rangle}.$$

In particular this means that for all k

$$\alpha_k'(1) = \alpha_{k+1}'(0).$$

Also, it follows from Eq.(38) that

$$\alpha_k'(t) - 1 = \frac{\langle \dot{\mathbf{y}}(t+k) - G(\psi^{\alpha_k(t)}(\overline{\mathbf{x}}_k)), F(\mathbf{y}(t+k)) \rangle}{\langle G(\psi^{\alpha_k(t)}(\overline{\mathbf{x}}_k)), F(\mathbf{y}(t+k)) \rangle} - \frac{\langle \psi^{\alpha_k(t)}(\overline{\mathbf{x}}_k) - \mathbf{y}(t+k), DF(\mathbf{y}(t+k))\dot{\mathbf{y}}(t+k) \rangle}{\langle G(\psi^{\alpha_k(t)}(\overline{\mathbf{x}}_k)), F(\mathbf{y}(t+k)) \rangle}$$

and so, using Eq.(37), for $0 \le t \le 1$ we estimate

$$\begin{aligned}
&|\alpha_k'(t) - 1| \\
&\le 4\Delta^{-1} \Big[\|\dot{\mathbf{y}}(t+k) - G(\psi^{\alpha_k(t)}(\overline{\mathbf{x}}_k))\| \\
&\qquad + M_1 \|\psi^{\alpha_k(t)}(\overline{\mathbf{x}}_k) - \mathbf{y}(t+k)\| \frac{\|\dot{\mathbf{y}}(t+k)\|}{\|F(\mathbf{y}(t+k))\|} \Big] \\
&\le 4\Delta^{-1} \big[\|\dot{\mathbf{y}}(t+k) - F(\mathbf{y}(t+k))\| + \|F(\mathbf{y}(t+k)) - F(\psi^{\alpha_k(t)}(\overline{\mathbf{x}}_k))\| \\
&\qquad + \|F(\psi^{\alpha_k(t)}(\overline{\mathbf{x}}_k)) - G(\psi^{\alpha_k(t)}(\overline{\mathbf{x}}_k))\| \\
&\qquad\qquad + M_1(1 + \delta(\Delta - M_1 d)^{-1}) \|\psi^{\alpha_k(t)}(\overline{\mathbf{x}}_k) - \mathbf{y}(t+k)\| \big] \\
&\le 4\Delta^{-1} \left[\delta + \sigma + M_1(2 + \delta(\Delta - M_1 d)^{-1}) \|\psi^{\alpha_k(t)}(\overline{\mathbf{x}}_k) - \mathbf{y}(t+k)\| \right].
\end{aligned} \tag{39}$$

Now we define $\alpha : I\!R \to I\!R$ by

$$\alpha(t) = \begin{cases} s_0 + \ldots + s_{k-1} + \alpha_k(t-k) & \text{if } k \le t \le k+1,\ k \ge 1 \\ \alpha_0(t) & \text{if } 0 \le t \le 1 \\ -s_k - \ldots - s_{-1} + \alpha_k(t-k) & \text{if } k \le t \le k+1,\ k \le -1 \end{cases}$$

and we define the solution

$$\mathbf{x}(t) = \psi^t(\overline{\mathbf{x}}_0).$$

Then it follows from the facts that each $\alpha_k(t)$ is C^1, $\alpha_k(0) = 0$, $\alpha_k(1) = s_k$ and $\alpha_k'(1) = \alpha_{k+1}'(0)$, that $\alpha(t)$ is a C^1 function with $\alpha(0) = 0$. Moreover, it follows from Eqs.(27), (28), (34), (35), (38) and (39) that for all t

$$\langle \mathbf{x}(\alpha(t)) - \mathbf{y}(t), F(\mathbf{y}(t)) \rangle = 0,$$

$$\|\mathbf{x}(\alpha(t)) - \mathbf{y}(t)\|$$
$$\leq e^{M_1}[8(M_0+\sigma)\Delta^{-1}+1][\sigma+\delta+(8(M_0+\sigma)\Delta^{-1}+1)(\varepsilon+d)], \tag{40}$$

$$\alpha'(t) = \frac{\langle \dot{\mathbf{y}}(t), F(\mathbf{y}(t))\rangle - \langle \mathbf{x}(\alpha(t)) - \mathbf{y}(t), DF(\mathbf{y}(t))\dot{\mathbf{y}}(t)\rangle}{\langle G(\mathbf{x}(\alpha(t))), F(\mathbf{y}(t))\rangle} \tag{41}$$

and

$$|\alpha'(t) - 1| \leq 4\Delta^{-1}\left[\delta+\sigma+M_1(2+\delta(\Delta - M_1 d)^{-1})\|\mathbf{x}(\alpha(t)) - \mathbf{y}(t)\|\right]. \tag{42}$$

So, with suitable choice of M, the existence part of the theorem is proved.

What remains to be proved is the uniqueness. First we suppose δ, σ, d are so small that the right side of inequality (42) is less than one. Then it follows that $\alpha(t)$ is a homeomorphism. Now note that

$$\operatorname{dist}(\mathbf{x}(t), S) \leq d + M(\sigma+\delta+d)$$

and so if δ, σ, d are sufficiently small it follows from Theorem 7.14 that the solution $\mathbf{x}(t)$ lies in a compact hyperbolic set S_O with exponents $3\alpha_1/4$, $3\alpha_2/4$ and with constants and bounds N^0, N^s, N^u on the projections depending only on F and S. Also note if we carry through the proof of Proposition 7.10 for Eq.(2) and S_O, the modulus of continuity $\omega(\varepsilon)$ would be replaced by $\omega(\varepsilon)+2\sigma$, where $\omega(\cdot)$ is the modulus of continuity of DF, and so we would need ε and σ sufficiently small depending only on F and S. Thus the expansivity constant for S_O derived in Theorem 7.11 can be chosen to depend only on F and S.

Now let $\mathbf{z}(t)$ be another solution of Eq.(2) and $\beta : \mathbb{R} \to \mathbb{R}$ a continuously differentiable function with $\beta(0) = 0$ such that for all t

$$\|\mathbf{z}(\beta(t)) - \mathbf{y}(t)\| \leq M(\delta+\sigma+d).$$

It follows that for all t

$$\|\mathbf{z}(\beta(t)) - \mathbf{x}(\alpha(t))\| \leq 2M(\delta+\sigma+d)$$

and hence for all t

$$\|\mathbf{z}(\gamma(t)) - \mathbf{x}(t)\| \leq 2M(\delta+\sigma+d),$$

where $\gamma = \beta \circ \alpha^{-1}$. Then if $2M(\delta+\sigma+d)$ does not exceed the expansivity constant for the hyperbolic set S_O, it follows from Theorem 7.11 that there exists a real number τ such that

$$\mathbf{x}(0) = \psi^{\tau}(\mathbf{z}(0)),$$

where

$$|\tau| \leq \text{constant} \cdot 2M(\delta+\sigma+d),$$

the constant depending only on F and S if δ, σ and d are sufficiently small. Now suppose we also assume that

$$\langle \mathbf{z}(\beta(t)) - \mathbf{y}(t), F(\mathbf{y}(t)) \rangle = 0 \tag{43}$$

for all t. In particular,

$$\langle \mathbf{z}(0) - \mathbf{y}(0), F(\mathbf{y}(0)) \rangle = 0.$$

Since $\beta(0) = 0$, we also have

$$\|\mathbf{z}(0) - \mathbf{y}(0)\| \leq M(\delta + \sigma + d).$$

However, note that

$$\langle \psi^{\tau}(\mathbf{z}(0)) - \mathbf{y}(0), F(\mathbf{y}(0)) \rangle = \langle \mathbf{x}(0) - \mathbf{y}(0), F(\mathbf{y}(0)) \rangle = 0.$$

Then, it follows from the uniqueness in Lemma 6.2 that $\tau = 0$ provided δ, σ and d are sufficiently small. Hence

$$\mathbf{x}(0) = \mathbf{z}(0)$$

and so $\mathbf{x}(t) = \mathbf{z}(t)$ for all t.

Next note from Eq.(41) that $\alpha(t)$ is a solution of the scalar differential equation

$$\alpha' = \frac{\langle \dot{\mathbf{y}}(t), F(\mathbf{y}(t)) \rangle - \langle \psi^{\alpha}(\mathbf{x}(0)) - \mathbf{y}(t), DF(\mathbf{y}(t))\dot{\mathbf{y}}(t) \rangle}{\langle G(\psi^{\alpha}(\mathbf{x}(0))), F(\mathbf{y}(t)) \rangle}$$

with $\alpha(0) = 0$. Now, since Eq.(43) holds for all t and $\mathbf{x}(0) = \mathbf{z}(0)$, and provided δ, σ and d are so small that

$$\langle G(\mathbf{z}(\beta(t)), F(\mathbf{y}(t)) \rangle > 0$$

for all t, $\beta(t)$ is a solution of the same initial value problem. So, by uniqueness,

$$\alpha(t) = \beta(t)$$

for all t. This completes the proof of the uniqueness and the proof of the theorem.

Actually the proof shows that the uniqueness is stronger than stated in the theorem. It was just sufficient to assume that $\beta(t)$ is continuous since if Eq.(43) holds for all t we saw in Lemma 7.9 that $\beta(t)$ is continuously differentiable.

9.3 MORE ON THE ROUGHNESS OF HYPERBOLIC SETS

Let S be a compact hyperbolic set for Eq.(1). It was shown in Section 7.6 that if O is a sufficiently "tight" neighbourhood of S and $G : U \to \mathbb{R}^n$ (U open and convex) is a C^1 vector field sufficiently close in the C^1 topology to F, then the maximal invariant set S_O for the system

$$\dot{\mathbf{x}} = G(\mathbf{x}) \tag{44}$$

in $\overline{O}$ is hyperbolic. In this section we want to use shadowing to prove that if S is an *isolated* invariant set for Eq.(1), then the flows on S and S_0 are *topologically conjugate*, a concept which we now define.

Definition 9.5. Let X, Y be metric spaces and let $\phi^t : X \to X$ and $\psi^t : Y \to Y$ be flows. Then ϕ^t and ψ^t are said to be *topologically conjugate* if there is a homeomorphism $h : X \to Y$ and a continuous function $\alpha : \mathbb{R} \times X \to \mathbb{R}$ with the cocycle property

$$\alpha(t+s, \mathbf{x}) = \alpha(s, \mathbf{x}) + \alpha(t, \phi^s(\mathbf{x}))$$

such that

$$h(\phi^t(\mathbf{x})) = \psi^{\alpha(t,\mathbf{x})}(h(\mathbf{x})).$$

Remark 9.6. It is easy to see that if the conditions of Definition 9.5 hold, then $\overline{\psi} : \mathbb{R} \times X \to Y$ defined by $\overline{\psi}(t, \mathbf{y}) = \psi^{\alpha(t,h^{-1}(\mathbf{y}))}(\mathbf{y})$ defines a flow on Y and that

$$h \circ \phi^t = \overline{\psi}^t \circ h$$

for all t.

Definition 9.5 is not as restrictive as may seem at first sight since Thomas [1985, Lemma 4] has shown that when X and Y are compact, $\phi^t : X \to X$ has no fixed points and there is a homeomorphism $h : X \to Y$ which maps the orbits of ϕ^t onto those of ψ^t, then ϕ^t and ψ^t are topologically conjugate in the sense of Definition 9.5 and the cocycle $\alpha(t, \mathbf{x})$ is unique. Moreover, for each fixed $\mathbf{x}$, $\alpha_{\mathbf{x}}(t) = \alpha(t, \mathbf{x})$ is a strictly increasing homeomorphism.

Our aim in this section is to prove the following theorem.

Theorem 9.7. *Let U be a convex open set in $\mathbb{R}^n$, let $F : U \to \mathbb{R}^n$ be a C^1 vector field and let S be an isolated compact hyperbolic set for Eq.(1). Then if O is a sufficiently tight open neighbourhood of S and $G : U \to \mathbb{R}^n$ is a C^1 vector field satisfying*

$$\sup_{\mathbf{x}\in U} \|G(\mathbf{x}) - F(\mathbf{x})\| + \sup_{\mathbf{x}\in U} \|DG(\mathbf{x}) - DF(\mathbf{x})\| \le \sigma \tag{45}$$

for σ sufficiently small, the maximal invariant set S_O for Eq.(44) in $\overline{O}$ is isolated hyperbolic and the flows $\phi^t : S \to S$, $\psi^t : S_O \to S_O$ corresponding to Eqs.(1) and (44) respectively are topologically conjugate with conjugacy $h : S \to S_O$ satisfying

$$\|h(\mathbf{x}) - \mathbf{x}\| \le M\sigma, \quad \langle h(\mathbf{x}) - \mathbf{x}, F(\mathbf{x}) \rangle = 0$$

and continuous cocycle $\alpha : \mathbb{R} \times S \to \mathbb{R}$ with continuous derivative $\frac{\partial \alpha}{\partial t}(t, \mathbf{x})$ satisfying

$$\left| \frac{\partial \alpha}{\partial t}(t, \mathbf{x}) - 1 \right| \le M\sigma,$$

where M is the constant in Theorem 9.4.

Proof. By Theorem 7.14 there are numbers $d_0 > 0$ and $\overline{\sigma}_0 > 0$ such that if $\text{dist}(\mathbf{x}, S) \le d_0$ for all $\mathbf{x}$ in $\overline{O}$ and σ in Eq.(45) does not exceed $\overline{\sigma}_0$, then the maximal invariant set S_O for Eq.(44) in $\overline{O}$ is hyperbolic. We assume these conditions hold in the rest of the proof.

Let δ_0, σ_0, M be the constants in the continuous shadowing Theorem 9.4 applied to F and S. Let $\mathbf{x} \in S$ and put $\mathbf{x}(t) = \phi^t(\mathbf{x})$. Then, by Theorem 9.4 with $\delta = d = 0$, if $\sigma \le \sigma_0$, there is a solution $\mathbf{z}(t)$ of Eq.(44) and a continuously differentiable function $\alpha : \mathbb{R} \to \mathbb{R}$ with $\alpha(0) = 0$ such that for all t

$$\|\mathbf{z}(\alpha(t)) - \mathbf{x}(t)\| \le M\sigma, \ |\alpha'(t) - 1| \le M\sigma \tag{46}$$

and

$$\langle \mathbf{z}(\alpha(t)) - \mathbf{x}(t), F(\mathbf{x}(t)) \rangle = 0. \tag{47}$$

Also $\mathbf{z}(t)$ is the unique solution of Eq.(44) and $\alpha(t)$ the unique continuously differentiable function $\alpha : \mathbb{R} \to \mathbb{R}$ with $\alpha(0) = 0$ for which Eq.(47) holds and

$$\|\mathbf{z}(\alpha(t)) - \mathbf{x}(t)\| \le M(\sigma_0 + \delta_0)$$

for all t. Then we define

$$h(\mathbf{x}) = \mathbf{z}(0), \ \ \alpha(t, \mathbf{x}) = \alpha(t).$$

Note that it follows from Eq.(46) that

$$\|h(\mathbf{x}) - \mathbf{x}\| \le M\sigma$$

and from (47) that

$$\langle h(\mathbf{x}) - \mathbf{x}, F(\mathbf{x}) \rangle = 0.$$

Next fixing the real number s, we define

$$\bar{\mathbf{z}}(t) = \mathbf{z}(t + \alpha(s)), \ \bar{\alpha}(t) = \alpha(t + s) - \alpha(s). \tag{48}$$

We see that

$$\bar{\mathbf{z}}(\bar{\alpha}(t)) - \mathbf{x}(t+s) = \mathbf{z}(\alpha(t+s)) - \mathbf{x}(t+s)$$

and so for all t

$$\|\bar{\mathbf{z}}(\bar{\alpha}(t)) - \mathbf{x}(t+s)\| \leq M\sigma \quad \text{and} \quad \langle \bar{\mathbf{z}}(\bar{\alpha}(t)) - \mathbf{x}(t+s), F(\mathbf{x}(t+s)) \rangle = 0.$$

By uniqueness, it follows that $\bar{\mathbf{z}}(t)$ and $\bar{\alpha}(t)$ correspond to $\mathbf{x}(t+s) = \phi^t(\phi^s(\mathbf{x}))$ and so

$$h(\mathbf{x}(s)) = \bar{\mathbf{z}}(0) = \mathbf{z}(\alpha(s)) \quad \text{and} \quad \bar{\alpha}(t) = \alpha(t, \phi^s(\mathbf{x})).$$

Thus we have established the identities

$$h(\phi^s(\mathbf{x})) = \psi^{\alpha(s,\mathbf{x})}(h(\mathbf{x})) \tag{49}$$

and

$$\alpha(t, \phi^s(\mathbf{x})) = \alpha(t+s, \mathbf{x}) - \alpha(s, \mathbf{x}),$$

the latter implying that $\alpha(t, \mathbf{x})$ is a cocycle.

What remains to be shown is that S_O is isolated, h is a homeomorphism of S onto S_O, and $\alpha(t, \mathbf{x})$ and $\frac{\partial \alpha}{\partial t}(t, \mathbf{x})$ are continuous. First provided σ is so small that $M\sigma < 1$, so that $\alpha_{\mathbf{x}}(t) = \alpha(t, \mathbf{x})$ is a homeomorphism (conf. Eq.(46)), and so that

$$\{\mathbf{x} : \operatorname{dist}(\mathbf{x}, S) \leq M\sigma\} \subset O,$$

it follows from the facts that

$$\|h(\phi^t(\mathbf{x})) - \phi^t(\mathbf{x})\| \leq M\sigma \quad \text{and} \quad h(\phi^t(\mathbf{x})) = \psi^{\alpha(t,\mathbf{x})}(h(\mathbf{x}))$$

that the solution of Eq.(44) starting at $h(\mathbf{x})$ lies entirely in O. Hence $h(\mathbf{x}) \in S_O$ for all $\mathbf{x}$ in S. Thus h maps S into S_O.

Next we show that h is one to one. Suppose there exist $\mathbf{x}$ and $\mathbf{z}$ in S with $h(\mathbf{x}) = h(\mathbf{z}) = \mathbf{y}$. Then $\beta = \alpha_{\mathbf{z}}^{-1} \circ \alpha_{\mathbf{x}}$ (where $\alpha_{\mathbf{x}}(t) = \alpha(t, \mathbf{x})$, etc.) is a C^1 function satisfying

$$\begin{aligned} &\|\phi^t(\mathbf{x}) - \phi^{\beta(t)}(\mathbf{z})\| \\ &\leq \|\phi^t(\mathbf{x}) - \psi^{\alpha(t,\mathbf{x})}(h(\mathbf{x}))\| + \|\psi^{\alpha(\beta(t),\mathbf{z})}(h(\mathbf{z})) - \phi^{\beta(t)}(\mathbf{z})\| \\ &= \|\phi^t(\mathbf{x}) - h(\phi^t(\mathbf{x}))\| + \|h(\phi^{\beta(t)}(\mathbf{z})) - \phi^{\beta(t)}(\mathbf{z})\| \\ &\leq 2M\sigma \end{aligned} \tag{50}$$

for all t using the fact that

$$\|h(\mathbf{x}) - \mathbf{x}\| \leq M\sigma.$$

So provided σ is so small that $2M\sigma$ does not exceed an expansivity constant for F on S, there exists τ such that

$$\mathbf{z} = \phi^{\tau}(\mathbf{x}).$$

Also it follows from Theorem 7.11 that

$$|\tau| \leq 8\Delta^{-1}M\sigma, \tag{51}$$

where

$$\Delta = \inf_{\mathbf{x}\in S} \|F(\mathbf{x})\|.$$

To prove that $\tau = 0$, set

$$g(t) = \langle \mathbf{y} - \phi^t(\mathbf{x}), F(\phi^t(\mathbf{x})) \rangle.$$

Note that

$$g(0) = \langle h(\mathbf{x}) - \mathbf{x}, F(\mathbf{x}) \rangle = 0$$

and also that

$$g(\tau) = \langle \mathbf{y} - \phi^{\tau}(\mathbf{x}), F(\phi^{\tau}(\mathbf{x})) \rangle = \langle h(\mathbf{z}) - \mathbf{z}, F(\mathbf{z}) \rangle = 0.$$

Next we estimate for $|t| \leq |\tau|$,

$$\begin{aligned} g'(t) &= -\|F(\phi^t(\mathbf{x}))\|^2 + \langle \mathbf{y} - \phi^t(\mathbf{x}), DF(\phi^t(\mathbf{x}))F(\phi^t(\mathbf{x})) \rangle \\ &\leq -\|F(\phi^t(\mathbf{x}))\| \, [\, \|F(\phi^t(\mathbf{x}))\| - M_1\|\mathbf{y} - \phi^t(\mathbf{x})\| \,] \\ &\leq -\|F(\phi^t(\mathbf{x}))\| \, [\, \|F(\phi^t(\mathbf{x}))\| - M_1(\|\mathbf{y} - \mathbf{x}\| + M_0|\tau|) \,] \\ &\leq -\Delta[\Delta - M_1 M(1 + 8\Delta^{-1}M_0)\sigma] \\ &< 0, \end{aligned}$$

provided

$$M_1 M(1 + 8\Delta^{-1}M_0)\sigma < \Delta,$$

where

$$M_0 = \sup_{\mathbf{x}\in U} \|F(\mathbf{x})\|, \; M_1 = \sup_{\mathbf{x}\in U} \|DF(\mathbf{x})\|.$$

This implies that $\tau = 0$. So $\mathbf{x} = \mathbf{z}$ and thus we have proved that h is one to one.

Now we prove h is surjective. We define

$$d = \max_{\mathbf{x}\in\overline{O}} \mathrm{dist}(\mathbf{x}, S).$$

Let $\mathbf{z} \in S_O$ and define $\mathbf{y}(t) = \psi^t(\mathbf{z})$. Then

$$\|\dot{\mathbf{y}}(t) - F(\mathbf{y}(t))\| \leq \sigma$$

for all t. We apply Theorem 9.4 to deduce that if $d \leq d_0$, $\sigma \leq \delta_0$ there exists a unique solution $\mathbf{x}(t)$ of Eq.(1) and a unique continuously differentiable real valued function $\alpha(t)$ with $\alpha(0) = 0$ such that for all t

$$\|\mathbf{x}(\alpha(t)) - \mathbf{y}(t)\| \leq M(\sigma + d), \tag{52}$$

$$|\alpha'(t) - 1| \leq M(\sigma + d) \tag{53}$$

and

$$\langle\, \mathbf{x}(\alpha(t)) - \mathbf{y}(t), F(\mathbf{y}(t)) \,\rangle = 0.$$

Now if the quantity on the right side of Eq.(53) is less than 1, $\alpha(t)$ is a homeomorphism of $\mathbb{R}$ onto itself. Note that

$$\mathrm{dist}(\mathbf{x}(\alpha(t)), S) \leq d + M(\sigma + d).$$

So if d and σ are sufficiently small, $\mathbf{x}(t) \in S$ for all t since S is isolated. Also we can rewrite inequality (52) as

$$\|\mathbf{y}(\alpha^{-1}(t)) - \mathbf{x}(t)\| \leq M(\sigma + d) \tag{54}$$

for all t.

Now, assuming

$$\sigma < \Delta,$$

we apply Lemma 6.2 to Eq.(44) with $\mathbf{x}(t)$ as $\mathbf{y}$, $\mathbf{y}(\alpha^{-1}(t))$ as $\mathbf{x}$,

$$\alpha = (\Delta - \sigma)/2(M_0 + \sigma)(M_1 + \sigma)$$

and $F(\mathbf{x}(t))$ as $\mathbf{v}$. Then, if

$$M(\sigma + d) < \min\{(\Delta - \sigma)/2(M_1 + \sigma),\ (\Delta - \sigma)^2/8(M_0 + \sigma)(M_1 + \sigma)\},$$

it follows from Lemma 6.2 that there exists $\tau = \tau(t)$ satisfying

$$|\tau(t)| \leq 4(\Delta - \sigma)^{-1}\|\mathbf{y}(\alpha^{-1}(t)) - \mathbf{x}(t)\| \ \ (< \alpha)$$

and

$$\langle\, \psi^\tau(\mathbf{y}(\alpha^{-1}(t))) - \mathbf{x}(t),\ F(\mathbf{x}(t)) \,\rangle = 0. \tag{55}$$

Moreover, $\tau(t)$ is the unique τ satisfying Eq.(55) in $|\tau| \leq \alpha$. We estimate that for all t,

$$\begin{aligned} \|\mathbf{y}(\alpha^{-1}(t)+\tau(t)) - \mathbf{x}(t)\| & \\ &\leq (M_0+\sigma)|\tau(t)| + \|\mathbf{y}(\alpha^{-1}(t)) - \mathbf{x}(t)\| \\ &\leq [4(M_0+\sigma)(\Delta-\sigma)^{-1}+1]\|\mathbf{y}(\alpha^{-1}(t)) - \mathbf{x}(t)\| \\ &\leq [4(M_0+\sigma)(\Delta-\sigma)^{-1}+1]M(\sigma+d) \end{aligned} \tag{56}$$

and note also that we have

$$\langle \mathbf{y}(\alpha^{-1}(t)+\tau(t)) - \mathbf{x}(t), F(\mathbf{x}(t)) \rangle = 0. \tag{57}$$

To show that $\tau(t)$ is continuously differentiable, define $g(\tau, t)$ by

$$g(\tau,t) = \langle \psi^\tau(\mathbf{y}(\alpha^{-1}(t))) - \mathbf{x}(t), F(\mathbf{x}(t)) \rangle.$$

Differentiating with respect to τ and using Eq.(56), we find for $\tau = \tau(t)$ that

$$\begin{aligned} \frac{\partial g}{\partial \tau}(\tau,t) &= \langle G(\psi^\tau(\mathbf{y}(\alpha^{-1}(t)))), F(\mathbf{x}(t)) \rangle \\ &\geq \|F(\mathbf{x}(t))\|(\|F(\mathbf{x}(t))\| - \|G(\psi^\tau(\mathbf{y}(\alpha^{-1}(t)))) - F(\mathbf{x}(t))\|) \\ &\geq \|F(\mathbf{x}(t))\|(\Delta - [\sigma + \|F(\psi^\tau(\mathbf{y}(\alpha^{-1}(t)))) - F(\mathbf{x}(t))\|]) \\ &\geq \|F(\mathbf{x}(t))\|\,(\Delta - [\sigma + M_1\|\mathbf{y}(\alpha^{-1}(t)+\tau) - \mathbf{x}(t)\|]) \\ &\geq \|F(\mathbf{x}(t))\|\,(\Delta - [\sigma + M_1(4(M_0+\sigma)(\Delta-\sigma)^{-1}+1)M(\sigma+d)]) \\ &> 0. \end{aligned}$$

Then it follows from the implicit function theorem and the uniqueness of $\tau(t)$ that $\tau(t)$ is C^1.

So provided

$$[4(M_0+\sigma)(\Delta-\sigma)^{-1}+1](\sigma+d) \leq \sigma_0+\delta_0,$$

it follows from Eqs.(56), (57), the fact that $\alpha^{-1}(t) + \tau(t)$ is continuously differentiable with $\alpha^{-1}(0) = 0$ and the uniqueness in Theorem 9.4 that

$$h(\mathbf{x}(0)) = \mathbf{y}(\tau(0)) = \psi^{\tau(0)}(\mathbf{z}).$$

But, by Eq.(49),

$$h(\phi^s(\mathbf{x}(0))) = \psi^{\alpha(s,\mathbf{x}(0))}(h(\mathbf{x}(0))) = \psi^{\alpha(s,\mathbf{x}(0))+\tau(0)}(\mathbf{z}).$$

So if we choose s so that $\alpha(s, \mathbf{x}(0)) = -\tau(0)$, we see that

$$h(\phi^s(\mathbf{x}(0))) = \mathbf{z}.$$

Hence $h : S \to S_O$ is surjective. Moreover, since $h(S) \subset O$, it follows that $S_O \subset O$ and so S_O is isolated.

Now we prove h is continuous. Let $\mathbf{x}$ and $\overline{\mathbf{x}}$ be in S and suppose $-T \le t \le T$, where T is to be chosen later. Then, by Eq.(49),

$$\begin{aligned}\|\psi^{\alpha(t,\overline{\mathbf{x}})}(h(\overline{\mathbf{x}})) - \phi^t(\mathbf{x})\| &= \|h(\phi^t(\overline{\mathbf{x}})) - \phi^t(\mathbf{x})\| \\ &\le \|h(\phi^t(\overline{\mathbf{x}})) - \phi^t(\overline{\mathbf{x}})\| + \|\phi^t(\overline{\mathbf{x}}) - \phi^t(\mathbf{x})\| \\ &\le M\sigma + \|\phi^t(\overline{\mathbf{x}}) - \phi^t(\mathbf{x})\|\end{aligned}$$

and

$$\|\psi^{\alpha(t,\mathbf{x})}(h(\mathbf{x})) - \phi^t(\mathbf{x})\| = \|h(\phi^t(\mathbf{x})) - \phi^t(\mathbf{x})\| \le M\sigma. \tag{58}$$

Next, with a view to applying Proposition 7.10, we apply Lemma 6.2 to Eq.(44) with $\psi^{\alpha(t,\overline{\mathbf{x}})}(h(\overline{\mathbf{x}}))$ as $\mathbf{x}$, $\phi^t(\mathbf{x})$ as $\mathbf{y}$, $F(\phi^t(\mathbf{x}))$ as $\mathbf{v}$ and $\alpha = (\Delta - \sigma)/2(M_0 + \sigma)(M_1 + \sigma)$. Then if σ and $\|\overline{\mathbf{x}} - \mathbf{x}\|$ are so small that

$$2(M_1 + \sigma)\left[M\sigma + \sup_{-T\le u\le T}\|\phi^u(\overline{\mathbf{x}}) - \phi^u(\mathbf{x})\|\right] \le \Delta - \sigma$$

and

$$8(M_0 + \sigma)(M_1 + \sigma)\left[M\sigma + \sup_{-T\le u\le T}\|\phi^u(\overline{\mathbf{x}}) - \phi^u(\mathbf{x})\|\right] < (\Delta - \sigma)^2,$$

there exists $\tau = \tau(t)$, unique in $|\tau| \le \alpha$, satisfying

$$\begin{aligned}&|\tau(t)| \\ &\quad\le 4(\Delta - \sigma)^{-1}\|F(\phi^t(\mathbf{x}))\|^{-1}\left|\langle\, \psi^{\alpha(t,\overline{\mathbf{x}})}(h(\overline{\mathbf{x}})) - \phi^t(\mathbf{x}),\, F(\phi^t(\mathbf{x}))\,\rangle\right|\end{aligned} \tag{59}$$

such that

$$\langle\, \psi^{\alpha(t,\overline{\mathbf{x}})+\tau(t)}(h(\overline{\mathbf{x}})) - \phi^t(\mathbf{x}),\, F(\phi^t(\mathbf{x}))\,\rangle = 0.$$

Note, using Eq.(59), that

$$\begin{aligned}&\|\psi^{\alpha(t,\overline{\mathbf{x}})+\tau(t)}(h(\overline{\mathbf{x}})) - \phi^t(\mathbf{x})\| \\ &\qquad\le \|\psi^{\alpha(t,\overline{\mathbf{x}})+\tau(t)}(h(\overline{\mathbf{x}})) - \psi^{\alpha(t,\overline{\mathbf{x}})}(h(\overline{\mathbf{x}}))\| \\ &\qquad\qquad + \|\psi^{\alpha(t,\overline{\mathbf{x}})}(h(\overline{\mathbf{x}})) - \phi^t(\mathbf{x})\| \\ &\qquad\le (M_0 + \sigma)|\tau(t)| + \|\psi^{\alpha(t,\overline{\mathbf{x}})}(h(\overline{\mathbf{x}})) - \phi^t(\mathbf{x})\| \\ &\qquad\le [1 + 4(M_0 + \sigma)(\Delta - \sigma)^{-1}](M\sigma + \|\phi^t(\overline{\mathbf{x}}) - \phi^t(\mathbf{x})\|).\end{aligned} \tag{60}$$

To see that $\tau(t)$ is continuous, we consider the function

$$g(t,\tau) = \langle \psi^{\alpha(t,\overline{\mathbf{x}})+\tau}(h(\overline{\mathbf{x}})) - \phi^t(\mathbf{x}), F(\phi^t(\mathbf{x})) \rangle.$$

This is a C^1 function with

$$\frac{\partial g}{\partial \tau}(t,\tau) = \langle G(\psi^{\alpha(t,\overline{\mathbf{x}})+\tau}(h(\overline{\mathbf{x}}))), F(\phi^t(\mathbf{x})) \rangle.$$

Then, when $\tau = \tau(t)$, we see using Eq.(60) that

$$\begin{aligned}
&\frac{\partial g}{\partial \tau}(t,\tau) \\
&\quad \geq \|F(\phi^t(\mathbf{x}))\| \left[\|F(\phi^t(\mathbf{x})\| - \|G(\psi^{\alpha(t,\overline{\mathbf{x}})+\tau}(h(\overline{\mathbf{x}}))) - F(\phi^t(\mathbf{x}))\|\right] \\
&\quad \geq \|F(\phi^t(\mathbf{x}))\| \left[\Delta - \sigma - M_1\|\psi^{\alpha(t,\overline{\mathbf{x}})+\tau}(h(\overline{\mathbf{x}})) - \phi^t(\mathbf{x})\|\right] \\
&\quad \geq \|F(\phi^t(\mathbf{x}))\| \left[\Delta - \sigma - M_1(1 + 4(M_0+\sigma)(\Delta-\sigma)^{-1}) \times \right. \\
&\qquad\qquad\qquad \left. (M\sigma + \|\phi^t(\overline{\mathbf{x}}) - \phi^t(\mathbf{x})\|)\right] \\
&\quad > 0,
\end{aligned}$$

by the conditions imposed on σ and $\|\overline{\mathbf{x}} - \mathbf{x}\|$ above. Then it follows from the implicit function theorem and the uniqueness of $\tau(t)$ that when $-T \leq t \leq T$, $\tau(t)$ is C^1 and hence certainly continuous.

Now, taking $\beta_1 = \alpha_1/2$ and $\beta_2 = \alpha_2/2$, we apply Proposition 7.10 with $\mathbf{y}(t) = \psi^t(h(\overline{\mathbf{x}}))$, $\mathbf{z}(t) = \psi^t(h(\mathbf{x}))$, $\alpha(t) = \alpha(t,\overline{\mathbf{x}}) + \tau(t)$, $\beta(t) = \alpha(t,\mathbf{x})$, $\mathbf{x}(t) = \phi^t(\mathbf{x})$, $a = -T$ and $b = T$. We assume σ is so small and $\overline{\mathbf{x}}$ so close to $\mathbf{x}$ that

$$\left[1 + 4(M_0+\sigma)(\Delta-\sigma)^{-1}\right]\left(M\sigma + \sup_{-T\leq t\leq T}\|\phi^t(\overline{\mathbf{x}}) - \phi^t(\mathbf{x})\|\right) \leq \varepsilon_0,$$

where ε_0 satisfies the conditions satisfied by ε in Proposition 7.10. Then, referring to Eqs.(58) and (60), we deduce when σ and $\|\overline{\mathbf{x}} - \mathbf{x}\|$ are sufficiently small (depending on F, S and T) that for $-T \leq t \leq T$

$$\|\psi^{\alpha(t,\overline{\mathbf{x}})+\tau(t)}(h(\overline{\mathbf{x}})) - \psi^{\alpha(t,\mathbf{x})}(h(\mathbf{x}))\| \leq \left[L_1 e^{-\alpha_1(t+T)/2} + L_2 e^{-\alpha_2(T-t)/2}\right]\varepsilon_0,$$

where L_1 and L_2 depend only on F and S. Taking $t = 0$, we see that

$$\begin{aligned}
\|h(\overline{\mathbf{x}}) - h(\mathbf{x})\| &\leq \|\psi^{\tau(0)}(h(\overline{\mathbf{x}})) - h(\overline{\mathbf{x}})\| + \|\psi^{\tau(0)}(h(\overline{\mathbf{x}})) - h(\mathbf{x})\| \\
&\leq (M_0+\sigma)|\tau(0)| + (L_1 e^{-\alpha_1 T/2} + L_2 e^{-\alpha_2 T/2})\varepsilon_0.
\end{aligned}$$

Now given $\varepsilon > 0$, we choose $T > 0$ so that

$$\left(L_1 e^{-\alpha_1 T/2} + L_2 e^{-\alpha_2 T/2}\right)\varepsilon_0 < \varepsilon/2.$$

Then, using Eq.(59),

$$\begin{aligned}\|h(\overline{\mathbf{x}}) - h(\mathbf{x})\| \\ &\leq 4(M_0+\sigma)\Delta^{-1}(\Delta-\sigma)^{-1}|\langle\, h(\overline{\mathbf{x}}) - \mathbf{x}, F(\mathbf{x})\,\rangle| \;+\; \varepsilon/2.\end{aligned}$$

Continuing, since $\langle\, h(\overline{\mathbf{x}}) - \overline{\mathbf{x}}, F(\overline{\mathbf{x}})\,\rangle = 0$,

$$\begin{aligned}\|h(\overline{\mathbf{x}}) - h(\mathbf{x})\| \\ &\leq 4(M_0+\sigma)\Delta^{-1}(\Delta-\sigma)^{-1}|\langle\, h(\overline{\mathbf{x}}) - \overline{\mathbf{x}}, F(\mathbf{x}) - F(\overline{\mathbf{x}})\,\rangle \\ &\qquad + \langle\, \overline{\mathbf{x}} - \mathbf{x}, F(\mathbf{x})\,\rangle| \;+\; \varepsilon/2 \\ &\leq 4(M_0+\sigma)\Delta^{-1}(\Delta-\sigma)^{-1}\left[M_1 M\sigma + M_0\right]\|\overline{\mathbf{x}} - \mathbf{x}\| \;+\; \varepsilon/2 \\ &< \varepsilon\end{aligned}$$

if $\|\overline{\mathbf{x}} - \mathbf{x}\|$ is small enough. Thus the continuity of h is established.

Now we prove h^{-1} is continuous. (Note this follows from compactness but we give a proof here that allows the possibility of extensions to the non-compact case.) Note first that if we write $\alpha_{\mathbf{x}}(t) = \alpha(t, \mathbf{x})$ and define for $\mathbf{z} \in S_O$

$$\gamma(t, \mathbf{z}) = \alpha^{-1}_{h^{-1}(\mathbf{z})}(t),$$

then it follows from Eq.(49) that

$$h^{-1}(\psi^t(\mathbf{z})) = \phi^{\gamma(t,\mathbf{z})}(h^{-1}(\mathbf{z})) \tag{61}$$

holds for all t and $\mathbf{z} \in S_O$. Next let $\mathbf{z}, \overline{\mathbf{z}}$ be in S_O and suppose $-T \leq t \leq T$, where T will be defined later. Set

$$\overline{\alpha}(t) = (\gamma_{\overline{\mathbf{z}}} \circ \gamma_{\mathbf{z}}^{-1})(t) = \gamma(s, \overline{\mathbf{z}}),$$

where

$$\gamma_{\mathbf{z}}(t) = \gamma(t, \mathbf{z}) \quad \text{and} \quad s = \gamma_{\mathbf{z}}^{-1}(t) = \alpha_{h^{-1}(\mathbf{z})}(t).$$

Note that, in view of Eq.(53), we have $-\overline{T} \leq s \leq \overline{T}$, where

$$\overline{T} = [1 + M(\sigma + d)]T.$$

Then

$$\|\phi^{\overline{\alpha}(t)}(h^{-1}(\overline{\mathbf{z}})) - \phi^t(h^{-1}(\mathbf{z}))\|$$

$$= \|\phi^{\gamma(s,\overline{\mathbf{z}})}(h^{-1}(\overline{\mathbf{z}})) - \phi^{\gamma(s,\mathbf{z})}(h^{-1}(\mathbf{z}))\|$$

$$= \|h^{-1}(\psi^s(\overline{\mathbf{z}})) - h^{-1}(\psi^s(\mathbf{z}))\| \text{ using Eq.(61)}$$

$$\leq \|h^{-1}(\psi^s(\overline{\mathbf{z}})) - \psi^s(\overline{\mathbf{z}})\| + \|\psi^s(\overline{\mathbf{z}}) - \psi^s(\mathbf{z})\| + \|h^{-1}(\psi^s(\mathbf{z})) - \psi^s(\mathbf{z})\|$$

$$\leq 2M\sigma + \|\psi^s(\overline{\mathbf{z}}) - \psi^s(\mathbf{z})\|.$$

Now, with a view to applying Proposition 7.10, we apply Lemma 6.2 to Eq.(1) with $\phi^{\overline{\alpha}(t)}(h^{-1}(\overline{\mathbf{z}}))$ as $\mathbf{x}$, $\phi^t(h^{-1}(\mathbf{z}))$ as $\mathbf{y}$, $F(\phi^t(h^{-1}(\mathbf{z})))$ as $\mathbf{v}$ and $\alpha = \Delta/2M_0M_1$. Then if σ and $\|\overline{\mathbf{z}} - \mathbf{z}\|$ are so small that

$$2M_1 \left[2M\sigma + \sup_{-\overline{T}\leq u\leq \overline{T}} \|\psi^u(\overline{\mathbf{z}}) - \psi^u(\mathbf{z})\|\right] \leq \Delta$$

and

$$8M_0M_1 \left[2M\sigma + \sup_{-\overline{T}\leq u\leq \overline{T}} \|\psi^u(\overline{\mathbf{z}}) - \psi^u(\mathbf{z})\|\right] < \Delta^2,$$

there exists $\tau = \tau(t)$, unique in $|\tau| \leq \alpha$, satisfying

$$|\tau(t)| \leq 4\Delta^{-1}\|F(\phi^t(h^{-1}(\mathbf{z})))\|^{-1} \times \\ \left|\langle \phi^{\overline{\alpha}(t)}(h^{-1}(\overline{\mathbf{z}})) - \phi^t(h^{-1}(\mathbf{z})), F(\phi^t(h^{-1}(\mathbf{z})))\rangle\right| \quad (62)$$

such that

$$\langle \phi^{\overline{\alpha}(t)+\tau(t)}(h^{-1}(\overline{\mathbf{z}})) - \phi^t(h^{-1}(\mathbf{z})), F(\phi^t(h^{-1}(\mathbf{z})))\rangle = 0.$$

Also we estimate

$$\|\phi^{\overline{\alpha}(t)+\tau(t)}(h^{-1}(\overline{\mathbf{z}})) - \phi^t(h^{-1}(\mathbf{z}))\|$$

$$\leq M_0|\tau(t)| + \|\phi^{\overline{\alpha}(t)}(h^{-1}(\overline{\mathbf{z}})) - \phi^t(h^{-1}(\mathbf{z}))\|$$

$$\leq (1 + 4M_0\Delta^{-1})\|\phi^{\overline{\alpha}(t)}(h^{-1}(\overline{\mathbf{z}})) - \phi^t(h^{-1}(\mathbf{z}))\| \quad (63)$$

$$\leq \left(1 + 4M_0\Delta^{-1}\right)\left[2M\sigma + \|\psi^s(\overline{\mathbf{z}}) - \psi^s(\mathbf{z})\|\right].$$

To show that $\tau(t)$ is continuous, consider the function

$$g(t,\tau) = \langle \phi^{\overline{\alpha}(t)+\tau}(h^{-1}(\overline{\mathbf{z}})) - \phi^t(h^{-1}(\mathbf{z})), F(\phi^t(h^{-1}(\mathbf{z})))\rangle.$$

This is a C^1 function and when $\tau = \tau(t)$

$$\frac{\partial g}{\partial \tau}(t,\tau)$$
$$= \langle F(\phi^{\overline{\alpha}(t)+\tau}(h^{-1}(\overline{\mathbf{z}}))), F(\phi^t(h^{-1}(\mathbf{z})))\rangle$$
$$\geq \|F(\phi^t(h^{-1}(\mathbf{z})))\| \left[\|F(\phi^t(h^{-1}(\mathbf{z})))\| - M_1\|\phi^{\overline{\alpha}(t)+\tau}(h^{-1}(\overline{\mathbf{z}})) - \phi^t(h^{-1}(\mathbf{z}))\|\right]$$
$$> 0$$

by the conditions imposed on σ and $\|\overline{\mathbf{z}} - \mathbf{z}\|$. Then it follows from the implicit function theorem and the uniqueness of $\tau(t)$ that $\tau(t)$ is C^1 and hence certainly continuous.

Now, taking $\beta_1 = \alpha_1/2$ and $\beta_2 = \alpha_2/2$, we apply Proposition 7.10 with $G = F$, $\mathbf{y}(t) = \phi^t(h^{-1}(\overline{\mathbf{z}}))$, $\mathbf{z}(t) = \mathbf{x}(t) = \phi^t(h^{-1}(\mathbf{z}))$, $\alpha(t) = \overline{\alpha}(t) + \tau(t)$, $\beta(t) = t$, $a = -T$ and $b = T$. We assume σ is so small and $\overline{\mathbf{z}}$ so close to $\mathbf{z}$ that

$$(1 + 4M_0\Delta^{-1})\left[2M\sigma + \sup_{-\overline{T}\leq u\leq \overline{T}} \|\psi^u(\overline{\mathbf{z}}) - \psi^u(\mathbf{z})\|\right] \leq \varepsilon_0,$$

where ε_0 satisfies the conditions satisfied by ε in Proposition 7.10. Then, using Eq.(63), we deduce that for $-T \leq t \leq T$

$$\|\phi^{\overline{\alpha}(t)+\tau(t)}(h^{-1}(\overline{\mathbf{z}})) - \phi^t(h^{-1}(\mathbf{z}))\| \leq \left[L_1e^{-\alpha_1(t+T)/2} + L_2e^{-\alpha_2(T-t)/2}\right]\varepsilon_0,$$

where L_1 and L_2 depend only on F and S. Then we see that

$$\begin{aligned}\|h^{-1}(\overline{\mathbf{z}}) - h^{-1}(\mathbf{z})\| &\leq \|\phi^{\tau(0)}(h^{-1}(\overline{\mathbf{z}})) - h^{-1}(\overline{\mathbf{z}})\| + \|\phi^{\tau(0)}(h^{-1}(\overline{\mathbf{z}})) - h^{-1}(\mathbf{z})\| \\ &\leq M_0|\tau(0)| + (L_1e^{-\alpha_1 T/2} + L_2e^{-\alpha_2 T/2})\varepsilon_0.\end{aligned}$$

Next given $\varepsilon > 0$, we choose $T > 0$ so that

$$\left(L_1e^{-\alpha_1 T/2} + L_2e^{-\alpha_2 T/2}\right)\varepsilon_0 < \varepsilon/4.$$

Then, using Eq.(62) with $t = 0$,

$$\begin{aligned}
\|h^{-1}(\overline{\mathbf{z}}) - h^{-1}(\mathbf{z})\| &< 4M_0\Delta^{-2}\left|\langle h^{-1}(\overline{\mathbf{z}}) - h^{-1}(\mathbf{z}), F(h^{-1}(\mathbf{z}))\rangle\right| + \frac{\varepsilon}{4} \\
&= 4M_0\Delta^{-2}|\langle \overline{\mathbf{z}} - \mathbf{z}, F(h^{-1}(\mathbf{z}))\rangle \\
&\qquad + \langle h^{-1}(\overline{\mathbf{z}}) - \overline{\mathbf{z}}, F(h^{-1}(\mathbf{z})) - F(h^{-1}(\overline{\mathbf{z}}))\rangle| + \frac{\varepsilon}{4} \\
&\qquad \text{since } \langle h(\mathbf{x}) - \mathbf{x}, F(\mathbf{x})\rangle = 0 \\
&\leq 4M_0\Delta^{-2}[M_0\|\overline{\mathbf{z}} - \mathbf{z}\| + M\sigma M_1\|h^{-1}(\mathbf{z}) - h^{-1}(\overline{\mathbf{z}})\|] + \frac{\varepsilon}{4}.
\end{aligned}$$

Now we assume σ is so small that

$$4MM_0M_1\Delta^{-2}\sigma \leq \frac{1}{2}.$$

Then

$$\|h^{-1}(\mathbf{z}) - h^{-1}(\overline{\mathbf{z}})\| \leq 8M_0^2\Delta^{-2}\|\overline{\mathbf{z}} - \mathbf{z}\| + \frac{\varepsilon}{2} < \varepsilon$$

if $\|\overline{\mathbf{z}} - \mathbf{z}\|$ is small enough. This establishes the continuity of h^{-1}.

Finally to prove that $\alpha(t, \mathbf{x})$ and $\frac{\partial\alpha}{\partial t}(t, \mathbf{x})$ are continuous, observe that it follows by differentiating Eq.(47) that $\alpha(t) = \alpha(t, \mathbf{x})$ is a solution of the equation

$$\alpha' = \frac{\langle F(\phi^t(\mathbf{x})), F(\phi^t(\mathbf{x}))\rangle - \langle \psi^\alpha(h(\mathbf{x})) - \phi^t(\mathbf{x}), DF(\phi^t(\mathbf{x}))F(\phi^t(\mathbf{x}))\rangle}{\langle G(\psi^\alpha(h(\mathbf{x}))), F(\phi^t(\mathbf{x})\rangle}$$

with initial condition $\alpha(0) = 0$. Since the right side and its derivative in α are continuous functions of $(t, \alpha, \mathbf{x})$, the continuity of $\alpha(t, \mathbf{x})$ and $\frac{\partial\alpha}{\partial t}(t, \mathbf{x})$ follows from standard theorems about continuous dependence on parameters. Thus the proof of the theorem is complete.

Remark 9.8. We can prove a topological conjugacy result even when S is not isolated. We just follow through the same proof omitting the proof of surjectivity. So h maps S into S_O and $h(S)$ is a hyperbolic set, being a subset of S_O.

9.4 ASYMPTOTIC PHASE FOR HYPERBOLIC SETS

Let $F : U \to I\!R^n$ be a C^1 vector field and let S be a compact hyperbolic set for Eq.(1). For the set S we can define its *stable manifold*

$$W^s(S) = \{\mathbf{x} \in U : \text{dist}(\phi^t(\mathbf{x}), S) \to 0 \text{ as } t \to \infty\}$$

and its *unstable manifold*

$$W^u(S) = \{\mathbf{x} \in U : \text{dist}(\phi^t(\mathbf{x}), S) \to 0 \text{ as } t \to -\infty\}.$$

We can also define the stable and unstable manifolds for points in S as follows. If $\mathbf{x} \in S$ its stable manifold is

$$W^s(\mathbf{x}) = \{\mathbf{y} \in U : \|\phi^t(\mathbf{y}) - \phi^t(\mathbf{x})\| \to 0 \text{ as } t \to \infty\}$$

and its unstable manifold is

$$W^s(\mathbf{x}) = \{\mathbf{y} \in U : \|\phi^t(\mathbf{y}) - \phi^t(\mathbf{x})\| \to 0 \text{ as } t \to -\infty\}.$$

It is obvious that

$$\cup_{\mathbf{x}\in S} W^s(\mathbf{x}) \subset W^s(S), \quad \cup_{\mathbf{x}\in S} W^u(\mathbf{x}) \subset W^u(S).$$

When S is an isolated invariant set, we still use shadowing to prove that the reverse inclusions hold. This means that an orbit which is asymptotic to the set S must be asymptotic to some orbit in S.

Now we state the main theorem of this section (conf. Hirsch, Palis, Pugh and Shub (1970)).

Theorem 9.9. *Let S be an isolated compact hyperbolic set for Eq.(1) with $F : U \to I\!R^n$ a C^1 vector field with U open and convex. Then*

$$\cup_{\mathbf{x}\in S} W^s(\mathbf{x}) = W^s(S), \quad \cup_{\mathbf{x}\in S} W^u(\mathbf{x}) = W^u(S).$$

Proof. We just prove the first since the proof of the second is analogous. All we need show is that

$$W^s(S) \subset \bigcup_{\mathbf{x}\in S} W^s(\mathbf{x})$$

since the other inclusion is obvious. So let $\mathbf{z} \in W^s(S)$. Then $\text{dist}(\phi^t(\mathbf{z}), S) \to 0$ as $t \to \infty$. Without loss of generality, we may suppose

$$\text{dist}(\phi^t(\mathbf{z}), S) \le d \text{ for } t \ge 0,$$

where d is a positive number to be chosen later.

For $k \geq 0$, define

$$\mathbf{z}_k = \phi^k(\mathbf{z})$$

and choose $\mathbf{y}_k \in S$ such that

$$\|\mathbf{y}_k - \mathbf{z}_k\| \leq d.$$

Then, using Gronwall's lemma,

$$\begin{aligned}\|\mathbf{y}_{k+1} - \phi^1(\mathbf{y}_k)\| &\leq \|\mathbf{y}_{k+1} - \mathbf{z}_{k+1}\| + \|\phi^1(\mathbf{z}_k) - \phi^1(\mathbf{y}_k)\| \\ &\leq d + e^{M_1}\|\mathbf{z}_k - \mathbf{y}_k\| \\ &\leq (1 + e^{M_1})d,\end{aligned}$$

where

$$M_1 = \sup_{\mathbf{x} \in U} \|DF(\mathbf{x})\|.$$

So if we define

$$\mathbf{y}_k = \phi^k(\mathbf{y}_0)$$

for $k < 0$, we see that $\{\mathbf{y}_k\}_{k=-\infty}^{\infty}$ is a discrete δ pseudo orbit in S of Eq.(1) with $\delta = (1 + e^{M_1})d$ and with associated times $h_k = 1$. We apply Theorem 9.3 with $h_{max} = h_{min} = 1$. Then if

$$(1 + e^{M_1})d \leq \delta_0,$$

there are sequences $\{\mathbf{x}_k\}_{k=-\infty}^{\infty}$ and $\{t_k\}_{k=-\infty}^{\infty}$ such that for all k

$$\mathbf{x}_{k+1} = \phi^{t_k}(\mathbf{x}_k)$$

and

$$\|\mathbf{x}_k - \mathbf{y}_k\| \leq M(1 + e^{M_1})d, \quad |t_k - 1| \leq M(1 + e^{M_1})d. \tag{64}$$

Since S is isolated, if d is sufficiently small, it follows that $\mathbf{x}_k \in S$ for all k.

So we have sequences $\{\mathbf{x}_k\}_{k=0}^{\infty}$, $\{t_k\}_{k=0}^{\infty}$, $\{\mathbf{z}_k\}_{k=0}^{\infty}$ such that for $k \geq 0$

$$\mathbf{x}_{k+1} = \phi^{t_k}(\mathbf{x}_k), \quad \mathbf{z}_{k+1} = \phi^1(\mathbf{z}_k),$$

and

$$\|\mathbf{x}_k - \mathbf{z}_k\| \leq [1 + M(1 + e^{M_1})]d, \quad |t_k - 1| \leq M(1 + e^{M_1})d.$$

Now we apply Lemma 6.2 with $\mathbf{z}_k$ as $\mathbf{x}$, $\mathbf{x}_k$ as $\mathbf{y}$, $F(\mathbf{x}_k)$ as $\mathbf{v}$ and $\alpha = \Delta/2M_0M_1$, where

$$M_0 = \sup_{\mathbf{x} \in U} \|F(\mathbf{x})\|, \; \Delta = \inf_{\mathbf{x} \in S} \|F(\mathbf{x})\|.$$

If

$$2M_1 \left[1 + M(1 + e^{M_1})\right] d \leq \Delta$$

and

$$8M_0M_1\left[1+M(1+e^{M_1})\right]d \le \Delta^2,$$

there exists s_k satisfying

$$|s_k| \le 4\Delta^{-1}\|\mathbf{x}_k - \mathbf{z}_k\| \le 4\Delta^{-1}\left[1+M(1+e^{M_1})\right]d$$

such that

$$\langle\, \phi^{s_k}(\mathbf{z}_k) - \mathbf{x}_k, F(\mathbf{x}_k)\,\rangle = 0.$$

Then if we define

$$\mathbf{w}_k = \phi^{s_k}(\mathbf{z}_k), \quad \tau_k = s_{k+1} + 1 - s_k$$

we see that for $k \ge 0$

$$\mathbf{w}_{k+1} = \phi^{\tau_k}(\mathbf{w}_k),$$

$$|\tau_k - 1| \le 8\Delta^{-1}[1+M(1+e^{M_1})]d, \tag{65}$$

$$\|\mathbf{w}_k - \mathbf{x}_k\| \le (1+4M_0\Delta^{-1})[1+M(1+e^{M_1})]d \tag{66}$$

and

$$\langle\, \mathbf{w}_k - \mathbf{x}_k, F(\mathbf{x}_k)\,\rangle = 0.$$

Next for each $k \ge 0$, we seek a function $\alpha_k : [0, t_k] \to \mathbb{R}$ such that

$$\langle\, \phi^{\alpha_k(t)}(\mathbf{w}_k) - \phi^t(\mathbf{x}_k), F(\phi^t(\mathbf{x}_k))\,\rangle = 0$$

for $0 \le t \le t_k$. Again we apply Lemma 6.2 with $\phi^t(\mathbf{w}_k)$ as $\mathbf{x}$, $\phi^t(\mathbf{x}_k)$ as $\mathbf{y}$, $F(\phi^t(\mathbf{x}_k))$ as $\mathbf{v}$ and $\alpha = \Delta/2M_0M_1$. Note that by Gronwall's lemma and using Eqs.(64) and (66), if $0 \le t \le t_k$,

$$\|\phi^t(\mathbf{w}_k) - \phi^t(\mathbf{x}_k)\| \le e^{M_1 t}\|\mathbf{w}_k - \mathbf{x}_k\| \le d_1,$$

where

$$d_1 = e^{M_1[1+M(1+e^{M_1})d]}\left(1+4M_0\Delta^{-1}\right)\left[1+M(1+e^{M_1})\right]d.$$

Hence if

$$d_1 < \min\{\Delta/2M_1, \Delta^2/8M_0M_1\},$$

it follows from Lemma 6.2 that there exists $\tau = \tau_k(t)$ satisfying

$$|\tau_k(t)| \le 4\Delta^{-1}\|\phi^t(\mathbf{w}_k) - \phi^t(\mathbf{x}_k)\| \ (< \alpha)$$

and

$$\langle\, \phi^{t+\tau}(\mathbf{w}_k) - \phi^t(\mathbf{x}_k), F(\phi^t(\mathbf{x}_k))\,\rangle = 0. \tag{67}$$

Moreover, $\tau_k(t)$ is the unique number τ satisfying Eq.(67) in $|\tau| \le \alpha$. So, by uniqueness, it follows that

$$\tau_k(0) = 0, \quad \tau_k(t_k) = \tau_k - t_k \tag{68}$$

provided, using Eqs.(64) and (65),

$$[M(1 + e^{M_1}) + 8\Delta^{-1}(1 + M(1 + e^{M_1}))]d \leq \Delta/2M_0M_1.$$

Next we prove that $\tau_k(t)$ is continuous on $[0, t_k]$. If we define

$$g(t, \tau) = \langle \phi^{t+\tau}(\mathbf{w}_k) - \phi^t(\mathbf{x}_k), F(\phi^t(\mathbf{x}_k)) \rangle,$$

we see that for $\tau = \tau_k(t)$

$$\begin{aligned}
\frac{\partial g}{\partial \tau}(t, \tau) &= \langle F(\phi^{t+\tau}(\mathbf{w}_k)), F(\phi^t(\mathbf{x}_k)) \rangle \\
&\geq \|F(\phi^t(\mathbf{x}_k))\| \, [\|F(\phi^t(\mathbf{x}_k))\| - M_1\|\phi^{t+\tau}(\mathbf{w}_k) - \phi^t(\mathbf{x}_k)\|] \\
&\geq \|F(\phi^t(\mathbf{x}))\| \, [\Delta - M_1(M_0|\tau| + \|\phi^t(\mathbf{w}_k) - \phi^t(\mathbf{x}_k)\|)] \\
&\geq \|F(\phi^t(\mathbf{x}))\| \, [\Delta - M_1(1 + 4M_0\Delta^{-1})\|\phi^t(\mathbf{w}_k) - \phi^t(\mathbf{x}_k)\|] \\
&\geq \|F(\phi^t(\mathbf{x}))\|[\Delta - M_1(1 + 4M_0\Delta^{-1})d_1] \\
&> 0.
\end{aligned}$$

Then it follows from the implicit function theorem and the uniqueness of $\tau_k(t)$ that $\tau_k(t)$ is continuous in $[0, t_k]$.

If we define

$$\alpha_k(t) = t + \tau_k(t) \quad \text{for} \quad 0 \leq t \leq t_k,$$

we see that $\alpha_k(t)$ is continuous and that for $0 \leq t \leq t_k$

$$\langle \phi^{\alpha_k(t)}(\mathbf{w}_k) - \phi^t(\mathbf{x}_k), F(\phi^t(\mathbf{x}_k)) \rangle = 0$$

and

$$\begin{aligned}
\|\phi^{\alpha_k(t)}(\mathbf{w}_k) - \phi^t(\mathbf{x}_k)\| &\leq \|\phi^{\alpha_k(t)}(\mathbf{w}_k) - \phi^t(\mathbf{w}_k)\| + \|\phi^t(\mathbf{w}_k) - \phi^t(\mathbf{x}_k)\| \\
&\leq M_0|\tau_k(t)| + \|\phi^t(\mathbf{w}_k) - \phi^t(\mathbf{x}_k)\| \\
&\leq (4M_0\Delta^{-1} + 1)\|\phi^t(\mathbf{w}_k) - \phi^t(\mathbf{x}_k)\| \\
&\leq (4M_0\Delta^{-1} + 1)d_1.
\end{aligned}$$

Also it follows from Eq.(68) that

$$\alpha_k(0) = 0, \quad \alpha_k(t_k) = \tau_k.$$

Now we define $\alpha : [0, \infty) \to I\!R$ by

$$\alpha(t) = \alpha_0(t) \text{ if } 0 \le t \le t_0$$

and

$$\alpha(t) = \tau_0 + \ldots + \tau_{k-1} + \alpha_k\left(t - \sum_{i=0}^{k-1} t_i\right) \text{ if } \sum_{i=0}^{k-1} t_i \le t \le \sum_{i=0}^{k} t_i, \; k \ge 1.$$

This is valid provided $\sum_{i=0}^{k} t_i \to \infty$ as $k \to \infty$, which is ensured (conf. Eq.(64)) by

$$M(1 + e^{M_1})d < 1.$$

It follows that $\alpha(t)$ is a continuous function with $\alpha(0) = 0$ such that for all $t \ge 0$

$$\langle\, \phi^{\alpha(t)}(\mathbf{w}_0) - \phi^t(\mathbf{x}_0), F(\phi^t(\mathbf{x}_0))\,\rangle = 0$$

and

$$\|\phi^{\alpha(t)}(\mathbf{w}_0) - \phi^t(\mathbf{x}_0)\| \le d_2,$$

where

$$d_2 = (4M_0\Delta^{-1} + 1)d_1. \tag{69}$$

Then it follows from Proposition 7.10 that, provided d is sufficiently small, there is a constant L_1 such that

$$\|\phi^{\alpha(t)}(\mathbf{w}_0) - \phi^t(\mathbf{x}_0)\| \le L_1 e^{-\alpha_1 t/2}\|\mathbf{w}_0 - \mathbf{x}_0\| \tag{70}$$

for $t \ge 0$.

Now it also follows from Lemma 7.9 that $\alpha(t)$ is continuously differentiable with

$$\alpha'(t) - 1 =$$

$$\frac{1}{\langle\, F(\phi^{\alpha(t)}(\mathbf{w}_0)), F(\phi^t(\mathbf{x}_0))\,\rangle}[\langle\, \phi^{\alpha(t)}(\mathbf{w}_0) - \phi^t(\mathbf{x}_0), DF(\phi^t(\mathbf{x}_0))F(\phi^t(\mathbf{x}_0))\,\rangle$$

$$- \langle\, F(\phi^{\alpha(t)}(\mathbf{w}_0)) - F(\phi^t(\mathbf{x}_0)), F(\phi^t(\mathbf{x}_0))\,\rangle].$$

Hence

$$\begin{aligned} |\alpha'(t) - 1| &\le \frac{2M_1\|\phi^{\alpha(t)}(\mathbf{w}_0) - \phi^t(\mathbf{x}_0))\|}{\|F(\phi^t(\mathbf{x}_0))\| - \|F(\phi^{\alpha(t)}(\mathbf{w}_0)) - F(\phi^t(\mathbf{x}_0))\|} \\ &\le 2M_1(\Delta - M_1 d_2)^{-1}\|\phi^{\alpha(t)}(\mathbf{w}_0) - \phi^t(\mathbf{x}_0)\|, \end{aligned}$$

where d_2 is the quantity defined in Eq.(69). Thus we see from this and Eq.(70) that

$$T = \int_0^\infty \alpha'(t) - 1\, dt$$

is well defined and

$$\begin{aligned}
|\alpha(t) - t - T| &= \left|\int_t^\infty \alpha'(s) - 1\, ds\right| \\
&\leq 2M_1(\Delta - M_1 d_2)^{-1} L_1 \|\mathbf{w}_0 - \mathbf{x}_0\| \int_t^\infty e^{-\alpha_1 s/2} ds \\
&= 2M_1(\Delta - M_1 d_2)^{-1} L_1 \|\mathbf{w}_0 - \mathbf{x}_0\| 2\alpha_1^{-1} e^{-\alpha_1 t/2}.
\end{aligned}$$

Then for $t \geq 0$,

$$\begin{aligned}
&\|\phi^{t+T}(\mathbf{w}_0) - \phi^t(\mathbf{x}_0)\| \\
&\leq \|\phi^{t+T}(\mathbf{w}_0) - \phi^{\alpha(t)}(\mathbf{w}_0)\| + \|\phi^{\alpha(t)}(\mathbf{w}_0) - \phi^t(\mathbf{x}_0)\| \\
&\leq M_0|\alpha(t) - t - T| + L_1 e^{-\alpha_1 t/2} \|\mathbf{w}_0 - \mathbf{x}_0\| \\
&\leq L_1 \left[1 + 4M_0 M_1 (\Delta - M_1 d_2)^{-1} \alpha_1^{-1}\right] e^{-\alpha_1 t/2} \|\mathbf{w}_0 - \mathbf{x}_0\|.
\end{aligned}$$

From this we see that

$$\begin{aligned}
\phi^t(\mathbf{z}_0) - \phi^{t-T-s_0}(\mathbf{x}_0) &= \phi^{t-s_0}(\mathbf{w}_0) - \phi^{t-T-s_0}(\mathbf{x}_0) \\
&= \phi^{t-s_0-T+T}(\mathbf{w}_0) - \phi^{t-s_0-T}(\mathbf{x}_0) \\
&\to \mathbf{0}
\end{aligned}$$

as $t \to \infty$. Thus

$$\mathbf{z} = \mathbf{z}_0 \in W^s(\phi^{-s_0-T}(\mathbf{x}_0))$$

and the proof of the theorem is complete.

10. SYMBOLIC DYNAMICS NEAR A TRANSVERSAL HOMOCLINIC ORBIT OF A SYSTEM OF ORDINARY DIFFERENTIAL EQUATIONS

We consider the autonomous system of ordinary differential equations

$$\dot{\mathbf{x}} = F(\mathbf{x}), \tag{1}$$

where $F : U \to \mathbb{R}^n$ is a C^1 vectorfield, the set U being open and convex. Denote by ϕ the corresponding flow. Suppose $\mathbf{u}(t)$ is a hyperbolic periodic orbit for Eq.(1) with minimal period $T > 0$ and let $\mathbf{p}_0$ be in the intersection of the stable manifold $W^s(\mathbf{u})$ and the unstable manifold $W^u(\mathbf{u})$. If $\mathbf{p}_0$ satisfies the transversality condition

$$T_{\mathbf{p}_0} W^s(\mathbf{u}) \cap T_{\mathbf{p}_0} W^u(\mathbf{u}) = \operatorname{span}\{F(\mathbf{p}_0)\},$$

then we know from Theorem 8.2 that the set

$$S = \{\mathbf{u}(t) : -\infty < t < \infty\} \cup \{\phi^t(\mathbf{p}_0) : -\infty < t < \infty\}$$

is hyperbolic. Now we state our main theorem, which uses symbolic dynamics to describe the solutions of Eq.(1) which remain in a neighbourhood of the set S.

Theorem 10.1. **(Silnikov)** *Let $\mathbf{u}(t)$ be a hyperbolic periodic orbit of Eq.(1) and let $\mathbf{p}_0$ be in the intersection of the stable manifold $W^s(\mathbf{u})$ and the unstable manifold $W^u(\mathbf{u})$ of $\mathbf{u}(t)$ with $\mathbf{p}_0 \neq \mathbf{u}(t)$ for all t. Suppose, moreover, this intersection is transversal, that is,*

$$T_{\mathbf{p}_0} W^s(\mathbf{u}) \cap T_{\mathbf{p}_0} W^u(\mathbf{u}) = \operatorname{span}\{F(\mathbf{p}_0)\}.$$

Then, for any sufficiently large positive integer L, there is an open neighbourhood O of the set

$$S = \{\mathbf{u}(t) : -\infty < t < \infty\} \cup \{\phi^t(\mathbf{p}_0) : -\infty < t < \infty\}$$

such that the solutions of Eq.(1) which remain in O for all time are in one to one correspondence with the set Y of bi-infinite sequences $\{a_k\}_{k=-\infty}^{\infty}$ with $a_k \in \{0, 1, 2\}$ described by the following properties:

(i) if $a_k = 0 \neq a_{k+1}$, then $a_{k+1} = 1$;

(ii) if $a_k = 1$, then $a_{k+1} = 2$;

(iii) if $a_k = 2$, then $a_{k+\ell} = 0$ for $1 \leq \ell \leq L$.

If the set Σ consists of the intersections the solutions remaining in O for all time make with a certain subset of the cross-section

$$C = \{\mathbf{x} \in \mathbb{R}^n : \langle \mathbf{x} - \mathbf{u}(0), \dot{\mathbf{u}}(0) \rangle = 0\},$$

a return map $G : \Sigma \to \Sigma$ can be defined so that this correspondence is given by a homeomorphism $\alpha : Y \to \Sigma$ satisfying

$$G \circ \alpha = \alpha \circ \sigma,$$

where $\sigma : Y \to Y$ is the shift defined by

$$\sigma(\{a_k\}_{k=-\infty}^{\infty}) = \{a_{k+1}\}_{k=-\infty}^{\infty}$$

and Y is given the product discrete topology.

Note that in the theorem each zero in the sequence $\{a_k\}_{k=-\infty}^{\infty}$ corresponds to a circuit near the periodic orbit, a two to a circuit near the homoclinic orbit and a one to a transition in between. In order to prove this theorem, we need to extend the discrete shadowing theorem proved in Chapter 9.

Theorem 10.2. *Let S be a hyperbolic set for Eq.(1) and let σ, h_{min}, h_{max} be positive numbers satisfying $\sigma \le 1$ and $h_{min} \le h_{max}$. Then there are positive numbers M and δ_0 depending only on F, S, σ, h_{min} and h_{max} such that if $\{\mathbf{y}_k\}_{k=-\infty}^{\infty}$ is a discrete δ pseudo orbit lying in S with $\delta \le \delta_0$ and with associated times $\{h_k\}_{k=-\infty}^{\infty}$ satisfying $h_{min} \le h_k \le h_{max}$, and if $\{\mathbf{v}_k\}_{k=-\infty}^{\infty}$ is a sequence of unit vectors satisfying*

$$\langle \mathbf{v}_k, F(\mathbf{y}_k) \rangle \ge \sigma \|F(\mathbf{y}_k)\|,$$

then there are unique sequences $\{\mathbf{x}_k\}_{k=-\infty}^{\infty}$ and $\{t_k\}_{k=-\infty}^{\infty}$ such that for all k

$$\mathbf{x}_{k+1} = \phi^{t_k}(\mathbf{x}_k),$$

$$\|\mathbf{x}_k - \mathbf{y}_k\| \le M\delta, \ |t_k - h_k| \le M\delta$$

and

$$\langle \mathbf{x}_k - \mathbf{y}_k, \mathbf{v}_k \rangle = 0.$$

Proof. When $\mathbf{v}_k = F(\mathbf{y}_k)/\|F(\mathbf{y}_k)\|$, the theorem follows from Theorem 9.3. Here we need only extend Theorem 9.3 to the slightly different situation here. Set

$$M_0 = \sup_{\mathbf{x} \in U} \|F(\mathbf{x})\|, \ \Delta = \inf_{\mathbf{x} \in S} \|F(\mathbf{x})\|, \ M_1 = \sup_{\mathbf{x} \in U} \|DF(\mathbf{x})\|.$$

Then let $\overline{M}$ and $\overline{\delta}_0$ be the positive numbers in Theorem 9.3 depending only on F, S, h_{min} and h_{max} and set

$$\alpha = \sigma\Delta/2M_0M_1, \ M = \max\{1 + 4M_0/\sigma\Delta, 1 + 8/\sigma\Delta\}\overline{M}.$$

Next let the positive number δ_0 not exceed any of the quantities

$$\overline{\delta}_0, \frac{\sigma\Delta}{2\overline{M}M_1}, \frac{\alpha\Delta\sigma}{4\overline{M}}, \frac{\Delta}{2M_1M}, \frac{\alpha\Delta}{4M}, \frac{\overline{M}\,\overline{\delta}_0}{M(1+4M_0/\Delta)}, \frac{\overline{M}\,\overline{\delta}_0}{M(1+8/\Delta)}$$

and also have the property that when $\text{dist}(\mathbf{x}, S) \leq \overline{M}\delta_0$, $\phi^t(\mathbf{x})$ is defined for $|t| \leq \alpha$.

By Theorem 9.3, if $\{\mathbf{y}_k\}_{k=-\infty}^{\infty}$ is a discrete δ pseudo orbit for Eq.(1) in S with $\delta \leq \delta_0$ and associated times $\{h_k\}_{k=-\infty}^{\infty}$ satisfying $h_{min} \leq h_k \leq h_{max}$, then there are unique sequences $\{\bar{\mathbf{x}}_k\}_{k=-\infty}^{\infty}$ and $\{\bar{t}_k\}_{k=-\infty}^{\infty}$ such that

$$\bar{\mathbf{x}}_{k+1} = \phi^{\bar{t}_k}(\bar{\mathbf{x}}_k),$$

$$\|\bar{\mathbf{x}}_k - \mathbf{y}_k\| \leq \overline{M}\delta, \ |\bar{t}_k - h_k| \leq \overline{M}\delta$$

and

$$\langle \bar{\mathbf{x}}_k - \mathbf{y}_k, F(\mathbf{y}_k) \rangle = 0.$$

We apply Lemma 6.2 to $\bar{\mathbf{x}}_k$, $\mathbf{y}_k$, $\mathbf{v}_k$ and α. Now if $|t| \leq \alpha$,

$$\|F(\bar{\mathbf{x}}_k) - F(\mathbf{y}_k)\| \leq M_1\overline{M}\delta \leq \sigma\Delta/2 \leq \frac{\sigma}{2}\|F(\mathbf{y}_k)\| \leq \frac{1}{2}|\langle F(\mathbf{y}_k), \mathbf{v}_k \rangle|,$$

$$\|DF(\phi^t(\mathbf{x}))F(\phi^t(\mathbf{x}))\| \leq M_0M_1 = \Delta\sigma/2\alpha \leq |\langle F(\mathbf{y}_k), \mathbf{v}_k \rangle|/2\alpha,$$

and

$$|\langle \bar{\mathbf{x}}_k - \mathbf{y}_k, \mathbf{v}_k \rangle| \leq \overline{M}\delta \leq \alpha\sigma\Delta/4 \leq \alpha|\langle F(\mathbf{y}_k), \mathbf{v}_k \rangle|/4.$$

So there is a t in

$$|t| \leq 4|\langle \bar{\mathbf{x}}_k - \mathbf{y}_k, \mathbf{v}_k \rangle|/\langle F(\mathbf{y}_k), \mathbf{v}_k \rangle| \leq 4\overline{M}\delta/\sigma\Delta$$

for which

$$\langle \phi^t(\bar{\mathbf{x}}_k) - \mathbf{y}_k, \mathbf{v}_k \rangle = 0$$

and which is unique in $|t| \leq \alpha$. Denote this value of t by $\tilde{t}_k$. Then we define

$$\mathbf{x}_k = \phi^{\tilde{t}_k}(\bar{\mathbf{x}}_k), \ t_k = \tilde{t}_{k+1} + \bar{t}_k - \tilde{t}_k.$$

We see that

$$\mathbf{x}_{k+1} = \phi^{t_k}(\mathbf{x}_k),$$

that

$$\begin{aligned} \|\mathbf{x}_k - \mathbf{y}_k\| &\leq \|\phi^{\tilde{t}_k}(\bar{\mathbf{x}}_k) - \bar{\mathbf{x}}_k\| + \|\bar{\mathbf{x}}_k - \mathbf{y}_k\| \\ &\leq M_0|\tilde{t}_k| + \overline{M}\delta \\ &\leq M_0 \cdot 4\overline{M}\delta/\sigma\Delta + \overline{M}\delta \\ &\leq M\delta, \end{aligned}$$

that

$$\begin{aligned}|t_k - h_k| &\le |\tilde{t}_{k+1}| + |\bar{t}_k - h_k| + |\tilde{t}_k| \\ &\le 4\overline{M}\delta/\sigma\Delta + \overline{M}\delta + 4\overline{M}\delta/\sigma\Delta \\ &\le M\delta\end{aligned}$$

and that

$$\langle \mathbf{x}_k - \mathbf{y}_k, \mathbf{v}_k \rangle = 0.$$

Now we prove that $\{\mathbf{x}_k\}_{k=-\infty}^{\infty}$ and $\{t_k\}_{k=-\infty}^{\infty}$ are unique. Let $\{\mathbf{z}_k\}_{k=-\infty}^{\infty}$ and $\{s_k\}_{k=-\infty}^{\infty}$ be another pair of sequences satisfying

$$\mathbf{z}_{k+1} = \phi^{s_k}(\mathbf{z}_k),$$

$$\|\mathbf{z}_k - \mathbf{y}_k\| \le M\delta, \; |s_k - h_k| \le M\delta$$

and

$$\langle \mathbf{z}_k - \mathbf{y}_k, \mathbf{v}_k \rangle = 0.$$

We apply Lemma 6.2 to $\mathbf{z}_k$, $\mathbf{y}_k$, $F(\mathbf{y}_k)/\|F(\mathbf{y}_k)\|$ and $\alpha = \sigma\Delta/2M_0M_1$. Now if $|t| \le \alpha$,

$$\|F(\mathbf{z}_k) - F(\mathbf{y}_k)\| \le M_1M\delta \le \frac{1}{2}\Delta \le \frac{1}{2}\langle F(\mathbf{y}_k), \frac{F(\mathbf{y}_k)}{\|F(\mathbf{y}_k)\|} \rangle,$$

$$\|DF(\phi^t(\mathbf{z}_k))F(\phi^t(\mathbf{z}_k))\| \le M_0M_1 \le \Delta/2\alpha \le \langle F(\mathbf{y}_k), \frac{F(\mathbf{y}_k)}{\|F(\mathbf{y}_k)\|} \rangle/2\alpha$$

and

$$\left|\langle \mathbf{z}_k - \mathbf{y}_k, \frac{F(\mathbf{y}_k)}{\|F(\mathbf{y}_k)\|} \rangle\right| \le M\delta \le \alpha\Delta/4 \le \alpha\langle F(\mathbf{y}_k), \frac{F(\mathbf{y}_k)}{\|F(\mathbf{y}_k)\|} \rangle/4.$$

So the equation

$$\langle \phi^t(\mathbf{z}_k) - \mathbf{y}_k, F(\mathbf{y}_k)\rangle = 0$$

holds for some t in

$$|t| \le 4|\langle \mathbf{z}_k - \mathbf{y}_k, F(\mathbf{y}_k) \rangle|/\|F(\mathbf{y}_k)\|^2 \le 4M\delta/\Delta.$$

Denote this value of t by $\tilde{s}_k$. Then we define

$$\bar{\mathbf{z}}_k = \phi^{\tilde{s}_k}(\mathbf{z}_k), \; \bar{s}_k = \tilde{s}_{k+1} + s_k - \tilde{s}_k.$$

We see that

$$\bar{\mathbf{z}}_{k+1} = \phi^{\bar{s}_k}(\bar{\mathbf{z}}_k),$$

that

$$\begin{aligned}\|\bar{\mathbf{z}}_k - \mathbf{y}_k\| &\le \|\phi^{\tilde{s}_k}(\mathbf{z}_k) - \mathbf{z}_k\| + \|\mathbf{z}_k - \mathbf{y}_k\| \\ &\le M_0|\tilde{s}_k| + M\delta \\ &\le M_0 \cdot 4M\delta/\Delta + M\delta \\ &\le \overline{M}\,\overline{\delta}_0,\end{aligned}$$

that

$$\begin{aligned}|\bar{s}_k - h_k| &\le |\tilde{s}_{k+1}| + |s_k - h_k| + |\tilde{s}_k| \\ &\le 4M\delta/\Delta + M\delta + 4M\delta/\Delta \\ &\le \overline{M}\,\overline{\delta}_0\end{aligned}$$

and that

$$\langle \bar{\mathbf{z}}_k - \mathbf{y}_k, F(\mathbf{y}_k) \rangle = 0.$$

By the uniqueness in Theorem 9.3,

$$\bar{\mathbf{z}}_k = \bar{\mathbf{x}}_k \text{ and } \bar{s}_k = \bar{t}_k$$

for all k. This means that

$$\langle \phi^{-\tilde{s}_k}(\bar{\mathbf{x}}_k) - \mathbf{y}_k, \mathbf{v}_k \rangle = \langle \phi^{-\tilde{s}_k}(\bar{\mathbf{z}}_k) - \mathbf{y}_k, \mathbf{v}_k \rangle = \langle \mathbf{z}_k - \mathbf{y}_k, \mathbf{v}_k \rangle = 0,$$

where

$$|-\tilde{s}_k| \le 4M\delta/\Delta \le \alpha.$$

By the uniqueness of the solution t of the equation $\langle \phi^t(\bar{\mathbf{x}}_k) - \mathbf{y}_k, \mathbf{v}_k \rangle = 0$ in $|t| \le \alpha$, it follows that $-\tilde{s}_k = \tilde{t}_k$. Hence

$$\mathbf{x}_k = \phi^{\tilde{t}_k}(\bar{\mathbf{x}}_k) = \phi^{-\tilde{s}_k}(\bar{\mathbf{z}}_k) = \mathbf{z}_k \quad \text{and} \quad t_k = \tilde{t}_{k+1} + \bar{t}_k - \tilde{t}_k = -\tilde{s}_{k+1} + \bar{s}_k + \tilde{s}_k = s_k.$$

Thus the uniqueness is proved and the proof of the theorem is complete.

Remark 10.3. In the application of Theorem 10.2 in the proof of Theorem 10.1, we have to consider δ_0 pseudo orbits $\{\mathbf{y}_k\}_{k=-\infty}^{\infty}$ in S with associated times $\{h_k\}_{k=-\infty}^{\infty}$ where $h_{min} \le h_k$ but for some of the k

$$h_{max} < h_k \le T_1,$$

with T_1 a fixed constant. We want to add a condition which will ensure there is *at most one* shadowing orbit. For convenience, we assume $h_{max} \ge 2h_{min}$, which is satisfied in our application.

We choose

$$\varepsilon_2 = \min\{\Delta/2M_1, \Delta^2/8M_0M_1, \Delta M\delta_0/8, \Delta M\delta_0/(4M_0 + \Delta)\},$$

with the notations as in Theorem 10.2 and its proof. We show *there is at most one pair of sequences* $\{\mathbf{x}_k\}_{k=-\infty}^{\infty}$ *and* $\{t_k\}_{k=-\infty}^{\infty}$ *such that for all k*

$$\mathbf{x}_{k+1} = \phi^{t_k}(\mathbf{x}_k),$$

$$\|\mathbf{x}_k - \mathbf{y}_k\| \leq M\delta_0, \text{ when } \quad h_{min} \leq h_k \leq h_{max},$$

$$\|\phi^t(\mathbf{x}_k) - \phi^t(\mathbf{y}_k)\| \leq \varepsilon_2 \quad \text{for} \quad 0 \leq t \leq h_k, \text{ when } \quad h_{max} < h_k \leq T_1,$$

$$\langle \mathbf{x}_k - \mathbf{y}_k, \mathbf{v}_k \rangle = 0$$

and

$$|t_k - h_k| \leq \begin{cases} M\delta_0, & \text{if } h_{min} \leq h_k \leq h_{max}, \\ M\delta_0/2, & \text{if } h_{max} < h_k \leq T_1. \end{cases}$$

To prove this, let $\{\mathbf{x}_k\}_{k=-\infty}^{\infty}$ and $\{t_k\}_{k=-\infty}^{\infty}$ be sequences with the stated properties. Consider a k for which $h_{max} < h_k \leq T_1$. Then, since $2h_{min} \leq h_{max}$, there is a positive integer m_k such that

$$h_k = m_k h_{min} + \ell_k,$$

where $h_{min} \leq \ell_k \leq h_{max}$. For $r = 1, \ldots, m_k$ write

$$\mathbf{q}_r = \phi^{r h_{min}}(\mathbf{y}_k)$$

and let $\mathbf{p}_r$ be the intersection the solution $\phi^t(\mathbf{x}_k)$ makes near $\mathbf{q}_r$ with the hyperplane through $\mathbf{q}_r$ and orthogonal to $F(\mathbf{q}_r)$. Note that

$$\|\phi^{r h_{min}}(\mathbf{x}_k) - \mathbf{q}_r\| \leq \varepsilon_2.$$

By applying Lemma 6.2 with $\mathbf{x} = \phi^{r h_{min}}(\mathbf{x}_k)$, $\mathbf{y} = \mathbf{q}_r$, $\mathbf{v} = F(\mathbf{q}_r)/\|F(\mathbf{q}_r)\|$ and $\alpha = \Delta/2M_0M_1$, we find that

$$\mathbf{p}_r = \phi^{s_k}(\mathbf{x}_k)$$

with

$$|s_r - r h_{min}| \leq 4\Delta^{-1}\varepsilon_2 \leq M\delta_0/2. \tag{2}$$

Now in our pseudo orbit we replace the point $\mathbf{y}_k$ by the segment

$$\mathbf{y}_k, \mathbf{q}_1, \ldots, \mathbf{q}_{m_k}$$

with associated times

$$h_{min}, h_{min}, \ldots, h_{min}, \ell_k, \tag{3}$$

thus obtaining a new δ_0 pseudo orbit with times between h_{min} and h_{max}. Correspondingly we replace the point $\mathbf{x}_k$ by the points

$$\mathbf{x}_k, \mathbf{p}_1, \ldots, \mathbf{p}_{m_k}$$

with corresponding times

$$s_1, s_2 - s_1, \ldots, s_{m_k} - s_{m_{k-1}}, t_k - s_{m_k}.$$

Using Eq.(2), we see that these times are within $M\delta_0$ of the times in Eq.(3). Also for $r = 1, \ldots, m_k$, we estimate

$$\begin{aligned} \|\mathbf{p}_r - \mathbf{q}_r\| &\leq \|\phi^{s_r - rh_{min}}(\phi^{rh_{min}}(\mathbf{x}_k)) - \phi^{rh_{min}}(\mathbf{x}_k)\| \\ &\qquad\qquad + \|\phi^{rh_{min}}(\mathbf{x}_k) - \phi^{rh_{min}}(\mathbf{y}_k)\| \\ &\leq M_0|s_r - rh_{min}| + \varepsilon_2 \\ &\leq (4M_0\Delta^{-1} + 1)\varepsilon_2 \\ &\leq M\delta_0. \end{aligned}$$

Of course, we take $F(\mathbf{q}_r)/\|F(\mathbf{q}_r)\|$ as the unit vector corresponding to the point $\mathbf{q}_r$ on the pseudo orbit. Now the uniqueness follows from Theorem 10.2.

Proof of Theorem 10.1. Let the constants M_0, M_1 and Δ be as in the proof of Theorem 10.2. The open set O in Theorem 10.1 will be the union of a tubular neighbourhood O_1 of the periodic orbit (the "periodic tube") and a tubular neighbourhood O_2 of the section of the homoclinic orbit which goes away from the periodic orbit (the "homoclinic tube"). A typical solution in O will make one journey around the homoclinic tube in between journeys around the periodic tube each consisting of several circuits. Segments of zeros in the sequence $\{a_k\}$ will give the number of circuits in journeys around the periodic tube. A one corresponds to a circuit around the periodic tube leading to a journey around the homoclinic tube which corresponds to a two.

Definition of the open set O: We will define disjoint open subsets V_0, V_1, V_2 and V_3 of the hyperplane C given in the statement of the theorem and use these subsets to define the periodic and homoclinic tubes. V_0, V_1, V_2 and V_3 will be neighbourhoods of disjoint subsets of the set consisting of $\mathbf{u}(0)$ and those points in which the homoclinic orbit intersects C. First choose positive numbers $\overline{M}$ and δ_1 such that

(A1) $\overline{M}$ and δ_1 can be taken as the constants in Theorem 10.2 applied to Eq.(1) and S with $h_{min} = 3T_0/4, h_{max} = 2T_0$ and $\sigma = \frac{1}{2}$, where T_0 is the period of $\mathbf{u}(t)$;

(A2) $\varepsilon_1 = \overline{M}\delta_1 < T_0/4$;

(A3) if $\mathbf{y} \in S$ and $\|\mathbf{x} - \mathbf{y}\| \leq \varepsilon_1$, then $\|F(\mathbf{x}) - F(\mathbf{y})\| \leq \Delta/4$.

Next we choose an open neighbourhood N of $\mathbf{u}(0)$ in C with the following properties:

(B1) $\mathrm{diam}(N) \le \min\{\varepsilon_1, \delta_1\}$;

(B2) the Poincaré map $P : N \to P(N) \subset C$ associated with the cross-section C is a well-defined C^1 diffeomorphism such that

$$\mathrm{diam}(P(N)) \le \min\{\varepsilon_1, \delta_1\}$$

and

$$P(\mathbf{x}) = \phi^{\tau_0(\mathbf{x})}(\mathbf{x}) \quad \text{with} \quad |\tau_0(\mathbf{x}) - T_0| \le \varepsilon_1/2;$$

also if for some $\mathbf{x} \in N$ and t in $-2\varepsilon_1 \le t \le T_0 + 3\varepsilon_1$,

$$\phi^t(\mathbf{x}) \in N \cup P(N),$$

then $t = 0$ or $\tau_0(\mathbf{x})$;

(B3) if $P^k(\mathbf{x})$ is defined for all integers k, then $\mathbf{x} = \mathbf{u}(0)$.

That N can be chosen so that (B2) is satisfied follows from Theorem 6.1; that it can be chosen so that (B3) is satisfied follows from the fact that $\mathbf{u}(0)$ is an isolated invariant set for P (see beginning of Section 2.6).

Consider the intersections of the homoclinic orbit $\mathbf{p}(t)$ with N. By (A3) and (B1), the tangent vector $\dot{\mathbf{p}}(t) = F(\mathbf{p}(t))$ is transverse to C when $\mathbf{p}(t) \in N$. So the set $\{t : \mathbf{p}(t) \in N\}$ is isolated. It follows from Lemma 6.2, the fact that

$$\mathrm{dist}(\mathbf{p}(t), \mathbf{u}) \to 0$$

as $|t| \to \infty$ and (B2) that the set $\{t : \mathbf{p}(t) \in N\}$ is an increasing sequence $\{t_k\}_{k=-\infty}^{\infty}$, where $t_k \to \pm\infty$ as $k \to \pm\infty$ and

$$\mathbf{p}(t_{k+1}) = P(\mathbf{p}(t_k))$$

if $|k|$ is sufficiently large. It is also clear that $\mathbf{p}(t_k) \to \mathbf{u}(0)$ as $k \to \pm\infty$. For short, we write

$$\mathbf{p}_k = \mathbf{p}(t_k).$$

Because of property (B3) of P, there exists an integer K^- such that

$$\mathbf{p}_{k+1} = P(\mathbf{p}_k) \quad \text{if} \quad k < K^-$$

but

$$P(\mathbf{p}_{K^-}) = \mathbf{v}_1 \ne \mathbf{p}_{K^-+1}.$$

It follows from (B2) that $\phi^t(\mathbf{p}_{K^-}) \notin N$ if $0 < t < \tau_0(\mathbf{p}_{K^-})$ and so if $\mathbf{v}_1$ were in N, we would have $\mathbf{v}_1 = \mathbf{p}_{K^-+1}$. So $\mathbf{v}_1 \notin N$. Next we take K^+ to be an integer greater than K^- such that $\mathbf{p}_{k+1} = P(\mathbf{p}_k)$ if $k \ge K^+$ and $\mathbf{v}_2 = \mathbf{p}_{K^+} = \phi^{T_1}(\mathbf{v}_1)$, where $T_1 \ge T_0$.

Making N smaller if necessary, we may assume that the homoclinic orbit does not intersect $\partial N \cap C$ so that, in particular, $\mathbf{v}_1$ is not in the closure $\overline{N}$ of N.

The homoclinic tube will be a neighbourhood of the segment of the homoclinic orbit from $\mathbf{v}_1$ to $\mathbf{p}_{K^+ +1}$.

Now consider the subsets I_0, $\{\mathbf{v}_2\}$ and J_0 of N where

$$I_0 = \{\ldots, \mathbf{p}_{K^- -1}, \mathbf{p}_{K^-}\} \cup \{\mathbf{u}(0)\} \cup \{\mathbf{p}_{K^+ +1}, \mathbf{p}_{K^+ +2}, \ldots\},$$

$$J_0 = \{\mathbf{p}_{K^- +1}, \ldots, \mathbf{p}_{K^+ -1}\}$$

and choose open neighbourhoods $\tilde{V}_2 \subset N$ of $\mathbf{v}_2 = \mathbf{p}_{K^+}$ and $V_3 \subset N$ of J_0 such that the closures of $\tilde{V}_2$ and V_3 are disjoint and do not intersect I_0.

Next we define a neighbourhood V_1 in C of $\mathbf{v}_1 = P(\mathbf{p}_{K^-}) \in P(N)\backslash\bar{N}$. Note, using (B2) and the definition of the $\mathbf{p}_k$, that in the time interval $[-2\varepsilon_1, T_1+2\varepsilon_1]$, the homoclinic solution $\phi^t(\mathbf{v}_1)$ never intersects itself and intersects N just at the $K^+ - K^- - 1$ points of J_0 and at $\mathbf{v}_2 = \mathbf{p}_{K^+} = \phi^{T_1}(\mathbf{v}_1)$ and these intersections are transversal. Then let $V_1 \subset C$ be an open ball with centre $\mathbf{v}_1$ such that

(C1) $V_1 \subset P(N)$, $N \cap V_1$ is empty, and if $\mathbf{x} \in V_1$ and $0 \leq t \leq T_1 + \varepsilon_1$, then $\phi^t(\mathbf{x})$ is defined and

$$\|\phi^t(\mathbf{x}) - \phi^t(\mathbf{v}_1)\| \leq \varepsilon_2,$$

where

$$\varepsilon_2 = \min\{\Delta/2M_1, \Delta^2/8M_0M_1, \Delta\varepsilon_1/8, (1+4M_0\Delta^{-1})^{-1}\varepsilon_1\};$$

(C2) if $\mathbf{x} \in V_1$, the solution $\phi^t(\mathbf{x})$ intersects N in the time interval $[-2\varepsilon_1, T_1+2\varepsilon_1]$ only at $K^+ - K^- - 1$ points in V_3 and at one point of $\tilde{V}_2$ and V_1 only at $\mathbf{x}$; moreover, if we write the point in $\tilde{V}_2$ as $Q(\mathbf{x})$, then (conf. Theorem 6.1) $Q : V_1 \to Q(V_1) \subset \tilde{V}_2$ is a C^1 diffeomorphism and $Q(\mathbf{x}) = \phi^{\tau_1(\mathbf{x})}(\mathbf{x})$ with $|\tau_1(\mathbf{x}) - T_1| \leq \varepsilon_1/2$.

Having chosen V_1, we choose an open neighbourhood $V_0 \subset N$ of I_0 such that $P(V_0) \subset N \cup V_1$ and $V_0 \cup P(V_0)$ does not intersect $\tilde{V}_2 \cup V_3$.

Next we replace the neighbourhood $\tilde{V}_2$ of $\mathbf{v}_2$ by a smaller open set V_2. To this end, choose a positive number $\delta \leq \delta_2$ satisfying

$$\varepsilon = \tilde{M}\delta < \min\{\varepsilon_1, \text{dist}(I_0, \partial V_0), \text{dist}(\mathbf{v}_1, \partial V_1), \text{dist}(\mathbf{v}_2, \partial\tilde{V}_2)\},$$

where the boundaries are relative to C and $\tilde{M}$, δ_2 are the numbers in Theorem 10.2 corresponding to F, S, $\sigma = \frac{1}{2}$, $h_{min} = 3T_0/4$ and $h_{max} = \max\{T_1, 2T_0\}$. Then choose the positive integer M so that

$$\|\mathbf{p}_{K^+ +M+1} - \mathbf{u}(0)\| \leq \delta, \quad \|\mathbf{p}_{K^- -M+1} - \mathbf{u}(0)\| \leq \delta.$$

We take $L = 2M$ (it is easy to see how this can be modified to allow L to be odd) and note that L can be as large as we like. Now define

$$V_2 = Q(V_1) \cap \bigcap_{k=1}^{2M} P^{-k}(V_0).$$

Finally define the open set O as $O_1 \cup O_2$, where O_1 is the "periodic tube" given by

$$O_1 = \{\phi^t(\mathbf{x}) : \mathbf{x} \in V_0, -\varepsilon_1 < t < \tau_0(\mathbf{x}) + \varepsilon_1\},$$

and O_2 is the "homoclinic tube" given by

$$O_2 = \{\phi^t(\mathbf{x}) : \mathbf{x} \in V_2, -\tau_1(Q^{-1}(\mathbf{x})) - \varepsilon_1 < t < \tau_0(\mathbf{x}) + \varepsilon_1\}.$$

Note that O_1 is a neighbourhood of the periodic orbit and O_2 is a neighbourhood of the segment of the homoclinic orbit from $\mathbf{v}_1 = P(\mathbf{p}_{K^-})$, where iterates of the Poincaré map first leave N, to $\mathbf{p}_{K^++1}$, which is in N and stays in N under iteration by P.

Solutions in O generate sequences $\{a_k\}_{k=-\infty}^{\infty}$. Consider a solution of Eq.(1) which remains in O for all time. We shall show that this solution intersects

$$W = V_0 \cup V_1 \cup V_2$$

in a sequence of points $\{\mathbf{w}_k\}_{k=-\infty}^{\infty}$, where $\mathbf{w}_{k+1} = \phi^{s_k}(\mathbf{w}_k)$ with $s_k \geq 3T_0/4$. Moreover,

$$\mathbf{w}_k \in V_0 \ \Rightarrow \ \mathbf{w}_{k+1} \in V_0 \cup V_1 \text{ and } \mathbf{w}_{k+1} = P(\mathbf{w}_k),$$

$$\mathbf{w}_k \in V_1 \ \Rightarrow \ \mathbf{w}_{k+1} \in V_2 \text{ and } \mathbf{w}_{k+1} = Q(\mathbf{w}_k)$$

and

$$\mathbf{w}_k \in V_2 \Rightarrow \mathbf{w}_{k+\ell} \in V_0 \text{ and } \mathbf{w}_{k+\ell} = P(\mathbf{w}_{k+\ell-1}) \text{ for } \ell = 1, \ldots, 2M.$$

In view of (A3), (B1) and (B2), and since $W \subset N \cup P(N)$, the vector field $F(\mathbf{x})$ is never parallel to C when $\mathbf{x} \in W$ and so the solution must intersect W in an isolated, hence countable, set. Note that the set of intersections with W cannot be empty since a solution through a point in O_1 intersects W in V_0 and a solution through a point in O_2 intersects W in V_2. Let $\mathbf{w}$ be such an intersection. Then $\mathbf{w} \in V_0$, V_1 or V_2. We consider these cases in turn.

Suppose $\mathbf{w} \in V_0$. If $\phi^{\varepsilon_1}(P(\mathbf{w})) \in O_1$, then $\phi^{\varepsilon_1}(P(\mathbf{w})) = \phi^t(\mathbf{x})$, where $\mathbf{x} \in V_0$ and $-\varepsilon_1 < t < \tau_0(\mathbf{x}) + \varepsilon_1$. This implies $P(\mathbf{w}) = \phi^{t-\varepsilon_1}(\mathbf{x})$, where $-2\varepsilon_1 < t - \varepsilon_1 < \tau_0(\mathbf{x})$. By (B2), $t - \varepsilon_1 = 0$ and so $P(\mathbf{w}) \in V_0$. Otherwise, $\phi^{\varepsilon_1}(P(\mathbf{w})) \in O_2$. This means that $\phi^{\varepsilon_1}(P(\mathbf{w})) = \phi^t(\mathbf{x})$, where $\mathbf{x} \in V_2$ and $-\tau_1(Q^{-1}(\mathbf{x})) - \varepsilon_1 < t < \tau_0(\mathbf{x}) + \varepsilon_1$. Now we cannot have $t - \varepsilon_1 > 0$ for then $\phi^{t-\varepsilon_1}(\mathbf{x}) = P(\mathbf{w}) \in P(V_0)$ with $0 < t - \varepsilon_1 < \tau_0(\mathbf{x})$, which is impossible by (B2). So $t - \varepsilon_1 \leq 0$ and then $P(\mathbf{w}) = \phi^{t-\varepsilon_1+\tau_1(\mathbf{z})}(\mathbf{z})$, where $\mathbf{z} = Q^{-1}(\mathbf{x}) \in V_1$ and $-2\varepsilon_1 < t - \varepsilon_1 + \tau_1(\mathbf{z}) \leq \tau_1(\mathbf{z}) \leq T_1 + \varepsilon/2$. Now $P(\mathbf{w}) \in P(V_0)$ and $P(V_0) \subset N \cup V_1$ and is disjoint from $V_2 \cup V_3$. From (C2) applied to $\mathbf{z}$, it follows that $P(\mathbf{w}) = \mathbf{z} \in V_1$. So, if $\mathbf{w} \in V_0$, then $P(\mathbf{w}) \in V_0$ or V_1 and, by (B2), $P(\mathbf{w})$ must be the next intersection of $\phi^t(\mathbf{w})$ with W.

Now if $\mathbf{w} \in V_1$, then $Q(\mathbf{w}) = \phi^{\tau_1(\mathbf{w})}(\mathbf{w}) \in \tilde{V}_2$. We show that, in fact, $Q(\mathbf{w}) \in V_2$. First, suppose $Q(\mathbf{w}) \in O_1$. Then $Q(\mathbf{w}) = \phi^t(\mathbf{x})$, where $\mathbf{x} \in V_0$ and $-\varepsilon_1 < t < \tau_0(\mathbf{x}) + \varepsilon_1$. Since $\mathbf{x}$ and $Q(\mathbf{w})$ are both in N, it follows from (B2) that $t = 0$ or $t = \tau_0(\mathbf{x})$. Since $V_0 \cap \tilde{V}_2$ is empty, $t = 0$ is not possible. So $Q(\mathbf{w}) = \phi^{\tau_0(\mathbf{x})}(\mathbf{x})$ and, therefore, $\mathbf{x} = \phi^{\tau_1(\mathbf{w})-\tau_0(\mathbf{x})}(\mathbf{w})$ where, using the fact that $T_1 \geq T_0$, and (B2) and (C2), we find that $-\varepsilon_1 \leq \tau_1(\mathbf{w}) - \tau_0(\mathbf{x}) < T_1 + \varepsilon_1$. However, by (C2), $\phi^t(\mathbf{w}) \notin V_0$ when $-\varepsilon_1 \leq t < T_1 + \varepsilon_1$. So $t = \tau_0(\mathbf{x})$ is not possible also. Thus $Q(\mathbf{w})$ must be in O_2. This means that $Q(\mathbf{w}) = \phi^t(\mathbf{x})$, where $\mathbf{x} \in V_2$ and $-\tau_1(Q^{-1}(\mathbf{x})) - \varepsilon_1 < t < \tau_0(\mathbf{x}) + \varepsilon_1$. If $t > 0$, we have $\mathbf{x} = \phi^{\tau_1(\mathbf{w})-t}(\mathbf{w})$, where $-2\varepsilon_1 < \tau_1(\mathbf{w}) - t < T_1 + \varepsilon_1$. By (C2), this implies that $Q(\mathbf{w}) = \mathbf{x}$. On the other hand, if $t \leq 0$, $Q(\mathbf{w}) = \phi^{t+\tau_1(\mathbf{z})}(\mathbf{z})$, where $\mathbf{z} = Q^{-1}(\mathbf{x})$ is in V_1 and $-\varepsilon_1 < t + \tau_1(\mathbf{z}) \leq T_1 + \varepsilon/2$. So, by (C2) again, $Q(\mathbf{w}) = Q(\mathbf{z}) = \mathbf{x}$ and thus, in either case, $Q(\mathbf{w})$ is in V_2. Finally, by (C2), $\phi^t(\mathbf{w})$ intersects $N \cup V_1$ (in which W is contained) only at $K^+ - K^- - 1$ points in V_3 in the time interval $(0, \tau_1(\mathbf{w}))$ and so $Q(\mathbf{w})$ is the next intersection with W.

Next suppose $\mathbf{w} \in V_2$. By definition of V_2, $P^k(\mathbf{w}) \in V_0$ for $k = 1, \ldots, 2M$. Hence, by (B2), the next $2M$ intersections after $\mathbf{w}$ are in V_0 and each one of them is the image under P of the previous intersection.

To show there is a previous intersection, suppose first that $\phi^{-\varepsilon_1}(\mathbf{w}) \in O_1$. Then $\mathbf{w} = \phi^{t+\varepsilon_1}(\mathbf{x})$, where $\mathbf{x} \in V_0$ and $0 < t + \varepsilon_1 < \tau_0(\mathbf{x}) + 2\varepsilon_1$. By (B2), this means $t + \varepsilon_1 = \tau_0(\mathbf{x})$ and so $\mathbf{x}$ would be a previous intersection. Otherwise, $\phi^{-\varepsilon_1}(\mathbf{w}) \in O_2$. Then $\mathbf{w} = \phi^{t+\varepsilon_1}(\mathbf{x})$, where $\mathbf{x} \in V_2$ and $-\tau_1(Q^{-1}(\mathbf{x})) < t + \varepsilon_1 < \tau_0(\mathbf{x}) + 2\varepsilon_1$. By (B2) and (C2), this means either $t + \varepsilon_1 = \tau_0(\mathbf{x})$ and so $\mathbf{x}$ is a previous intersection again or $t + \epsilon_1 = 0$ in which case $\mathbf{w} \in V_2$ and $Q^{-1}(\mathbf{w})$ is a previous intersection. So there must be a previous intersection.

We may take our solution in the form $\phi^t(\mathbf{w})$, where $\mathbf{w} \in W$ and then we assign to it the sequence $\{a_k\}_{k=-\infty}^{\infty}$, where $a_k \in \{0, 1, 2\}$ with

$$a_k = i \quad \text{when} \quad \mathbf{w}_k \in V_i,$$

where we take $\mathbf{w}_0 = \mathbf{w}$. Clearly the sequence $\{a_k\}_{k=-\infty}^{\infty}$ is of the required type. A zero in the sequence $\{a_k\}_{k=-\infty}^{\infty}$ corresponds to a journey around the periodic tube. $1, 2$ always occur together and correspond respectively to a transition journey around the periodic tube and a journey around the homoclinic tube. The periodic orbit $\phi^t(\mathbf{u}(0))$ generates the sequence consisting entirely of zeros and the homoclinic orbit $\phi^t(\mathbf{v}_1)$ generates the sequence $\{a_k\}_{k=-\infty}^{\infty}$ with $a_0 = 1$, $a_1 = 2$ and $a_k = 0$ for all other k.

A sequence $\{a_k\}_{k=-\infty}^{\infty}$ corresponds to at most one solution. We show that there is at most one solution $\phi^t(\mathbf{w}_0)$ (with $\mathbf{w}_0$ in W) of Eq.(1) staying in O for all time and which generates a given sequence $\{a_k\}_{k=-\infty}^{\infty}$ as above. Let $\{a_k\}_{k=-\infty}^{\infty}$ be a sequence of the type given in the theorem. We define a δ_1 pseudo orbit $\{\mathbf{y}_k\}_{k=-\infty}^{\infty}$ of Eq.(1) in S as follows:

$$\mathbf{y}_k = \begin{cases} \mathbf{u}(0), & \text{if } a_k = 0, \\ \mathbf{v}_1, & \text{if } a_k = 1, \\ \mathbf{v}_2, & \text{if } a_k = 2, \end{cases}$$

with corresponding times $\{h_k\}_{k=-\infty}^{\infty}$:

$$h_k = \begin{cases} T_0, & \text{if } a_k = 0, \\ T_1, & \text{if } a_k = 1, \\ \tau_0(\mathbf{v}_2), & \text{if } a_k = 2. \end{cases}$$

Consider a solution $\phi^t(\mathbf{w}_0)$ (with $\mathbf{w}_0 \in W$) in O which generates $\{a_k\}_{k=-\infty}^{\infty}$ corresponding to the sequence $\{\mathbf{w}_k\}_{k=-\infty}^{\infty}$ of intersections with W. We show that the sequence $\{\mathbf{w}_k\}_{k=-\infty}^{\infty}$ shadows the δ_1 pseudo orbit just defined, in the sense of Remark 10.3 (with $\overline{M}$, δ_1, $3T_0/4$, $2T_0$ playing the roles of M, δ_0, h_{min}, h_{max} respectively). If $\mathbf{w}_k$ is in V_0 or V_2 and hence in N, we see from (B1) that $\mathbf{w}_k$ is within distance $\varepsilon_1 = \overline{M}\delta_1$ of $\mathbf{y}_k$ ($= \mathbf{u}(0)$ or $\mathbf{v}_2$ respectively) and from (B2) that the corresponding time $\tau_0(\mathbf{w}_k)$ to $\mathbf{w}_{k+1} = P(\mathbf{w}_k)$ is within ε_1 of $h_k = T_0$ or $\tau_0(\mathbf{v}_2)$ respectively. If $\mathbf{w}_k$ is in V_1, we see from (C1) that for $0 \leq t \leq T_1$

$$\|\phi^t(\mathbf{w}_k) - \phi^t(\mathbf{y}_k)\| = \|\phi^t(\mathbf{w}_k) - \phi^t(\mathbf{v}_1)\| \leq \varepsilon_2$$

and that the corresponding time $\tau_1(\mathbf{w}_k)$ to $\mathbf{w}_{k+1} = Q(\mathbf{w}_k)$ is within $\varepsilon_1/2$ of $h_k = T_1$. Also in all cases,

$$\langle \mathbf{w}_k - \mathbf{y}_k, \dot{\mathbf{u}}(0) \rangle = 0,$$

where we note that

$$\langle F(\mathbf{y}_k), \frac{\dot{\mathbf{u}}(0)}{\|\dot{\mathbf{u}}(0)\|} \rangle \geq \frac{1}{2}\|F(\mathbf{y}_k)\|,$$

since $\mathbf{y}_k \in N \cup P(N)$ implies $\|\mathbf{y}_k - \mathbf{u}(0)\| \leq \varepsilon_1$ and so it follows from (A3) and Eq.(8) in Chapter 2 that

$$\begin{aligned} \langle \frac{F(\mathbf{y}_k)}{\|F(\mathbf{y}_k)\|}, \frac{F(\mathbf{u}(0))}{\|F(\mathbf{u}(0))\|} \rangle &\geq 1 - \left\| \frac{F(\mathbf{y}_k)}{\|F(\mathbf{y}_k)\|} - \frac{F(\mathbf{u}(0))}{\|F(\mathbf{u}(0))\|} \right\| \\ &\geq 1 - \frac{2\|F(\mathbf{y}_k) - F(\mathbf{u}(0))\|}{\|F(\mathbf{u}(0))\|} \\ &\geq 1 - 2\frac{(\Delta/4)}{\Delta} \\ &= \frac{1}{2}. \end{aligned}$$

Then it follows from (A1), the properties of ε_2 and Remark 10.3 that the sequences of points $\{\mathbf{w}_k\}_{k=-\infty}^{\infty}$ and corresponding times are unique. Thus the solution $\phi^t(\mathbf{w}_0)$ generating $\{a_k\}_{k=-\infty}^{\infty}$ is unique.

Each sequence $\{a_k\}_{k=-\infty}^{\infty}$ is generated by a solution. We show, given a sequence $\{a_k\}_{k=-\infty}^{\infty}$ of the type given in the theorem, that there is a solution $\phi^t(\mathbf{w}_0)$ of

Eq.(1) staying in O for all time which generates it. To this end, we define a δ pseudo orbit $\{\mathbf{y}_k\}_{k=-\infty}^{\infty}$ of Eq.(1) in S. Corresponding to a finite segment of at least $L = 2M$ zeros between two segments $1, 2$, we take the same number of points

$$\mathbf{p}_{K^++1}, \ldots, \mathbf{p}_{K^++M}, \mathbf{u}(0), \ldots, \mathbf{u}(0), \mathbf{p}_{K^--M+1}, \ldots, \mathbf{p}_{K^-},$$

with corresponding h_k's

$$\tau_0(\mathbf{p}_{K^++1}), \ldots, \tau_0(\mathbf{p}_{K^++M}), T_0, \ldots, T_0, \tau_0(\mathbf{p}_{K^--M+1}), \ldots, \tau_0(\mathbf{p}_{K^-});$$

if the segment of zeros is semi-infinite, there are infinitely many $\mathbf{u}(0)$'s and no $\mathbf{p}_{K^--M+1}, \ldots, \mathbf{p}_{K^-}$ if $1, 2$ is on the left, and no $\mathbf{p}_{K^++1}, \ldots, \mathbf{p}_{K^++M}$ if $1, 2$ is on the right; if the segment of zeros is bi-infinite, there are only $\mathbf{u}(0)$'s. If $a_k = 1$, we take $\mathbf{y}_k = \mathbf{v}_1$ and $h_k = T_1$ and if $a_k = 2$, we take $\mathbf{y}_k = \mathbf{v}_2$ and $h_k = \tau_0(\mathbf{v}_2)$.

In view of our choice of δ and ε, by Theorem 10.2 there are unique sequences $\{\mathbf{w}_k\}_{k=-\infty}^{\infty}$ and $\{t_k\}_{k=-\infty}^{\infty}$ such that

$$\mathbf{w}_{k+1} = \phi^{t_k}(\mathbf{w}_k),$$

$$\|\mathbf{w}_k - \mathbf{y}_k\| \leq \varepsilon, \quad |t_k - h_k| \leq \varepsilon$$

and

$$\langle \mathbf{w}_k - \mathbf{y}_k, \dot{\mathbf{u}}(0) \rangle = 0.$$

We show that the solution $\phi^t(\mathbf{w}_0)$ generates the sequence $\{a_k\}_{k=-\infty}^{\infty}$.

Note first that since $\mathbf{y}_k \in C$ and $\langle \mathbf{w}_k - \mathbf{y}_k, \dot{\mathbf{u}}(0) \rangle = 0$, it follows that $\mathbf{w}_k \in C$.

If $a_k = 0$ so that $\mathbf{y}_k$ is in the set

$$\{\mathbf{p}_{K^++1}, \ldots, \mathbf{p}_{K^++M}, \mathbf{u}(0), \mathbf{p}_{K^--M+1}, \ldots, \mathbf{p}_{K^-}\}$$

and hence in I_0, then since $\varepsilon < \text{dist}(I_0, \partial V_0)$ we have $\mathbf{w}_k \in V_0$. Also $h_k = \tau_0(\mathbf{y}_k)$ and so by (B2)

$$|t_k - T_0| \leq |t_k - h_k| + |\tau_0(\mathbf{y}_k) - T_0| \leq 3\varepsilon_1/2.$$

From (B2) again, it follows that $t_k = \tau_0(\mathbf{w}_k)$, $\mathbf{w}_{k+1} = P(\mathbf{w}_k)$ and $\phi^t(\mathbf{w}_k) \notin W$ if $0 < t < t_k$. Also it is clear that $\phi^t(\mathbf{w}_k) \in O_1$ for $0 \leq t \leq t_k$. This reasoning applies even when $\mathbf{y}_k = \mathbf{p}_{K^-}$ for then $a_{k+1} = 1$ and, as shown below, $\mathbf{w}_{k+1} \in V_1 \subset P(N)$.

Suppose now that $a_{k-1} = 1$ and $a_k = 2$. Then since $\varepsilon < \text{dist}(\mathbf{v}_1, \partial V_1)$, $\mathbf{w}_{k-1} \in V_1$ and

$$|t_{k-1} - T_1| = |t_{k-1} - h_{k-1}| \leq \varepsilon < \varepsilon_1. \tag{4}$$

Since $a_k = 2$, $\mathbf{y}_k = \mathbf{v}_2$ and so because $\varepsilon < \text{dist}(\mathbf{v}_2, \partial\tilde{V}_2)$, $\mathbf{w}_k \in \tilde{V}_2$. Note that $\mathbf{w}_k = \phi^{t_{k-1}}(\mathbf{w}_{k-1})$, where $\mathbf{w}_{k-1} \in V_1$. Then it follows from Eq.(4) and (C2) that $\mathbf{w}_k \in Q(V_1)$, that in fact $\mathbf{w}_k = Q(\mathbf{w}_{k-1})$ and that $\phi^t(\mathbf{w}_{k-1}) \notin W$ if $0 < t < t_{k-1}$. Next, since $h_k = \tau_0(\mathbf{y}_k)$, it follows as in the case $a_k = 0$ that

$\mathbf{w}_{k+1} = P(\mathbf{w}_k)$ and that $\phi^t(\mathbf{w}_k) \notin W$ if $0 < t < t_k$. Also, since $a_{k+\ell} = 0$ for $\ell = 1, \ldots, L = 2M$, $\mathbf{w}_{k+\ell} \in V_0$ and $\mathbf{w}_{k+\ell+1} = P(\mathbf{w}_{k+\ell})$ for $\ell = 1, \ldots, L$. These facts imply that $\mathbf{w}_k \in V_2$, that $\phi^t(\mathbf{w}_{k-1}) \in O_2$ if $0 \le t \le t_{k-1}$ and that $\phi^t(\mathbf{w}_k) \in O_2$ if $0 \le t \le t_k$.

So we can see that the solution $\phi^t(\mathbf{w}_0)$ lies in O for all t and intersects W in the sequence $\{\mathbf{w}_k\}_{k=-\infty}^{\infty}$ and that $\mathbf{w}_k \in V_i$ if and only if $a_k = i$ (for $i = 0, 1, 2$). So the solution does indeed generate the sequence $\{a_k\}_{k=-\infty}^{\infty}$.

Definition of the return map. Finally we define the set Σ and the return map $G : \Sigma \to \Sigma$. We take $\Sigma \subset W$ as the set of intersections with W of those solutions of Eq.(1) staying in O. As we have seen above, there is a unique correspondence between such solutions $\phi^t(\mathbf{w})$ with $\mathbf{w} \in \Sigma$ and sequences $\{a_k\}_{k=-\infty}^{\infty}$ in Y. We define $G(\mathbf{w})$ as the next intersection after $\mathbf{w}$ of the solution $\phi^t(\mathbf{w})$ with W and the mapping $\alpha : Y \to \Sigma$ by

$$\alpha(\{a_k\}_{k=-\infty}^{\infty}) = \mathbf{w}.$$

Clearly,

$$\alpha(\{a_{k+1}\}_{k=-\infty}^{\infty}) = G(\mathbf{w})$$

and so

$$\alpha \circ \sigma = G \circ \alpha.$$

Also it is clear that α is bijective.

To show that α is a homeomorphism, first we show α is continuous at $\{a_k\}_{k=-\infty}^{\infty} \in Y$. Suppose not. Then there exists a positive number γ and a sequence $a^{(p)} = \{a_k^{(p)}\}_{k=-\infty}^{\infty}$ in Y converging to $a = \{a_k\}_{k=-\infty}^{\infty}$ as $p \to \infty$ such that for all p

$$\|\alpha(a^{(p)}) - \alpha(a)\| \ge \gamma. \tag{5}$$

In the previous subsection, we used $a^{(p)}$ (resp. a) to define a δ pseudo orbit $\{\mathbf{y}_k^{(p)}\}_{k=-\infty}^{\infty}$ (resp. $\{\mathbf{y}_k\}_{k=-\infty}^{\infty}$) with associated times $\{h_k^{(p)}\}_{k=-\infty}^{\infty}$ (resp. $\{h_k\}_{k=-\infty}^{\infty}$) which was ε-shadowed uniquely by a sequence of points $\{\mathbf{w}_k^{(p)}\}_{k=-\infty}^{\infty}$ (resp. $\{\mathbf{w}_k\}_{k=-\infty}^{\infty}$) on a true orbit with

$$\mathbf{w}_{k+1}^{(p)} = \phi^{t_k^{(p)}}(\mathbf{w}_k^{(p)}) \quad (\text{resp. } \mathbf{w}_{k+1} = \phi^{t_k}(\mathbf{w}_k))$$

such that

$$\langle \mathbf{w}_k^{(p)} - \mathbf{y}_k^{(p)}, \dot{\mathbf{u}}(0) \rangle = 0 \quad (\text{resp.} \langle \mathbf{w}_k - \mathbf{y}_k, \dot{\mathbf{u}}(0) \rangle = 0).$$

In fact,

$$\alpha(a^{(p)}) = \mathbf{w}_0^{(p)} \quad \text{and} \quad \alpha(a) = \mathbf{w}_0.$$

Since $a^{(p)}$ converges to a, $a_k^{(p)} = a_k$ for $|k| \le K_p$ where $K_p \to \infty$ as $p \to \infty$. From the manner in which the pseudo orbits were defined we can see that $\mathbf{y}_k^{(p)} = \mathbf{y}_k$ and $h_k^{(p)} = h_k$ for $|k| \le L_p$, where $L_p = K_p - M \to \infty$ as $p \to \infty$. So

$$\|\mathbf{w}_k^{(p)} - \mathbf{y}_k\| \le \varepsilon, \quad |t_k^{(p)} - h_k| \le \varepsilon \quad \text{and} \quad \langle \mathbf{w}_k^{(p)} - \mathbf{y}_k, \dot{\mathbf{u}}(0) \rangle = 0, \tag{6}$$

if $|k| \leq L_p$. By compactness, taking a subsequence if necessary, we may assume that as p tends to ∞, $\lim \mathbf{w}_k^{(p)} = \mathbf{z}_k$ and $\lim t_k^{(p)} = s_k$ for all k. Letting p tend to ∞ in Eq.(6), we obtain

$$\|\mathbf{z}_k - \mathbf{y}_k\| \leq \varepsilon, \qquad |s_k - h_k| \leq \varepsilon \quad \text{and} \quad \langle\, \mathbf{z}_k - \mathbf{y}_k, \dot{\mathbf{u}}(0)\,\rangle = 0$$

for all k. By uniqueness of the shadowing orbit, $\mathbf{z}_k = \mathbf{w}_k$ for all k and so $\mathbf{w}_0 = \lim_{p\to\infty} \mathbf{w}_0^{(p)}$. This contradicts Eq.(5). Hence α is continuous and so, since Y is compact, α is a homeomorphism. (Note it would be possible to prove the continuity of α more directly using an expansiveness argument, as was done in the diffeomorphism case in Chapter 5.) This completes the proof of the theorem.

We remark here that this theorem could be proved using the theory for not necessarily invertible maps developed by Steinlein and Walther [1990] (conf. also Hale and Lin [1986]). Their theory would be applied to the map $f : N \cup V_1 \to N \cup P(N)$ defined by

$$f(\mathbf{x}) = \begin{cases} P(\mathbf{x}) & \text{if } \mathbf{x} \in N, \\ Q(\mathbf{x}) & \text{if } \mathbf{x} \in V_1. \end{cases}$$

Notice that $\mathbf{u}(0)$ is a hyperbolic fixed point of this map and the sequence

$$\{\ldots, \mathbf{p}_{K^- -1}, \mathbf{p}_{K^-}, \mathbf{v}_1 = P(\mathbf{p}_{K^-}), \mathbf{v}_2 = Q(\mathbf{v}_1) = \mathbf{p}_{K^+}, \mathbf{p}_{K^+ +1}, \ldots\}$$

is a transversal homoclinic orbit. However this map is not simply the Poincaré map associated with the periodic orbit. Also it does not appear to be one to one and so the theory for diffeomorphisms cannot be applied.

11. NUMERICAL SHADOWING

In this chapter our object is to demonstrate how shadowing ideas can be used to verify the accuracy of numerical simulations of dynamical systems and also how they can be used to rigorously establish the existence of periodic orbits and of chaotic behaviour.

11.1 FINITE TIME SHADOWING FOR MAPS

Let $f : \mathbb{R}^n \to \mathbb{R}^n$ be a C^2 map, which could possibly be the period map associated with a periodic system of differential equations as in Example 1.2. When we compute an orbit of f, we begin with a point $\mathbf{y}_0$ and compute a finite sequence $\{\mathbf{y}_k\}_{k=0}^N$ where for $k = 0, \ldots, N-1$, $\mathbf{y}_{k+1}$ is the computed value of $f(\mathbf{y}_k)$. Since the computer rounds all its calculations, $\mathbf{y}_{k+1}$ is only approximately equal to $f(\mathbf{y}_k)$. In fact, there is a small positive number δ such that

$$\|\mathbf{y}_{k+1} - f(\mathbf{y}_k)\| \le \delta$$

for $k = 0, \ldots, N-1$. So we make the following definition.

Definition 11.1. Let $f : \mathbb{R}^n \to \mathbb{R}^n$ be a C^2 map. A sequence $\{\mathbf{y}_k\}_{k=0}^N$ of points in $\mathbb{R}^n$ is said to be a δ *pseudo orbit* of f if

$$\|\mathbf{y}_{k+1} - f(\mathbf{y}_k)\| \le \delta \quad \text{for} \quad k = 0, \ldots, N-1.$$

If N is large, it may turn out that the true orbit $\{f^k(\mathbf{y}_0)\}_{k=0}^N$ beginning at $\mathbf{y}_0$ diverges very far from the computed orbit. The best we may be able to do is find a true orbit $\{\mathbf{x}_k\}_{k=0}^N$, that is, $\mathbf{x}_{k+1} = f(\mathbf{x}_k)$ for $k = 0, \ldots, N-1$, which is near the computed orbit. So we make the following definition.

Definition 11.2. Let $f : \mathbb{R}^n \to \mathbb{R}^n$ be a C^2 map. The δ pseudo orbit $\{\mathbf{y}_k\}_{k=0}^N$ of f is said to be ε-shadowed by a true orbit $\{\mathbf{x}_k\}_{k=0}^N$, that is, $\mathbf{x}_{k+1} = f(\mathbf{x}_k)$ for $k = 0, \ldots, N-1$, if

$$\|\mathbf{x}_k - \mathbf{y}_k\| \le \varepsilon \quad \text{for} \quad k = 0, \ldots, N.$$

We now prove the following theorem, which gives conditions under which a pseudo orbit of a map can be shadowed by a true orbit. This version of the theorem comes from Coomes, Koçak and Palmer [1996], which was a modification of results from Chow and Palmer [1991, 1992]. (See also Hadeler [1996].) Note that the first contribution to this problem was made by Hammel, Yorke and

Grebogi [1987, 1988]. Sauer and Yorke [1990, 1991] made a further contribution which also uses shadowing ideas.

Theorem 11.3. *Let $f : \mathbb{R}^n \to \mathbb{R}^n$ be a C^2 map and $\{\mathbf{y}_k\}_{k=0}^N$ be a δ pseudo orbit. For a given right inverse L^{-1} of the linear operator $L : (\mathbb{R}^n)^{N+1} \to (\mathbb{R}^n)^N$, defined for $\mathbf{u} = \{\mathbf{u}_k\}_{k=0}^N \in (\mathbb{R}^n)^{N+1}$ by*

$$(L\mathbf{u})_k = \mathbf{u}_{k+1} - Df(\mathbf{y}_k)\mathbf{u}_k \quad \text{for} \quad k = 0, \ldots, N-1,$$

set

$$\varepsilon = 2\|L^{-1}\|\delta,$$

where the norm of L^{-1} is the operator norm with respect to the supremum norm on $(\mathbb{R}^n)^{N+1}$ and $(\mathbb{R}^n)^N$. Next, let

$$M = \sup\{\|D^2 f(\mathbf{x})\| : \mathbf{x} \in \mathbb{R}^n,\ \|\mathbf{x} - \mathbf{y}_k\| \leq \varepsilon \quad \text{for some} \quad k = 0, \ldots N\}.$$

Then if

$$2M\|L^{-1}\|^2\delta \leq 1,$$

the δ pseudo orbit $\{\mathbf{y}_k\}_{k=0}^N$ of f is ε- shadowed by a true orbit $\{\mathbf{x}_k\}_{k=0}^N$ of f.

To prove this theorem, we use the following lemma.

Lemma 11.4. *Let X and Y be finite dimensional normed spaces, let O be an open subset of X and let $\mathcal{G} : O \to Y$ be a C^2 map. Suppose for some $\mathbf{y} \in O$, we have $\|\mathcal{G}(\mathbf{y})\| \leq \delta$ and the derivative $D\mathcal{G}(\mathbf{y})$ has a right inverse $\mathcal{K}$ with $\|\mathcal{K}\| \leq K$. Set*

$$\varepsilon = 2K\delta$$

and

$$M = \sup\{\|D^2\mathcal{G}(\mathbf{x})\| : \|\mathbf{x} - \mathbf{y}\| \leq \varepsilon\}.$$

Then if the closed ball about $\mathbf{y}$ with radius ε is contained in O and

$$MK\varepsilon = 2MK^2\delta \leq 1,$$

the equation

$$\mathcal{G}(\mathbf{x}) = \mathbf{0}$$

has a solution $\mathbf{x}$ with $\|\mathbf{x} - \mathbf{y}\| \leq \varepsilon$.

Proof. Define the continuous operator $F : O \to X$ by

$$F(\mathbf{x}) = \mathbf{y} - \mathcal{K}[\mathcal{G}(\mathbf{x}) - D\mathcal{G}(\mathbf{y})(\mathbf{x} - \mathbf{y})].$$

Clearly if $F(\mathbf{x}) = \mathbf{x}$, then $\mathcal{G}(\mathbf{x}) = \mathbf{0}$.

Moreover, if $\|\mathbf{x}-\mathbf{y}\| \leq \varepsilon$,

$$\begin{aligned}\|F(\mathbf{x})-\mathbf{y}\| &\leq \|\mathcal{K}\|\,\|\mathcal{G}(\mathbf{x})-\mathcal{G}(\mathbf{y})-D\mathcal{G}(\mathbf{y})(\mathbf{x}-\mathbf{y})+\mathcal{G}(\mathbf{y})\| \\ &\leq K\left[\frac{1}{2}M\|\mathbf{x}-\mathbf{y}\|^2+\delta\right] \\ &\leq KM\varepsilon\frac{\varepsilon}{2}+K\delta \\ &\leq \frac{\varepsilon}{2}+\frac{\varepsilon}{2} \\ &= \varepsilon.\end{aligned}$$

Now the lemma follows from Brouwer's fixed point theorem.

Proof of Theorem 11.3. We apply Lemma 11.4 to $X = (\mathbb{R}^n)^{N+1}$ and $Y = (\mathbb{R}^n)^N$ with the norms

$$\|\mathbf{x}\| = \|\{\mathbf{x}_k\}_{k=0}^N\| = \max_{0\leq k\leq N} \|\mathbf{x}_k\|$$

for $\mathbf{x} \in X$ and

$$\|\mathbf{g}\| = \|\{\mathbf{g}_k\}_{k=0}^{N-1}\| = \max_{0\leq k\leq N-1} \|\mathbf{g}_k\|$$

for $\mathbf{g} \in Y$. We define the C^2 function $\mathcal{G} : X \to Y$ for $\mathbf{x} = \{\mathbf{x}_k\}_{k=0}^N \in X$ by

$$[\mathcal{G}(\mathbf{x})]_k = \mathbf{x}_{k+1} - f(\mathbf{x}_k) \quad \text{for} \quad k = 0,\ldots,N-1.$$

If $\mathbf{y} = \{\mathbf{y}_k\}_{k=0}^N$ is a δ pseudo orbit of f, we see that

$$\|\mathcal{G}(\mathbf{y})\| \leq \delta.$$

Clearly $D\mathcal{G}(\mathbf{y}) = L$. Also note that if $\mathbf{x} = \{\mathbf{x}_k\}_{k=0}^N$, $\mathbf{u} = \{\mathbf{u}_k\}_{k=0}^N$ and $\mathbf{v} = \{\mathbf{v}_k\}_{k=0}^N$ are in X,

$$[D^2\mathcal{G}(\mathbf{x})\mathbf{u}\mathbf{v}]_k = -D^2f(\mathbf{x}_k)\mathbf{u}_k\mathbf{v}_k \quad \text{for} \quad k = 0,\ldots,N-1$$

and so if $\|\mathbf{x}-\mathbf{y}\| \leq \varepsilon = 2\|L^{-1}\|\delta$,

$$\|D^2\mathcal{G}(\mathbf{x})\| \leq M.$$

Hence, by Lemma 11.4 with $K = \|L^{-1}\|$, if

$$2M\|L^{-1}\|^2\delta \leq 1,$$

the equation

$$\mathcal{G}(\mathbf{x}) = 0$$

has a solution $\mathbf{x}$ with $\|\mathbf{x} - \mathbf{y}\| \leq \varepsilon$. Thus the theorem is proved.

Implementation of the theorem: Let $\{\mathbf{y}_k\}_{k=0}^N$ be a δ pseudo orbit of the map f. In order to find a right inverse for the operator L, we triangularise the matrices $Df(\mathbf{y}_k)$ as follows. We begin with an orthogonal matrix S_0 and then we define orthogonal matrices S_k and upper triangular matrices A_k with positive diagonal entries by performing a sequence of QR factorisations recursively:

$$Df(\mathbf{y}_k)S_k = S_{k+1}A_k \quad \text{for} \quad k = 0, \ldots, N-1.$$

Our aim is to find a right inverse for L, that is, given $\mathbf{g} = \{\mathbf{g}_k\}_{k=0}^{N-1} \in (\mathbb{R}^n)^N$, we have to solve the equation

$$L\mathbf{u} = \mathbf{g}$$

for $\mathbf{u} = \{\mathbf{u}_k\}_{k=0}^N \in (\mathbb{R}^n)^{N+1}$. This equation can be written as the difference equation

$$\mathbf{u}_{k+1} = Df(\mathbf{y}_k)\mathbf{u}_k + \mathbf{g}_k \quad \text{for} \quad k = 0, \ldots, N-1. \tag{1}$$

If we make the transformation

$$\mathbf{u}_k = S_k\mathbf{v}_k \quad \text{for} \quad k = 0, \ldots, N,$$

then Eq.(1) becomes

$$\mathbf{v}_{k+1} = A_k\mathbf{v}_k + S_{k+1}^*\mathbf{g}_k \quad \text{for} \quad k = 0, \ldots, N-1.$$

We define the operator T by

$$(T\mathbf{u})_k = \mathbf{u}_{k+1} - A_k\mathbf{u}_k \quad \text{for} \quad k = 0, \ldots, N-1,$$

where now we restrict to the two-dimensional case so that we may write

$$\mathbf{u}_k = \begin{bmatrix} v_k \\ w_k \end{bmatrix} \quad \text{and} \quad A_k = \begin{bmatrix} a_k & b_k \\ 0 & c_k \end{bmatrix}.$$

For $\mathbf{g} = \{\mathbf{g}_k\}_{k=0}^{N-1} \in (\mathbb{R}^2)^N$, we solve the difference equation

$$\mathbf{u}_{k+1} = A_k\mathbf{u}_k + \mathbf{g}_k \quad \text{for} \quad k = 0, \ldots, N-1$$

which, in components, reads

$$\begin{aligned} v_{k+1} &= a_k v_k + b_k w_k + \mathbf{g}_k^{(1)} \quad \text{for} \quad k = 0, \ldots, N-1 \\ w_{k+1} &= c_k w_k + \mathbf{g}_k^{(2)} \quad \text{for} \quad k = 0, \ldots, N-1. \end{aligned} \tag{2}$$

We construct the right inverse T^{-1} of T obtained by solving the second equation in Eq.(2) forwards starting with $w_0 = 0$ and then the first equation backwards starting with $v_N = 0$. (The rationale behind this procedure is explained at the end of this section.) If $\|\mathbf{g}\| \leq 1$, we get

$$|w_k| \leq \tilde{w}_k,$$

where the $\tilde{w}_k$'s are given by the recursion

$$\begin{aligned}\tilde{w}_0 &= 0,\\ \tilde{w}_{k+1} &= |c_k|\tilde{w}_k + 1 \quad \text{for} \quad k = 0, \ldots, N-1,\end{aligned}$$

and

$$|v_k| \leq \tilde{v}_k,$$

where the $\tilde{v}_k$'s are given by the backwards recursion

$$\begin{aligned}\tilde{v}_N &= 0,\\ \tilde{v}_k &= |a_k^{-1}|(\tilde{v}_{k+1} + |b_k|\tilde{w}_k + 1) \quad \text{for} \quad k = N-1, \ldots, 0.\end{aligned}$$

Then, as we are using the Euclidean norm in $\mathbb{R}^2$,

$$\|T^{-1}\| \leq \max_{0 \leq k \leq N} \sqrt{\tilde{v}_k^2 + \tilde{w}_k^2}.$$

Finally if we define $\bar{\mathbf{g}}$ by

$$\bar{\mathbf{g}}_k = S_{k+1}^* \mathbf{g}_k \quad \text{for} \quad k = 0, \ldots, N-1$$

we see that

$$(L^{-1}\mathbf{g})_k = S_k(T^{-1}\bar{\mathbf{g}})_k \quad \text{for} \quad k = 0, \ldots, N$$

defines a right inverse for L and

$$\|L^{-1}\| \leq \|T^{-1}\|.$$

Example: Consider the Tinkerbell map (Nusse and Yorke [1994, p.219])

$$f(x, y) = (x^2 - y^2 + c_1 x + c_2 y,\ 2xy + c_3 x + c_4 y)$$

where we take $c_1 = 0.9$, $c_2 = -0.6013$, $c_3 = 2$, $c_4 = 0.5$ and compute an orbit starting at $(-0.72, -0.64)$. This computed orbit stays in the region $R = \{(x, y) : -1.23312 \leq x \leq 0.461052, -1.54655 \leq y \leq 0.544313\}$. Let $\bar{f}(x, y)$ be the computed value of $f(x, y)$. Then if (x, y) is in the region R, the techniques in Wilkinson [1963, 1965] show on the machine used that, with the Euclidean norm in $\mathbb{R}^2$,

$$\|\bar{f}(x, y) - f(x, y)\| \leq \delta = 7.5 \times 10^{-13}.$$

The computed orbit is $\{(x_k, y_k)\}_{k=0}^{N}$ with $(x_0, y_0) = (-0.72, -0.64)$ and $N = 100000$. The matrices

$$Df(x_k, y_k) = \begin{bmatrix} 2x_k + c_1 & -2y_k + c_2 \\ 2y_k + c_3 & 2x_k + c_4 \end{bmatrix}$$

are calculated. Then the matrices S_k and A_k are determined and it is found, using the method given above, that

$$\|L^{-1}\| \leq \|T^{-1}\| \leq 479483.$$

Now

$$\|D^2 f(x, y)\| \leq 2.$$

Hence we may take $M = 2$. Then

$$2M\|L^{-1}\|^2\delta \leq .69 \leq 1.$$

Therefore our theorem allows us to conclude the existence of a true orbit which ε-shadows the computed one with

$$\varepsilon = 2\|L^{-1}\|\delta \leq 7.2 \times 10^{-7}.$$

We have ignored the effects of roundoff error in the calculation of $Df(\mathbf{y}_k)$, S_k and A_k. These can be handled as in Coomes, Koçak and Palmer [1996]. They do not have a significant effect on the results. Also in the calculation of the upper bound for $\|T^{-1}\|$ it may be necessary to use a truncation as was used in Coomes, Koçak and Palmer [1996].

Motivation for the choice of right inverse: Here we describe the rationale behind the choice of right inverse for T. If f is a chaotic map and $\{\mathbf{y}_k\}_{k=0}^{\infty}$ is a δ pseudo orbit, in the two-dimensional case we would expect the difference equation

$$\mathbf{u}_{k+1} = Df(\mathbf{y}_k)\mathbf{u}_k \tag{3}$$

to have an exponential dichotomy with a projection of rank 1 (actually for chaotic systems which arise in practice, this only seems to be approximately true). When we make a triangularizing transformation

$$\mathbf{u}_k = S_k \mathbf{v}_k,$$

we obtain the difference equation

$$\mathbf{v}_{k+1} = A_k \mathbf{v}_k, \tag{4}$$

where

$$A_k = \begin{bmatrix} a_k & b_k \\ 0 & c_k \end{bmatrix},$$

with a_k and c_k both positive.

By the arguments of Chapter 3 (or see Palmer [1988, p.269]), the nullspace for the projection corresponding to the dichotomy of Eq.(3) can be spanned by any vector not in the (uniquely determined) stable subspace. Let $\mathbf{u}_k$ be the solution of Eq.(3) such that

$$\mathbf{u}_k = S_k \begin{bmatrix} a_{k-1} \cdots a_0 \\ 0 \end{bmatrix}.$$

If $\mathbf{u}_k$ is in the stable subspace, there are positive constants K_1, λ_1 with $\lambda_1 < 1$ such that

$$\|\mathbf{u}_k\| \leq K_1 \lambda_1^{k-m} \|\mathbf{u}_m\|$$

for $0 \leq m \leq k$, in which case

$$a_{k-1} \cdots a_m \leq K_1 \lambda_1^{k-m} \tag{5}$$

for $0 \leq m \leq k$. Otherwise, there are positive constants K_2, λ_2 with $\lambda_2 < 1$ such that

$$\|\mathbf{u}_k\| \leq K_2 \lambda_2^{m-k} \|\mathbf{u}_m\|$$

for $0 \leq k \leq m$, in which case

$$a_k^{-1} \cdots a_{m-1}^{-1} \leq K_2 \lambda_2^{m-k} \tag{6}$$

for $0 \leq k \leq m$. By considering the equations adjoint to Eqs.(3) and (4), we can arrive at a similar conclusion for the c_k's.

For a suitably small positive number β, we make the transformation

$$\mathbf{v}_k = \begin{bmatrix} 1 & 0 \\ 0 & \beta \end{bmatrix} \mathbf{w}_k \tag{7}$$

in Eq.(4), obtaining the new equation

$$\mathbf{w}_k = \begin{bmatrix} a_k & \beta b_k \\ 0 & c_k \end{bmatrix}. \tag{8}$$

Now the diagonal system

$$\mathbf{w}_k = \begin{bmatrix} a_k & 0 \\ 0 & c_k \end{bmatrix} \tag{9}$$

has an exponential dichotomy with projection

$$\begin{bmatrix} 1 & 0 \\ 0 & 1 \end{bmatrix}, \begin{bmatrix} 0 & 0 \\ 0 & 0 \end{bmatrix}, \begin{bmatrix} 1 & 0 \\ 0 & 0 \end{bmatrix} \text{ or } \begin{bmatrix} 0 & 0 \\ 0 & 1 \end{bmatrix},$$

depending on whether the a_k's and c_k's both satisfy the same inequalities like Eq.(5) or Eq.(6) or different ones. Also, if β is sufficiently small, it follows from the roughness theorem that Eqs.(8) and (9) have exponential dichotomies with projections of the same rank. However, since Eq.(8) is obtained from Eq.(4) by the transformation (7), it has an exponential dichotomy with projection of rank 1. Hence the a_k's and the c_k's must satisfy different ones of inequalities (5) and (6).

Again let $\mathbf{u}_k$ be the solution of Eq.(3) such that

$$\mathbf{u}_k = S_k \begin{bmatrix} a_{k-1} \cdots a_0 \\ 0 \end{bmatrix}.$$

In fact, $\mathbf{u}_0$ is the first column of S_0. Since S_0 is chosen arbitrarily, we would not expect $\mathbf{u}_0$ to be in the stable subspace and even if it were, because of numerical error, $\mathbf{u}_1$ would almost certainly not be. Hence, in all probability, $\mathbf{u}_k$ will not be in the stable subspace. Hence we would expect the a_k's to satisfy inequality (6) and so the c_k's would satisfy inequality (5) (with a_k replaced by c_k).

Then it is not hard to see that the solutions of Eq.(2), which are bounded for $k \geq 0$, are given by

$$w_k = c_{k-1} \cdots c_0 w_0 + \sum_{m=0}^{k-1} c_{k-1} \cdots c_{m+1} \mathbf{g}_m^{(2)}$$

and

$$v_k = -\sum_{m=k}^{\infty} a_k^{-1} \cdots a_m^{-1} (b_m w_m + \mathbf{g}_m^{(1)}),$$

where w_0 is arbitrary. In the case of finite sequences $\{\mathbf{y}_k\}_{k=0}^{N}$, we truncate the last series and take $w_0 = 0$ (since we want our solutions to be as small as possible so that $\|T^{-1}\|$ and hence $\|L^{-1}\|$ are as small as possible) to get

$$w_k = \sum_{m=0}^{k-1} c_{k-1} \cdots c_{m+1} \mathbf{g}_m^{(2)}, \; v_k = -\sum_{m=k}^{N-1} a_k^{-1} \cdots a_m^{-1} (b_m w_m + \mathbf{g}_m^{(1)}).$$

This is just the solution of Eq.(2) with $w_0 = 0$ and $v_N = 0$.

11.2 PERIODIC SHADOWING FOR MAPS

Periodic orbits play an important role in the dynamics of maps: for example, they can give important information about the structure of chaotic attractors (conf. Auerbach, Cvitanovich, Eckmann, Gunaratne and Procaccia [1987], Grebogi, Ott and Yorke [1987]). In this section, which is based on Coomes, Koçak and Palmer [1996], we give a method for verifying that a computed orbit which looks periodic is, in fact, periodic. For related work, see Schwartz [1983], Ocken [1995] and Osipenko and Komarchev [1995].

For a C^2 map $f : \mathbb{R}^n \to \mathbb{R}^n$, a formal definition of an approximate periodic orbit is given now.

Definition 11.5. A sequence $\{\mathbf{y}_k\}_{k=0}^N$ is said to be a δ *pseudo periodic orbit* of f if

$$\|\mathbf{y}_{k+1} - f(\mathbf{y}_k)\| \leq \delta \quad \text{for} \quad k = 0, \ldots, N-1$$

and

$$\|\mathbf{y}_0 - f(\mathbf{y}_N)\| \leq \delta.$$

If $\{\mathbf{y}_k\}_{k=0}^N$ is a δ pseudo periodic orbit, we define the linear operator $L : (\mathbb{R}^n)^{N+1} \to (\mathbb{R}^n)^{N+1}$ for $\mathbf{u} = \{\mathbf{u}_k\}_{k=0}^N$ by

$$(L\mathbf{u})_k = \begin{cases} \mathbf{u}_{k+1} - Df(\mathbf{y}_k)\mathbf{u}_k & \text{for } k = 0, \ldots, N-1 \\ \mathbf{u}_0 - Df(\mathbf{y}_k)\mathbf{u}_N & \text{for } k = N. \end{cases}$$

Our main theorem is as follows (conf. Coomes, Koçak and Coomes [1996]).

Theorem 11.6. *Suppose $f : \mathbb{R}^n \to \mathbb{R}^n$ is a C^2 map. Let $\{\mathbf{y}_k\}_{k=0}^N$ be a δ pseudo periodic orbit of f and suppose that the operator L is invertible. Set*

$$\varepsilon = 2\|L^{-1}\|\delta,$$

where the norm of L^{-1} is the operator norm with respect to the supremum norm on $(\mathbb{R}^n)^{N+1}$, and

$$M = \sup\{\|D^2 f(\mathbf{x})\| : \mathbf{x} \in \mathbb{R}^n,\ \|\mathbf{x} - \mathbf{y}_k\| \leq \varepsilon \quad \text{for} \quad \text{some} \quad k = 0, \ldots, N\}.$$

Then if

$$2M\|L^{-1}\|^2\delta < 1,$$

there is a unique periodic orbit $\{\mathbf{x}_k\}_{k=0}^N$, that is,

$$\mathbf{x}_{k+1} = f(\mathbf{x}_k) \quad \text{for} \quad k = 0, \ldots, N-1 \quad \text{and} \quad \mathbf{x}_0 = f(\mathbf{x}_N),$$

such that

$$\|\mathbf{x}_k - \mathbf{y}_k\| \leq \varepsilon \quad \text{for} \quad k = 0, \ldots, N.$$

Proof. We need to find a solution of the difference equation

$$\mathbf{x}_{k+1} = f(\mathbf{x}_k) \quad \text{for} \quad k = 0, \ldots, N-1$$

such that

$$\mathbf{x}_0 = f(\mathbf{x}_N)$$

and

$$\|\mathbf{x}_k - \mathbf{y}_k\| \le \varepsilon \quad \text{for} \quad k = 0, \dots, N.$$

We define a C^2 map $\mathcal{G} : (\mathbb{R}^n)^{N+1} \to (\mathbb{R}^n)^{N+1}$ as follows: if $\mathbf{x} = \{\mathbf{x}_k\}_{k=0}^N \in (\mathbb{R}^n)^{N+1}$, then

$$[\mathcal{G}(\mathbf{x})]_k = \begin{cases} \mathbf{x}_{k+1} - f(\mathbf{x}_k) & \text{for} \quad k = 0, \dots, N-1, \\ \mathbf{x}_0 - f(\mathbf{x}_N) & \text{for} \quad k = N. \end{cases}$$

Then the theorem will be proved if we can show that the equation

$$\mathcal{G}(\mathbf{x}) = \mathbf{0}$$

has a unique solution $\mathbf{x} = \{\mathbf{x}_k\}_{k=0}^N$ with $\|\mathbf{x}_k - \mathbf{y}_k\| \le \varepsilon$ for $k = 0, \dots, N$. To prove this we use the following lemma, which is a C^2 version of Lemma 4.4.

Lemma 11.7. *Let X, Y be Banach spaces, $O \subset X$ an open set and $\mathcal{G} : O \to Y$ a C^2 function. Suppose $\mathbf{y}$ is an element of O for which*

$$\|\mathcal{G}(\mathbf{y})\| \le \delta$$

and $D\mathcal{G}(\mathbf{y}) = L$ is invertible with $\|L^{-1}\| \le K$. Set

$$\varepsilon = 2K\delta$$

and

$$M = \sup\{\|D^2\mathcal{G}(\mathbf{x})\| : \mathbf{x} \in O, \ \|\mathbf{x} - \mathbf{y}\| \le \varepsilon\}.$$

Then if the closed ball of radius ε around $\mathbf{y}$ is in O and

$$MK\varepsilon = 2MK^2\delta < 1,$$

there is a unique solution $\mathbf{x}$ of the equation

$$\mathcal{G}(\mathbf{x}) = \mathbf{0}$$

satisfying $\|\mathbf{x} - \mathbf{y}\| \le \varepsilon$.

Proof. Define the operator $F : O \to X$ by

$$F(\mathbf{x}) = \mathbf{y} - L^{-1}[\mathcal{G}(\mathbf{x}) - D\mathcal{G}(\mathbf{y})(\mathbf{x} - \mathbf{y})].$$

Clearly $\mathcal{G}(\mathbf{x}) = \mathbf{0}$ if and only if $F(\mathbf{x}) = \mathbf{x}$. Moreover, if $\|\mathbf{x} - \mathbf{y}\| \le \varepsilon$,

$$\begin{aligned} \|F(\mathbf{x}) - \mathbf{y}\| &\le \|L^{-1}\|\|\mathcal{G}(\mathbf{x}) - \mathcal{G}(\mathbf{y}) - D\mathcal{G}(\mathbf{y})(\mathbf{x} - \mathbf{y}) + \mathcal{G}(\mathbf{y})\| \\ &\le K\left[\frac{1}{2}M\|\mathbf{x} - \mathbf{y}\|^2 + \delta\right] \\ &\le MK\varepsilon \cdot \frac{1}{2}\varepsilon + \frac{1}{2}\varepsilon \\ &< \varepsilon. \end{aligned}$$

Furthermore, if $\|\mathbf{x}-\mathbf{y}\| \le \varepsilon$ and $\|\mathbf{z}-\mathbf{y}\| \le \varepsilon$, then

$$\|F(\mathbf{x})-F(\mathbf{z})\| \le \|L^{-1}\|\|\mathcal{G}(\mathbf{x})-\mathcal{G}(\mathbf{z})-D\mathcal{G}(\mathbf{y})(\mathbf{x}-\mathbf{z})\| \le KM\varepsilon\|\mathbf{x}-\mathbf{z}\|.$$

Since $KM\varepsilon < 1$, F is a contraction on the closed ball of radius ε, centre $\mathbf{y}$, and thus the lemma follows from the contraction mapping principle.

To prove the theorem we apply Lemma 11.7 to the map $\mathcal{G}: X \to X$ defined before the lemma with $X = (\mathbb{R}^n)^{N+1}$, where we use the norm

$$\|\mathbf{x}\| = \|\{\mathbf{x}_k\}_{k=0}^N\| = \max_{0\le k\le N} \|\mathbf{x}_k\|.$$

First note that if $\mathbf{y} = \{\mathbf{y}_k\}_{k=0}^N$ is our δ pseudo orbit, then

$$\|\mathcal{G}(\mathbf{y})\| \le \delta.$$

We calculate

$$[D\mathcal{G}(\mathbf{y})\mathbf{u}]_k = \begin{cases} \mathbf{u}_{k+1} - Df(\mathbf{y}_k)\mathbf{u}_k & \text{for} \quad k = 0,\dots,N-1 \\ \mathbf{u}_0 - Df(\mathbf{y}_k)\mathbf{u}_N & \text{for} \quad k = N. \end{cases}$$

So $D\mathcal{G}(\mathbf{y}) = L$. Next, if $\mathbf{x} = \{\mathbf{x}_k\}_{k=0}^N$, $\mathbf{u} = \{\mathbf{u}_k\}_{k=0}^N$ and $\mathbf{v} = \{\mathbf{v}_k\}_{k=0}^N$ are in X

$$[D^2\mathcal{G}(\mathbf{x})\mathbf{u}\mathbf{v}]_k = -D^2 f(\mathbf{x}_k)\mathbf{u}_k\mathbf{v}_k \quad \text{for} \quad k = 0,\dots,N.$$

Hence

$$\sup\{\|D^2\mathcal{G}(\mathbf{x})\| : \|\mathbf{x}-\mathbf{y}\| \le \varepsilon\} \le M,$$

with M as in the statement of the theorem and $\varepsilon = 2\|L^{-1}\|\delta$. So the theorem follows using Lemma 11.7 with $K = \|L^{-1}\|$.

Implementation of the theorem: Let $\{\mathbf{y}_k\}_{k=0}^N$ be a δ pseudo periodic orbit of a C^2 map $f: \mathbb{R}^n \to \mathbb{R}^n$. In order to find an inverse for L, we triangularise the matrices $Df(\mathbf{y}_k)$ as follows. We begin with an orthogonal matrix S_0 and then we define orthogonal matrices S_k and upper triangular matrices A_k by performing a sequence of QR factorisations recursively:

$$Df(\mathbf{y}_k)S_k = S_{k+1}A_k \quad \text{for} \quad k = 0,\dots,N.$$

We find that, for a chaotic map, regardless of the choice of S_0, up to sign the columns of S_{N+1} turn out to be the same (in the machine — the reason for this is given at the end of this section). Then we take S_{N+1} as the new S_0 and repeat the procedure. Instead of taking A_N as what was found from the recursion, we take it as the matrix $S_0^* Df(\mathbf{y}_N)S_N$, which is still upper triangular since S_0 and S_{N+1} are the same except perhaps for the sign of the columns.

Our aim is to find the inverse of L, that is, given $\mathbf{g} = \{\mathbf{g}_k\}_{k=0}^N \in (\mathbb{R}^n)^{N+1}$, we have to solve the equation

$$L\mathbf{u} = \mathbf{g}$$

for $\mathbf{u} = \{\mathbf{u}_k\}_{k=0}^N \in (\mathbb{R}^n)^{N+1}$. This equation can be written as the difference equation

$$\begin{aligned} \mathbf{u}_{k+1} &= Df(\mathbf{y}_k)\mathbf{u}_k + \mathbf{g}_k \quad \text{for} \quad k = 0, \ldots, N-1 \\ \mathbf{u}_0 &= Df(\mathbf{y}_N)\mathbf{u}_N + \mathbf{g}_N. \end{aligned} \tag{10}$$

We make the transformation

$$\mathbf{u}_k = S_k \mathbf{v}_k \quad \text{for} \quad k = 0, \ldots, N.$$

Then Eq.(10) becomes

$$\begin{aligned} \mathbf{u}_{k+1} &= A_k \mathbf{v}_k + S_{k+1}^* \mathbf{g}_k \quad \text{for} \quad k = 0, \ldots, N-1 \\ \mathbf{u}_0 &= A_N \mathbf{u}_N + S_0^* \mathbf{g}_N. \end{aligned}$$

Consider the corresponding operator T given for $\mathbf{u} = \{\mathbf{u}_k\}_{k=0}^N$ by

$$(T\mathbf{u})_k = \begin{cases} \mathbf{u}_{k+1} - A_k \mathbf{u}_k & \text{for } k = 0, \ldots, N-1, \\ \mathbf{u}_0 - A_N \mathbf{u}_N & \text{for } k = N. \end{cases}$$

Now we restrict to the two dimensional case so that A_k has the form

$$A_k = \begin{bmatrix} a_k & b_k \\ 0 & c_k \end{bmatrix} \quad \text{for} \quad k = 0, \ldots, N.$$

Then if T^{-1} exists, $\mathbf{u}_k = (T^{-1}\mathbf{g})_k$ is a solution of the difference equation

$$\begin{aligned} \mathbf{u}_{k+1} &= A_k \mathbf{u}_k + \mathbf{g}_k \quad \text{for} \quad k = 0, \ldots, N-1 \\ \mathbf{u}_0 &= A_N \mathbf{u}_N + \mathbf{g}_N. \end{aligned}$$

In components $\mathbf{u}_k = \begin{bmatrix} v_k \\ w_k \end{bmatrix}$ these equations read

$$\begin{aligned} v_{k+1} &= a_k v_k + b_k w_k + \mathbf{g}_k^{(1)} \quad \text{for} \quad k = 0, \ldots, N-1 \\ v_0 &= a_N v_N + b_N w_N + \mathbf{g}_N^{(1)} \end{aligned} \tag{11}$$

and

$$\begin{aligned} w_{k+1} &= c_k w_k + \mathbf{g}_k^{(2)} \quad \text{for} \quad k = 0, \ldots, N-1 \\ w_0 &= c_N w_N + \mathbf{g}_N^{(2)}. \end{aligned} \tag{12}$$

As with finite time shadowing, we expect the a_k's to be expanding and the c_k's to be contracting. So we solve Eq.(12) forwards to find that

$$w_0 = (1 - c_N \cdots c_0)^{-1} \sum_{m=0}^{N} c_N \cdots c_{m+1} \mathbf{g}_m^{(2)},$$

where we would expect that

$$|c_N \cdots c_0| < 1. \tag{13}$$

Hence if $\mathbf{g}$ is a unit vector,

$$|w_0| \leq (1 - |c_N \cdots c_0|)^{-1} \sum_{m=0}^{N} |c_N \cdots c_{m+1}|.$$

Then if we set

$$\tilde{w}_0 = (1 - |c_N \cdots c_0|)^{-1} \sum_{m=0}^{N} |c_N \cdots c_{m+1}|$$

and use the recursion

$$\tilde{w}_{k+1} = |c_k|\tilde{w}_k + 1 \quad \text{for} \quad k = 0, \ldots, N-1,$$

we obtain upper bounds $\tilde{w}_k$ for $|w_k|$. Next we solve Eq.(11) backwards to find

$$v_0 = -(1 - a_0^{-1} \cdots a_N^{-1})^{-1} \sum_{m=0}^{N} a_0^{-1} \cdots a_m^{-1}(b_m w_m + \mathbf{g}_m^{(1)}),$$

where we would expect that

$$|a_0^{-1} \cdots a_N^{-1}| < 1. \tag{14}$$

So if $\mathbf{g}$ is a unit vector,

$$|v_0| \leq (1 - |a_0^{-1} \cdots a_N^{-1}|)^{-1} \sum_{m=0}^{N} |a_0^{-1} \cdots a_m^{-1}|(|b_m|\tilde{w}_m + 1).$$

Then we set

$$\tilde{v}_0 = (1 - |a_0^{-1} \cdots a_N^{-1}|)^{-1} \sum_{m=0}^{N} |a_0^{-1} \cdots a_m^{-1}|(|b_m|\tilde{w}_m + 1)$$

and use the backwards recursion

$$\begin{aligned} \tilde{v}_N &= |a_N^{-1}|(\tilde{v}_0 + |b_N|\tilde{w}_N + 1), \\ \tilde{v}_k &= |a_k^{-1}|(\tilde{v}_{k+1} + |b_k|\tilde{w}_k + 1) \quad \text{for} \quad k = N-1, \ldots, 1 \end{aligned}$$

to calculate upper bounds $\tilde{v}_k$ for $|v_k|$. Hence if the conditions in Eqs.(13) and (14) hold, T is invertible and

$$\|T^{-1}\| \leq \max_{0 \leq k \leq N} \sqrt{(\tilde{v}_k)^2 + (\tilde{w}_k)^2}.$$

Then we see that if the conditions in Eqs.(13) and (14) hold, L is invertible and L^{-1} is given by

$$(L^{-1}\mathbf{g})_k = S_k(T^{-1}\bar{\mathbf{g}})_k,$$

where

$$\bar{\mathbf{g}}_k = \begin{cases} S_{k+1}^*\mathbf{g}_k & \text{for } k = 0, \dots, N-1, \\ S_0^*\mathbf{g}_N & \text{for } k = N. \end{cases}$$

Also,

$$\|L^{-1}\| \le \|T^{-1}\| \le \max_{0\le k\le N} \sqrt{(\tilde{v}_k)^2 + (\tilde{w}_k)^2}.$$

Example: Consider the Hénon map $f : I\!R^2 \to I\!R^2$ given by

$$f(x, y) = (1 - ax^2 + y, -bx)$$

with $a = 1.4$ and $b = -0.3$. By a method described in Coomes, Koçak and Palmer [1998] we find a δ pseudo periodic orbit $\{\mathbf{y}_k\}_{k=0}^{100002}$ with initial data

$$(0.53201104688235434, -0.22573297271273646)$$

and

$$\delta = 6.52 \times 10^{-15}.$$

We check that the conditions in Eqs.(13) and (14) hold and hence obtain an upper bound for $\|L^{-1}\|$. We find that

$$2M\|L^{-1}\|^2\delta = 4a\|L^{-1}\|^2\delta \le 0.000475$$

and

$$\varepsilon = 2\|L^{-1}\|\delta \le 1.49 \times 10^{-9}.$$

Now the theorem implies the existence of a true periodic orbit of period 100003 near the computed one.

Motivation for the choice of S_0: As at the end of the previous section, we assume Eq.(3) has an exponential dichotomy on $[0, N+1]$ with a projection of rank 1 (we restrict to the two-dimensional case). Note that if $\mathbf{e}_k$ is the first column of S_k, then it follows from the relation $Df(\mathbf{y}_k)S_k = S_{k+1}A_k$ that

$$Df(\mathbf{y}_k)\mathbf{e}_k = a_k\mathbf{e}_{k+1}.$$

Hence

$$\mathbf{e}_k = \frac{\mathbf{u}_k}{a_{k-1}\cdots a_0},$$

where $\mathbf{u}_k$ is a solution of Eq.(3). According to the reasoning at the end of the previous section, we may assume $\mathbf{u}_k$ is not in the stable subspace. Let P_k be the projection with range the stable subspace and nullspace the one-dimensional subspace spanned by $\mathbf{u}_k$. It follows from Proposition 2.3 in Palmer [1988] that Eq.(3) has an exponential dichotomy with projection P_k. Corresponding to another choice of S_0, we obtain another solution $\overline{\mathbf{u}}_k$ and another projection $\overline{P}_k$. Then if $\Phi(k, m)$ is the transition matrix for Eq.(3), we obtain

$$\begin{aligned}
\|P_k - \overline{P}_k\| &= \|\Phi(k,0)(P_0 - \overline{P}_0)\Phi(0,k)\| \quad \text{by invariance} \\
&= \|\Phi(k,0)(P_0 - P_0\overline{P}_0)\Phi(0,k)\| \quad \text{since} \quad \mathcal{R}(P_0) = \mathcal{R}(\overline{P}_0) \\
&= \|\Phi(k,0)P_0(I - \overline{P}_0)\Phi(0,k)\| \\
&= \|\Phi(k,0)P_0.\Phi(0,k)(I - \overline{P}_k)\| \quad \text{by invariance} \\
&\to 0 \text{ as } k \to \infty.
\end{aligned}$$

Hence the subspaces spanned by $\mathbf{u}_k$ and $\overline{\mathbf{u}}_k$ approach each other as $k \to \infty$. Thus, for k large, in the computer the first column of S_k is unique up to sign and hence, because of orthogonality, both columns are unique up to sign. This means that for N large, the columns of S_{N+1} are, up to sign, independent of the choice of S_0.

11.3 FINITE TIME SHADOWING FOR DIFFERENTIAL EQUATIONS

In this section we show that computed solutions of chaotic autonomous systems of differential equations are shadowed for long times by true solutions. Note, as demonstrated in Coomes, Koçak and Palmer [1995a], the article on which this section is based, one does not obtain good results by simply applying the "map method" of Section 11.1 to the time-h map. One has to take into account the fact that there is no hyperbolicity in the direction of the vector field.

We begin with the definition of a pseudo orbit of an autonomous system

$$\dot{\mathbf{x}} = F(\mathbf{x}), \tag{15}$$

where $F : \mathbb{R}^n \to \mathbb{R}^n$ is a C^2 vector field. We denote by ϕ the associated flow.

Definition 11.8. For a given positive number δ, a sequence of points $\{\mathbf{y}_k\}_{k=0}^N$, with $F(\mathbf{y}_k) \neq \mathbf{0}$ for all k, is said to be a δ ***pseudo orbit*** of Eq.(15) if there is an associated sequence $\{h_k\}_{k=0}^{N-1}$ of positive times such that

$$\|\mathbf{y}_{k+1} - \phi^{h_k}(\mathbf{y}_k)\| \leq \delta \quad \text{for} \quad k = 0, \ldots N-1.$$

When we compute a solution of Eq.(15), we would obtain such a pseudo orbit. However, unlike for the maps, here the major contribution to δ usually comes from the discretization error associated with the numerical method used. Again our aim is to find a true orbit near the computed orbit. First we introduce notations and constructions needed for the statement of our theorem.

So let $\{\mathbf{y}_k\}_{k=0}^N$ be a δ pseudo orbit of Eq.(15) with associated times $\{h_k\}_{k=0}^{N-1}$. For $k = 0, \ldots N$, let S_k be an $n \times (n-1)$ matrix chosen so that its columns form an orthonormal basis for the subspace orthogonal to $F(\mathbf{y}_k)$. Now we let

$$A_k = S_{k+1}^* D\phi^{h_k}(\mathbf{y}_k) S_k \quad \text{for} \quad k = 0, \ldots N-1.$$

Geometrically, A_k is $D\phi^{h_k}(\mathbf{y}_k)$ restricted to the subspace orthogonal to $F(\mathbf{y}_k)$ and projected to the subspace orthogonal to $F(\mathbf{y}_{k+1})$.

Next we define a linear operator $L : (\mathbb{R}^{(n-1)})^{(N+1)} \to (\mathbb{R}^{(n-1)})^N$ in the following way. If $\{\xi_k\}_{k=0}^N$ is in $(\mathbb{R}^{(n-1)})^{(N+1)}$, then we take $L\xi = \{(L\xi)_k\}_{k=0}^{N-1}$ to be

$$(L\xi)_k = \xi_{k+1} - A_k \xi_k \quad \text{for} \quad k = 0, \ldots, N-1.$$

The operator L has right inverses and we choose one such right inverse L^{-1}.

Finally we define several constants. Let U be a convex set containing $\{\mathbf{y}_k\}_{k=0}^N$ in its interior. Then we define

$$M_0 = \sup_{\mathbf{x} \in U} \|F(\mathbf{x})\|, \quad M_1 = \sup_{\mathbf{x} \in U} \|DF(\mathbf{x})\|, \quad M_2 = \sup_{\mathbf{x} \in U} \|D^2F(\mathbf{x})\|.$$

Next we define

$$h_{min} = \min_{0 \le k \le N-1} h_k, \quad h_{max} = \max_{0 \le k \le N-1} h_k.$$

Then we choose a positive number $\varepsilon_0 \le h_{min}$ such that for $k = 0, \ldots, N-1$ and $\|\mathbf{x} - \mathbf{y}_k\| \le \varepsilon_0$ the solution $\phi^t(\mathbf{x})$ is defined for $0 \le t \le h_k + \varepsilon_0$ and is in U. Finally, we define

$$\Delta = \inf_{0 \le k \le N} \|F(\mathbf{y}_k)\|, \quad \overline{M}_0 = \sup_{0 \le k \le N} \|F(\mathbf{y}_k)\|, \quad \overline{M}_1 = \sup_{0 \le k \le N} \|DF(\mathbf{y}_k)\|,$$

and

$$\theta = \sup_{0 \le k \le N} \|D\phi^{h_k}(\mathbf{y}_k)\|.$$

Now we can state our main theorem (conf. Coomes, Koçak and Palmer [1995a]).

Theorem 11.9. *Let $\{\mathbf{y}_k\}_{k=0}^N$ be a δ pseudo orbit of the autonomous system (15), let*

$$C = \max\{\Delta^{-1}(\theta\|L^{-1}\| + 1), \|L^{-1}\|\}$$

and let

$$\overline{M} = (\overline{M}_0 + M_1\nu\delta)(\overline{M}_1 + M_2\nu\delta) + 2(\overline{M}_1 + M_2\nu\delta)e^{M_1(h_{max}+\varepsilon_0)}$$

$$+ M_2(h_{max} + \varepsilon_0)e^{2M_1(h_{max}+\varepsilon_0)},$$

where

$$\nu = 2C(e^{M_1(h_{max}+\varepsilon_0)} + M_0)(1 - CM_1\delta)^{-1} + 1.$$

If δ satisfies the inequalities

$$\text{(i)} \quad CM_1\delta < 1,$$

$$\text{(ii)} \quad 2C(1 - CM_1\delta)^{-1}\delta < \varepsilon_0, \quad \textit{and}$$

$$\text{(iii)} \quad 2\overline{M}C^2(1 - CM_1\delta)^{-2}\delta \leq 1,$$

then there exists a sequence of points $\{\mathbf{x}_k\}_{k=0}^{N}$ and a sequence of positive times $\{t_k\}_{k=0}^{N-1}$ such that

$$\mathbf{x}_{k+1} = \phi^{t_k}(\mathbf{x}_k) \quad \text{for} \quad k = 0, \dots, N-1$$

$$\|\mathbf{x}_k - \mathbf{y}_k\| \leq \varepsilon, \quad \text{for} \quad k = 0, \dots, N$$

and

$$|t_k - h_k| \leq \varepsilon \quad \text{for} \quad k = 0, \dots, N-1,$$

where

$$\varepsilon = 2C(1 - CM_1\delta)^{-1}\delta.$$

Proof. We begin with the δ pseudo orbit $\{\mathbf{y}_k\}_{k=0}^{N}$ of Eq.(15) and look for the points $\mathbf{x}_k$ in the hyperplane $\mathcal{H}_k$ through $\mathbf{y}_k$ and normal to $F(\mathbf{y}_k)$. Note that $\mathbf{x}_k \in \mathcal{H}_k$ if and only if there is a vector $\mathbf{z} \in \mathbb{R}^{n-1}$ such that $\mathbf{x}_k = \mathbf{y}_k + S_k\mathbf{z}$. So our problem then is to find a sequence of times $\{t_k\}_{k=0}^{N-1}$ and a sequence of vectors $\{\mathbf{z}_k\}_{k=0}^{N}$ in $\mathbb{R}^{n-1}$ such that

$$\mathbf{y}_{k+1} + S_{k+1}\mathbf{z}_{k+1} = \phi^{t_k}(\mathbf{y}_k + S_k\mathbf{z}_k), \quad k = 0, \dots, N-1.$$

We next introduce the space $X = \mathbb{R}^N \times (\mathbb{R}^{n-1})^{N+1}$ with norm

$$\|(\{t_k\}_{k=0}^{N-1}, \{\mathbf{z}_k\}_{k=0}^{N})\| = \max\left\{\max_{0\leq k\leq N-1} |t_k|, \max_{0\leq k\leq N} \|\mathbf{z}_k\|\right\},$$

and the space $Y = (\mathbb{R}^n)^N$ with norm

$$\|\{\mathbf{g}_k\}_{k=0}^{N-1}\| = \max_{0\leq k\leq N-1} \|\mathbf{g}_k\|.$$

Now, we let O be the open set in X consisting of those $\mathbf{v} = (\{t_k\}_{k=0}^{N-1}, \{\mathbf{z}_k\}_{k=0}^{N})$ with $|t_k - h_k| < \varepsilon_0$ and $\|\mathbf{z}_k\| < \varepsilon_0$, and introduce the function $\mathcal{G} : O \to Y$ given by

$$[\mathcal{G}(\mathbf{v})]_k = \mathbf{y}_{k+1} + S_{k+1}\mathbf{z}_{k+1} - \phi^{t_k}(\mathbf{y}_k + S_k\mathbf{z}_k), \quad k = 0, \dots, N-1.$$

Then the theorem will be proved if we can find a solution $\mathbf{v} = (\{t_k\}_{k=0}^{N-1}, \{\mathbf{z}_k\}_{k=0}^{N})$ of the equation

$$\mathcal{G}(\mathbf{v}) = 0 \tag{16}$$

in the closed ball of radius $\varepsilon = 2C(1 - CM_1\delta)^{-1}\delta$ about $\mathbf{v}_0 = (\{h_k\}_{k=0}^{N-1}, \mathbf{0})$. We use Lemma 11.4 to solve this equation with $\mathbf{y} = \mathbf{v}_0$.

Construction of right inverse for $D\mathcal{G}(\mathbf{v}_0)$: Note that for

$$\mathbf{u} = (\{\tau_k\}_{k=0}^{N-1}, \{\xi_k\}_{k=0}^{N}) \in X,$$

the derivative of $\mathcal{G}$ at $\mathbf{v}_0$ is given by

$$[D\mathcal{G}(\mathbf{v}_0)\mathbf{u}]_k = -\tau_k F(\phi^{h_k}(\mathbf{y}_k)) + S_{k+1}\xi_{k+1} - D\phi^{h_k}(\mathbf{y}_k)S_k\xi_k, \quad k = 0, \ldots, N-1.$$

We approximate $D\mathcal{G}(\mathbf{v}_0)$ by the linear operator $T : X \to Y$ defined by:

$$[T\mathbf{u}]_k = -\tau_k F(\mathbf{y}_{k+1}) + S_{k+1}\xi_{k+1} - D\phi^{h_k}(\mathbf{y}_k)S_k\xi_k, \quad k = 0, \ldots, N-1.$$

In order to find a right inverse for T, for given $\mathbf{g} = \{\mathbf{g}_k\}_{k=0}^{N-1}$ in Y, we must solve the set of equations

$$[T\mathbf{u}]_k = \mathbf{g}_k, \quad k = 0, \ldots, N-1, \tag{17}$$

for $\mathbf{u}$. Since, for each k, the matrix

$$[\, F(\mathbf{y}_{k+1})/\|F(\mathbf{y}_{k+1})\| \quad S_{k+1} \,]$$

is orthogonal, this set of equations is equivalent to the following two sets of equations, one set obtained by premultiplying the kth member in Eq.(17) by $F(\mathbf{y}_{k+1})^*$, and the other set obtained by premultiplying the kth member in Eq.(17) by S_{k+1}^*:

$$\begin{aligned} -\|F(\mathbf{y}_{k+1})\|^2\tau_k - F(\mathbf{y}_{k+1})^* D\phi^{h_k}(\mathbf{y}_k)S_k\xi_k &= F(\mathbf{y}_{k+1})^*\mathbf{g}_k, \\ \xi_{k+1} - A_k\xi_k &= S_{k+1}^*\mathbf{g}_k, \end{aligned}$$

for $k = 0, \ldots, N-1$. If we write $\bar{\mathbf{g}} = \{S_{k+1}^*\mathbf{g}_k\}_{k=0}^{N-1}$, then we see that a solution of the second set is

$$\xi_k = (L^{-1}\bar{\mathbf{g}})_k, \quad k = 0, \ldots, N-1. \tag{18}$$

We substitute this into the first set and obtain

$$\tau_k = -\frac{F(\mathbf{y}_{k+1})^*}{\|F(\mathbf{y}_{k+1})\|^2}\{D\phi^{h_k}(\mathbf{y}_k)S_k(L^{-1}\bar{\mathbf{g}})_k + \mathbf{g}_k\}, \quad k = 0, \ldots, N-1. \tag{19}$$

Then we define our right inverse of T by

$$T^{-1}\mathbf{g} = (\{\tau_k\}_{k=0}^{N-1}, \{\xi_k\}_{k=0}^{N}), \tag{20}$$

with ξ_k and τ_k given in Eqs. (18) and (19) respectively. For future use we note here that Eqs.(18)-(20) imply the inequality

$$\|T^{-1}\| \leq C,$$

where C is given in the statement of the theorem.

Next note that

$$\begin{aligned}
\|T^{-1}(D\mathcal{G}(\mathbf{v}_0) - T)\| &\leq \|T^{-1}\|\|D\mathcal{G}(\mathbf{v}_0) - T\| \\
&\leq \|T^{-1}\| \sup_{0\leq k\leq N-1} \|F(\phi^{h_k}(\mathbf{y}_k)) - F(\mathbf{y}_{k+1})\| \\
&\leq \|T^{-1}\|M_1 \cdot \sup_{0\leq k\leq N-1} \|\phi^{h_k}(\mathbf{y}_k) - \mathbf{y}_{k+1}\| \\
&\leq CM_1\delta \\
&< 1.
\end{aligned}$$

Then it follows from standard operator theory that the operator $\mathcal{K}$ defined by

$$\mathcal{K} = [I + T^{-1}(D\mathcal{G}(\mathbf{v}_0) - T)]^{-1}T^{-1}$$

is a right inverse of $D\mathcal{G}(\mathbf{v}_0)$ and also that

$$\|\mathcal{K}\| \leq C(1 - CM_1\delta)^{-1}. \tag{21}$$

Now we complete the verification of the hypotheses of Lemma 11.4 with $\mathbf{y} = \mathbf{v}_0$ and $K = C(1 - CM_1\delta)^{-1}$. First note that

$$\|\mathcal{G}(\mathbf{v}_0)\| = \sup_{0\leq k\leq N-1} \|\mathbf{y}_{k+1} - \phi^{h_k}(\mathbf{y}_k)\| \leq \delta \tag{22}$$

and since

$$\varepsilon = 2C(1 - CM_1\delta)^{-1}\delta < \varepsilon_0,$$

the closed ball of radius ε around $\mathbf{v}_0$ is contained in O.

Next we obtain a bound for $\|D^2\mathcal{G}(\mathbf{v})\|$. If $\mathbf{v} = (\{t_k\}_{k=0}^{N-1}, \{\mathbf{z}_k\}_{k=0}^{N})$, $\mathbf{u} = (\{s_k\}_{k=0}^{N-1}, \{\mathbf{w}_k\}_{k=0}^{N})$ and $\bar{\mathbf{u}} = (\{\tau_k\}_{k=0}^{N-1}, \{\xi_k\}_{k=0}^{N})$, one calculates that

$$\begin{aligned}
[D^2\mathcal{G}(\mathbf{v})\mathbf{u}\bar{\mathbf{u}}]_k = \; & -s_k\tau_k DF(\phi^{t_k}(\mathbf{y}_k + S_k\mathbf{z}_k))F(\phi^{t_k}(\mathbf{y}_k + S_k\mathbf{z}_k)) \\
& -s_k DF(\phi^{t_k}(\mathbf{y}_k + S_k\mathbf{z}_k))D\phi^{t_k}(\mathbf{y}_k + S_k\mathbf{z}_k)S_k\xi_k \\
& -\tau_k DF(\phi^{t_k}(\mathbf{y}_k + S_k\mathbf{z}_k))D\phi^{t_k}(\mathbf{y}_k + S_k\mathbf{z}_k)S_k\mathbf{w}_k \\
& -D^2\phi^{t_k}(\mathbf{y}_k + S_k\mathbf{z}_k)(S_k\mathbf{w}_k)(S_k\xi_k).
\end{aligned} \tag{23}$$

We also have the following estimates:

$$\|D\phi^t(\mathbf{x})\| \le e^{M_1 t}, \quad \|D^2\phi^t(\mathbf{x})\| \le M_2 t e^{2M_1 t}, \tag{24}$$

when $0 \le t \le h_k + \varepsilon_0$ and $\|\mathbf{x} - \mathbf{y}_k\| < \varepsilon_0$ for some k. The first follows from Gronwall's lemma and the second follows from the variation of constants formula using the fact that for vectors ξ and η the function $\mathbf{w}(t) = D^2\phi^t(\mathbf{x})\xi\,\eta$ is the solution of the inhomogeneous linear equation

$$\dot{\mathbf{w}} = DF(\phi^t(\mathbf{x}))\mathbf{w} + D^2F(\phi^t(\mathbf{x}))\, D\phi^t(\mathbf{x})\xi\, D\phi^t(\mathbf{x})\eta$$

with $\mathbf{w}(0) = \mathbf{0}$. Then note that if $\mathbf{v} = (\{t_k\}_{k=0}^{N-1}, \{\mathbf{z}_k\}_{k=0}^{N})$ is in the ball of radius ε about $\mathbf{v}_0$, for $0 \le k \le N-1$

$$\begin{aligned}\|\phi^{t_k}(\mathbf{y}_k + S_k\mathbf{z}_k) - \mathbf{y}_{k+1}\| &\le \|\phi^{t_k}(\mathbf{y}_k + S_k\mathbf{z}_k) - \phi^{t_k}(\mathbf{y}_k)\| \\ &\quad + \|\phi^{t_k}(\mathbf{y}_k) - \phi^{h_k}(\mathbf{y}_k)\| + \|\phi^{h_k}(\mathbf{y}_k) - \mathbf{y}_{k+1}\| \\ &\le e^{M_1(h_k+\varepsilon_0)}\|S_k\mathbf{z}_k\| + M_0|t_k - h_k| + \delta \\ &\le (e^{M_1(h_{max}+\varepsilon_0)} + M_0)\varepsilon + \delta \\ &= \nu\delta.\end{aligned}$$

It follows that for $k = 0, \ldots, N-1$

$$\begin{aligned}&\|F(\phi^{t_k}(\mathbf{y}_k + S_k\mathbf{z}_k))\| \\ &\quad \le \|F(\mathbf{y}_{k+1})\| + \|F(\phi^{t_k}(\mathbf{y}_k + S_k\mathbf{z}_k)) - F(\mathbf{y}_{k+1})\| \\ &\quad \le \overline{M}_0 + M_1\nu\delta,\end{aligned} \tag{25}$$

and

$$\begin{aligned}&\|DF(\phi^{t_k}(\mathbf{y}_k + S_k\mathbf{z}_k))\| \\ &\quad \le \|DF(\mathbf{y}_{k+1})\| + \|DF(\phi^{t_k}(\mathbf{y}_k + S_k\mathbf{z}_k)) - DF(\mathbf{y}_{k+1})\| \\ &\quad \le \overline{M}_1 + M_2\nu\delta.\end{aligned} \tag{26}$$

Now from Eqs.(23), (24), (25) and (26), one sees that for $\|\mathbf{v} - \mathbf{v}_0\| \le \varepsilon$,

$$\|D^2\mathcal{G}(\mathbf{v})\| \le \overline{M}.$$

Finally we note from hypothesis (iii) of the theorem that

$$2\overline{M}C^2(1 - CM_1\delta)^{-2}\delta \le 1.$$

This completes the verification of the conditions of Lemma 11.4 and thus we may assert that Eq.(16) has a solution $\mathbf{v}$ which corresponds to a true orbit having the desired properties. Thus the proof of Theorem 11.9 is complete.

Implementation of the Theorem: The purpose of Theorem 11.9 is to establish the existence of a true solution near a computed approximate solution of the initial value problem

$$\dot{\mathbf{x}} = F(\mathbf{x}), \quad \mathbf{x}(0) = \mathbf{y}_0.$$

We assume this approximate solution has been obtained by a standard one-step method. So we have a sequence $\{\mathbf{y}_k\}_{k=0}^{N}$ with associated time steps $\{h_k\}_{k=0}^{N-1}$ such that

$$||\mathbf{y}_{k+1} - \phi^{h_k}(\mathbf{y}_k)|| \leq \delta \quad \text{for} \quad k = 0, \ldots, N-1.$$

We also compute $D\phi^{h_k}(\mathbf{y}_k)$ by applying the same method to the enlarged initial value problem

$$\dot{\mathbf{x}} = F(\mathbf{x}), \quad \dot{X} = DF(\mathbf{x})X, \quad \mathbf{x}(0) = \mathbf{y}_k, \quad X(0) = I. \tag{27}$$

To compute the S_k's and A_k's, we proceed as follows. First we choose S_0 as any matrix such that

$$\left[\frac{F(\mathbf{y}_0)}{||F(\mathbf{y}_0)||} \quad S_0 \right]$$

is orthogonal. We generate the remaining S_k's recursively by performing the QR factorizations

$$\left[\frac{F(\mathbf{y}_{k+1})}{||F(\mathbf{y}_{k+1})||} \quad D\phi^{h_k}(\mathbf{y}_k)S_k \right] = \left[\frac{F(\mathbf{y}_{k+1})}{||F(\mathbf{y}_{k+1})||} \quad S_{k+1} \right] \begin{bmatrix} \cdots & \cdots \\ 0 & A_k \end{bmatrix}$$

for $k = 0, \ldots, N-1$, where

$$A_k = S_{k+1}^* D\phi^{h_k}(\mathbf{y}_k)S_k.$$

Note, of course, that the A_k are upper triangular so that a right inverse L^{-1} for L and an upper bound for $||L^{-1}||$ can be found as in the map case (conf. Section 11.1). The rest of the procedure is demonstrated in the following example.

Example: We apply Theorem 11.9 to the Lorenz equations

$$\begin{aligned} \dot{x} &= \sigma(y - x) \\ \dot{y} &= \rho x - y - xz \\ \dot{z} &= xy - \beta z \end{aligned} \tag{28}$$

where σ, ρ and β are parameters. We use the classic parameter values $\sigma = 10$, $\rho = 28$ and $\beta = 8/3$.

Using the Lyapunov function

$$V(x, y, z) = \rho x^2 + \sigma y^2 + \sigma(z - 2\rho)^2,$$

it can be shown that the set

$$U = \{(x, y, z) : \rho x^2 + \sigma y^2 + \sigma(z - 2\rho)^2 \leq \sigma\rho^2\beta^2/(\beta - 1)\}$$

is forward invariant under the flow of the Lorenz equations for $\sigma \geq 1$, $\rho > 0$ and $\beta > 1$. The pseudo orbit we calculate lies inside U. We find that

$$M_0 \leq 5547, \;\; M_1 \leq 88, \;\; M_2 = \sqrt{2}.$$

We take $\varepsilon_0 = .00001$ and check at each step that $\text{dist}(\mathbf{y}_k, \partial U) > \varepsilon_0$ so that if $\|\mathbf{x} - \mathbf{y}_k\| \leq \varepsilon_0$ then the solution $\phi^t(\mathbf{x})$ lies in U for $t \geq 0$.

The pseudo orbit $\{\mathbf{y}_k\}_{k=0}^N$ of the Lorenz equations and the matrices $\{D\phi^{h_k}(\mathbf{y}_k)\}_{k=0}^{N-1}$ are generated by applying a Taylor series method of order 31 to the initial value problem (27) for $k = 0, \ldots, N-1$, where $N = 87,140,000$, $F(\mathbf{x})$ is as in Eq.(28) and

$$h_k = 9.765625 \times 10^{-3}.$$

Taking into account both discretization and round-off error, we find that

$$\delta \leq 1.98 \times 10^{-13}.$$

Also we find that

$$\Delta \geq 2.83, \;\; \overline{M}_0 \leq 437, \;\; \overline{M}_1 \leq 40.8, \;\; \theta \leq 1.17, \;\; \overline{M} \leq 18030.$$

Next we calculate the matrices S_k and A_k as described above, taking into account roundoff error as in Coomes, Koçak and Palmer [1995a], and find that

$$\|L^{-1}\| \leq 6477, \;\; C \leq 6477.$$

Then we check that

$$CM_1\delta \leq .00000012 < 1,$$

$$\varepsilon = 2C(1 - CM_1\delta)^{-1}\delta \leq .0000000026 < \varepsilon_0 = .00001$$

and

$$2\overline{M}C^2(1 - CM_1\delta)^{-1}\delta \leq .3 < 1.$$

So Theorem 11.9 enables us to conclude that the pseudo orbit $\{\mathbf{y}_k\}_{k=0}^N$ is shadowed by points $\{\mathbf{x}_k\}_{k=0}^N$ on a true orbit such that

$$\mathbf{x}_{k+1} = \phi^{t_k}(\mathbf{x}_k) \quad \text{for} \quad k = 0, \ldots, N-1,$$

$$\|\mathbf{x}_k - \mathbf{y}_k\| \leq \varepsilon \leq .0000000026 \quad \text{for} \quad k = 0, \ldots, N$$

and

$$|t_k - h_k| \leq \varepsilon \leq .0000000026 \quad \text{for} \quad k = 0, \ldots, N-1.$$

11.4 PERIODIC SHADOWING IN DIFFERENTIAL EQUATIONS

In this section we show how a numerically computed apparent periodic orbit of an autonomous system of differential equations can be rigorously verified to approximate a true periodic orbit. The method we give is particularly suited for long periodic orbits of chaotic systems. This section is based on Coomes, Koçak and Palmer [1997]. See Sinai and Vul [1980] and Franke and Selgrade [1979] for related work.

We again consider the C^2 autonomous system Eq.(15) with associated flow ϕ. First we give formal expression to the idea of an approximate periodic orbit.

Definition 11.10. For a given positive number δ, a sequence of points $\{\mathbf{y}_k\}_{k=0}^N$, with $F(\mathbf{y}_k) \neq \mathbf{0}$ for all k, is said to be a δ *pseudo periodic orbit* of Eq.(15) if there is an associated sequence $\{h_k\}_{k=0}^N$ of positive times such that

$$\|\mathbf{y}_{k+1} - \phi^{h_k}(\mathbf{y}_k)\| \leq \delta \quad \text{for} \quad k = 0, \dots, N-1,$$

and

$$\|\mathbf{y}_0 - \phi^{h_N}(\mathbf{y}_N)\| \leq \delta.$$

In order to decide if a δ pseudo periodic orbit $\{\mathbf{y}_k\}_{k=0}^N$ is near a true periodic orbit, we need to compute other quantities. For $k = 0, \dots, N$, let S_k be an $n \times (n-1)$ matrix chosen so that the matrix

$$[\,F(\mathbf{y}_k)/\|F(\mathbf{y}_k)\| \quad S_k\,]$$

is orthogonal. Then we set

$$A_k = \begin{cases} S_{k+1}^* D\phi^{h_k}(\mathbf{y}_k) S_k & \text{for } k = 0, \dots, N-1, \\ S_0^* D\phi^{h_N}(\mathbf{y}_N) S_N & \text{for } k = N. \end{cases} \tag{29}$$

We define the linear operator $L : (I\!R^{n-1})^{N+1} \to (I\!R^{n-1})^{N+1}$ as follows: if $\xi = \{\xi_k\}_{k=0}^N$ then

$$(L\xi)_k = \begin{cases} \xi_{k+1} - A_k \xi_k & \text{for } k = 0, \dots, N-1, \\ \xi_0 - A_N \xi_N & \text{for } k = N. \end{cases}$$

Let U be a convex set containing $\{\mathbf{y}_k\}_{k=0}^N$ in its interior. For such a U, we define

$$M_0 = \sup_{\mathbf{x} \in U} \|F(\mathbf{x})\|, \quad M_1 = \sup_{\mathbf{x} \in U} \|DF(\mathbf{x})\|, \quad M_2 = \sup_{\mathbf{x} \in U} \|D^2 F(\mathbf{x})\|.$$

Then we define

$$h_{min} = \min_{0 \leq k \leq N} h_k, \quad h_{max} = \max_{0 \leq k \leq N} h_k.$$

Next, we choose a positive number $\varepsilon_0 \leq h_{min}$ such that for $k = 0, \ldots, N$ and $\|\mathbf{x} - \mathbf{y}_k\| \leq \varepsilon_0$ the solution $\phi^t(\mathbf{x})$ is defined and remains in U for $0 \leq t \leq h_k + \varepsilon_0$. Finally, we define

$$\Delta = \inf_{0 \leq k \leq N} \|F(\mathbf{y}_k)\|, \quad \overline{M}_0 = \sup_{0 \leq k \leq N} \|F(\mathbf{y}_k)\|, \quad \overline{M}_1 = \sup_{0 \leq k \leq N} \|DF(\mathbf{y}_k)\|,$$

and

$$\Theta = \sup_{0 \leq k \leq N} \|D\phi^{h_k}(\mathbf{y}_k)\|.$$

Theorem 11.11. *Let $\{\mathbf{y}_k\}_{k=0}^N$ be a δ pseudo periodic orbit of the autonomous system* (15) *such that the operator L is invertible. Let*

$$C = \max\{\Delta^{-1}(\Theta\|L^{-1}\| + 1), \|L^{-1}\|\},$$

and

$$\begin{aligned} \overline{M} &= (\overline{M}_0 + M_1\nu\delta)(\overline{M}_1 + M_2\nu\delta) + 2(\overline{M}_1 + M_2\nu\delta)e^{M_1(h_{max}+\varepsilon_0)} \\ &\quad + M_2(h_{max} + \varepsilon_0)e^{2M_1(h_{max}+\varepsilon_0)}, \end{aligned}$$

where

$$\nu = 2C(e^{M_1(h_{max}+\varepsilon_0)} + M_0)(1 - CM_1\delta)^{-1} + 1.$$

If these quantities together with δ, Δ and ε_0 satisfy the inequalities

(i) $CM_1\delta < 1,$

(ii) $2C(1 - CM_1\delta)^{-1}\delta < \varepsilon_0,$ *and*

(iii) $2\overline{M}C^2(1 - CM_1\delta)^{-2}\delta < 1,$

then there exists a true periodic orbit $\{\mathbf{x}_k\}_{k=0}^N$, that is, there are times $\{t_k\}_{k=0}^N$ for which $\mathbf{x}_{k+1} = \phi^{t_k}(\mathbf{x}_k)$ for $k = 0, \ldots, N-1$ and $\mathbf{x}_0 = \phi^{t_N}(\mathbf{x}_N)$, such that

$$\|\mathbf{x}_k - \mathbf{y}_k\| \leq \varepsilon, \quad |t_k - h_k| \leq \varepsilon,$$

for $k = 0, \ldots, N$, where

$$\varepsilon = 2C(1 - CM_1\delta)^{-1}\delta.$$

Proof. We look for the points $\mathbf{x}_k$ in the hyperplanes $\mathcal{H}_k$ through $\mathbf{y}_k$ and normal to $F(\mathbf{y}_k)$. So we find a sequence of times $\{t_k\}_{k=0}^N$ and a sequence of points $\{\mathbf{x}_k\}_{k=0}^N$ with $\mathbf{x}_k \in \mathcal{H}_k$ and near $\mathbf{y}_k$ such that

$$\mathbf{x}_{k+1} = \phi^{t_k}(\mathbf{x}_k) \quad \text{for} \quad k = 0, \ldots, N-1 \quad \text{and} \quad \mathbf{x}_0 = \phi^{t_N}(\mathbf{x}_N).$$

It is clear from the definition of S_k that $\mathbf{x}_k \in \mathcal{H}_k$ can be uniquely represented in the form $\mathbf{x}_k = \mathbf{y}_k + S_k\mathbf{z}_k$, where $\mathbf{z}_k \in \mathbb{R}^{n-1}$. So we must solve the equations

$$\begin{aligned} \mathbf{y}_{k+1} + S_{k+1}\mathbf{z}_{k+1} &= \phi^{t_k}(\mathbf{y}_k + S_k\mathbf{z}_k), \quad k = 0, \ldots N-1, \\ \mathbf{y}_0 + S_0\mathbf{z}_0 &= \phi^{t_N}(\mathbf{y}_N + S_N\mathbf{z}_N) \end{aligned}$$

for the sequences $\{t_k\}_{k=0}^N$ and $\{\mathbf{z}_k\}_{k=0}^N$. Next we introduce the space $X = \mathbb{R}^{N+1} \times (\mathbb{R}^{n-1})^{N+1}$ with norm

$$\|(\{t_k\}_{k=0}^N, \{\mathbf{z}_k\}_{k=0}^N)\| = \max\left\{\max_{0\le k\le N} |t_k|, \max_{0\le k\le N} \|\mathbf{z}_k\|\right\}$$

and the space $Y = (\mathbb{R}^n)^{N+1}$ with norm

$$\|\{\mathbf{g}_k\}_{k=0}^N\| = \max_{0\le k\le N} \|\mathbf{g}_k\|.$$

Now we let O be the open set in X consisting of those $\mathbf{v} = (\{t_k\}_{k=0}^N, \{\mathbf{z}_k\}_{k=0}^N)$ with $|t_k - h_k| < \varepsilon_0$ and $\|\mathbf{z}_k\| < \varepsilon_0$, and introduce the function $\mathcal{G} : O \to Y$ given by

$$[\mathcal{G}(\mathbf{v})]_k = \begin{cases} \mathbf{y}_{k+1} + S_{k+1}\mathbf{z}_{k+1} - \phi^{t_k}(\mathbf{y}_k + S_k\mathbf{z}_k), & \text{for} \quad k = 0, \ldots N-1 \\ \mathbf{y}_0 + S_0\mathbf{z}_0 - \phi^{t_N}(\mathbf{y}_N + S_N\mathbf{z}_N) & \text{for} \quad k = N. \end{cases}$$

Then the theorem will be proved if we can find a solution $\mathbf{v} = (\{t_k\}_{k=0}^N, \{\mathbf{z}_k\}_{k=0}^N)$ of the equation

$$\mathcal{G}(\mathbf{v}) = \mathbf{0}, \tag{30}$$

in the closed ball of radius ε about $\mathbf{v}_0 = (\{h_k\}_{k=0}^N, \mathbf{0})$. We use Lemma 11.7 with $\mathbf{y} = \mathbf{v}_0$ to solve this equation.

Construction of inverse for $D\mathcal{G}(\mathbf{v}_0)$: Note that for $\mathbf{u} = (\{\tau_k\}_{k=0}^N, \{\xi_k\}_{k=0}^N) \in X$, the derivative of $\mathcal{G}$ at $\mathbf{v}_0$ is given by

$$[D\mathcal{G}(\mathbf{v}_0)\mathbf{u}]_k =$$

$$\begin{cases} -\tau_k F(\phi^{h_k}(\mathbf{y}_k)) + S_{k+1}\xi_{k+1} - D\phi^{h_k}(\mathbf{y}_k)S_k\xi_k & \text{for} \quad k = 0, \ldots, N-1, \\ -\tau_N F(\phi^{h_N}(\mathbf{y}_N)) + S_0\xi_0 - D\phi^{h_N}(\mathbf{y}_N)S_N\xi_N & \text{for} \quad k = N. \end{cases}$$

We approximate $D\mathcal{G}(\mathbf{v}_0)$ by the linear operator $T : X \to Y$ defined by:

$$[T\mathbf{u}]_k = \begin{cases} -\tau_k F(\mathbf{y}_{k+1}) + S_{k+1}\xi_{k+1} - D\phi^{h_k}(\mathbf{y}_k)S_k\xi_k & \text{for} \quad k = 0, \ldots, N-1, \\ -\tau_N F(\mathbf{y}_0) + S_0\xi_0 - D\phi^{h_N}(\mathbf{y}_N)S_N\xi_N & \text{for} \quad k = N. \end{cases}$$

In order to find the inverse for T, for given $\mathbf{g} = \{\mathbf{g}_k\}_{k=0}^N$ in $(\mathbb{R}^n)^{N+1}$, we must solve the set of equations

$$[T\mathbf{u}]_k = \mathbf{g}_k, \quad k = 0, \ldots, N, \tag{31}$$

for $\mathbf{u}$. Since, for each k, the matrix

$$[F(\mathbf{y}_k)/\|F(\mathbf{y}_k)\| \quad S_k]$$

is orthogonal, this set of equations is equivalent to the following two sets of equations, one set obtained by premultiplying the kth member in Eq.(31) by $F(\mathbf{y}_{k+1})^*$, and the other set obtained by premultiplying the kth member in Eq.(31) by S_{k+1}^* (with $N+1$ understood as 0):

$$\begin{aligned}
-\|F(\mathbf{y}_{k+1})\|^2\tau_k - F(\mathbf{y}_{k+1})^* D\phi^{h_k}(\mathbf{y}_k)S_k\xi_k &= F(\mathbf{y}_{k+1})^*\mathbf{g}_k,\ k=0,\ldots,N-1, \\
-\|F(\mathbf{y}_0)\|^2\tau_k - F(\mathbf{y}_0)^* D\phi^{h_N}(\mathbf{y}_N)S_N\xi_N &= F(\mathbf{y}_0)^*\mathbf{g}_N, \\
\xi_{k+1} - A_k\xi_k &= S_{k+1}^*\mathbf{g}_k, \quad k=0,\ldots,N-1, \\
\xi_0 - A_N\xi_N &= S_0^*\mathbf{g}_N.
\end{aligned}$$

If we write $\bar{\mathbf{g}} = \{S_{k+1}^*\mathbf{g}_k\}_{k=0}^N$ (with $N+1$ understood as 0), then we see that a solution of the second set is

$$\xi_k = (L^{-1}\bar{\mathbf{g}})_k, \quad k=0,\ldots,N. \tag{32}$$

We substitute this into the first set and obtain

$$\tau_k = \begin{cases} -\dfrac{F(\mathbf{y}_{k+1})^*}{\|F(\mathbf{y}_{k+1})\|^2}\{D\phi^{h_k}(\mathbf{y}_k)S_k(L^{-1}\bar{\mathbf{g}})_k + \mathbf{g}_k\} & \text{for } k=0,\ldots,N-1, \\ -\dfrac{F(\mathbf{y}_0)^*}{\|F(\mathbf{y}_0)\|^2}\{D\phi^{h_N}(\mathbf{y}_N)S_N(L^{-1}\bar{\mathbf{g}})_N + \mathbf{g}_N\} & \text{for } k=N. \end{cases} \tag{33}$$

Then the inverse of T is given by

$$T^{-1}\mathbf{g} = (\{\tau_k\}_{k=0}^N, \{\xi_k\}_{k=0}^N), \tag{34}$$

with ξ_k and τ_k given in Eqs.(32) and (33) respectively. For future use we note here that

$$\|T^{-1}\| \le C,$$

where C is given in the statement of the theorem.

Next note that for $k=0,\ldots,N-1$,

$$\begin{aligned}
\|[(D\mathcal{G}(\mathbf{v}_0) - T)\mathbf{u}]_k\| &\le \|F(\phi^{h_k}(\mathbf{y}_k)) - F(\mathbf{y}_{k+1})\|\|\mathbf{u}\| \\
&\le M_1\|\phi^{h_k}(\mathbf{y}_k) - \mathbf{y}_{k+1}\|\|\mathbf{u}\| \\
&\le M_1\delta\|\mathbf{u}\|.
\end{aligned}$$

When $k=N$ a similar inequality holds and thus

$$\|D\mathcal{G}(\mathbf{v}_0) - T\| \le M_1\delta.$$

Since

$$\|T^{-1}\|M_1\delta \le CM_1\delta < 1,$$

it follows from standard operator theory that the operator $\mathcal{K}$ defined by

$$\mathcal{K} = [I + T^{-1}(D\mathcal{G}(\mathbf{v}_0) - T)]^{-1}T^{-1}$$

is the inverse of $D\mathcal{G}(\mathbf{v}_0)$ and also that

$$\|\mathcal{K}\| \le C(1 - CM_1\delta)^{-1}. \tag{35}$$

Now we complete the verification of the hypotheses of Lemma 11.7 with $\mathbf{y} = \mathbf{v}_0$ and $K = C(1 - CM_1\delta)^{-1}$. First

$$\|\mathcal{G}(\mathbf{v}_0)\| \le \delta \tag{36}$$

and since

$$\varepsilon = 2C(1 - CM_1\delta)^{-1}\delta < \varepsilon_0,$$

the closed ball of radius ε around $\mathbf{v}_0$ is contained in O.

Next we obtain a bound for $\|D^2\mathcal{G}(\mathbf{v})\|$. If $\mathbf{v} = (\{t_k\}_{k=0}^N, \{\mathbf{z}_k\}_{k=0}^N)$, $\mathbf{u} = (\{s_k\}_{k=0}^N, \{\mathbf{w}_k\}_{k=0}^N)$ and $\bar{\mathbf{u}} = (\{\tau_k\}_{k=0}^N, \{\xi_k\}_{k=0}^N)$, one calculates for $k = 0, \ldots, N$

$$\begin{aligned}[D^2\mathcal{G}(\mathbf{v})\mathbf{u}\bar{\mathbf{u}}]_k = \; & -s_k\tau_k DF(\phi^{t_k}(\mathbf{y}_k + S_k\mathbf{z}_k))F(\phi^{t_k}(\mathbf{y}_k + S_k\mathbf{z}_k)) \\ & -s_k DF(\phi^{t_k}(\mathbf{y}_k + S_k\mathbf{z}_k))D\phi^{t_k}(\mathbf{y}_k + S_k\mathbf{z}_k)S_k\xi_k \\ & -\tau_k DF(\phi^{t_k}(\mathbf{y}_k + S_k\mathbf{z}_k))D\phi^{t_k}(\mathbf{y}_k + S_k\mathbf{z}_k)S_k\mathbf{w}_k \\ & -D^2\phi^{t_k}(\mathbf{y}_k + S_k\mathbf{z}_k)(S_k\mathbf{w}_k)(S_k\xi_k).\end{aligned} \tag{37}$$

As in Eq.(24), we also have the following estimates:

$$\|D\phi^t(\mathbf{x})\| \le e^{M_1 t}, \quad \|D^2\phi^t(\mathbf{x})\| \le M_2 t e^{2M_1 t}, \tag{38}$$

when $0 \le t \le h_k + \varepsilon_0$ and $\|\mathbf{x} - \mathbf{y}_k\| < \varepsilon_0$ for some k. Then note that if $\mathbf{v} = (\{t_k\}_{k=0}^N, \{\mathbf{z}_k\}_{k=0}^N)$ is in the ball of radius ε about $\mathbf{v}_0$, for $k = 0, \ldots, N-1$

$$\begin{aligned}\|\phi^{t_k}(\mathbf{y}_k + S_k\mathbf{z}_k) - \mathbf{y}_{k+1}\| & \le \|\phi^{t_k}(\mathbf{y}_k + S_k\mathbf{z}_k) - \phi^{t_k}(\mathbf{y}_k)\| \\ & \quad + \|\phi^{t_k}(\mathbf{y}_k) - \phi^{h_k}(\mathbf{y}_k)\| + \|\phi^{h_k}(\mathbf{y}_k) - \mathbf{y}_{k+1}\| \\ & \le e^{M_1(h_k + \varepsilon_0)}\|S_k\mathbf{z}_k\| + M_0|t_k - h_k| + \delta \\ & \le (e^{M_1(h_{max} + \varepsilon_0)} + M_0)\varepsilon + \delta \\ & = \nu\delta.\end{aligned}$$

It follows that for $k = 0, \dots, N-1$

$$\begin{aligned} &\|F(\phi^{t_k}(\mathbf{y}_k + S_k\mathbf{z}_k))\| \\ &\qquad \le \|F(\mathbf{y}_{k+1})\| + \|F(\phi^{t_k}(\mathbf{y}_k + S_k\mathbf{z}_k)) - F(\mathbf{y}_{k+1})\| \qquad (39) \\ &\qquad \le \overline{M}_0 + M_1\nu\delta, \end{aligned}$$

and

$$\begin{aligned} &\|DF(\phi^{t_k}(\mathbf{y}_k + S_k\mathbf{z}_k))\| \\ &\qquad \le \|DF(\mathbf{y}_{k+1})\| + \|DF(\phi^{t_k}(\mathbf{y}_k + S_k\mathbf{z}_k)) - DF(\mathbf{y}_{k+1})\| \qquad (40) \\ &\qquad \le \overline{M}_1 + M_2\nu\delta. \end{aligned}$$

When $k = N$ we get the same estimates as above by simply replacing every occurrence of $\mathbf{y}_{k+1}$ by $\mathbf{y}_0$. Thus from Eqs.(37), (38), (39) and (40), one sees that for $\|\mathbf{v} - \mathbf{v}_0\| \le \varepsilon$,

$$\|D^2\mathcal{G}(\mathbf{v})\| \le \overline{M}.$$

Finally, we also know from hypothesis (iii) of the theorem that

$$2\overline{M}C^2(1 - CM_1\delta)^{-2}\delta < 1.$$

This completes the verification of the conditions of Lemma 11.7 and thus we may assert that Eq.(30) has a solution $\mathbf{v}$ which corresponds to a true orbit having the desired properties. Thus the proof of Theorem 11.11 is complete.

Implementation of Theorem 11.11: The purpose of Theorem 11.11 is to establish the existence of a true periodic solution near a computed approximate periodic solution of Eq.(15). So we assume we have a sequence $\{\mathbf{y}_k\}_{k=0}^N$ with associated time steps $\{h_k\}_{k=0}^N$ such that

$$\|\mathbf{y}_{k+1} - \phi^{h_k}(\mathbf{y}_k)\| \le \delta \quad \text{for} \quad k = 0, \dots, N-1$$

and

$$\|\mathbf{y}_0 - \phi^{h_N}(\mathbf{y}_N)\| \le \delta.$$

This sequence may have been obtained, for example, by the Newton's method described in Coomes, Koçak and Palmer [1997]. We also compute $D\phi^{h_k}(\mathbf{y}_k)$ by applying a standard one-step method to the initial value problem

$$\dot{\mathbf{x}} = F(\mathbf{x}), \quad \dot{X} = DF(\mathbf{x})X, \quad \mathbf{x}(0) = \mathbf{y}_k, \quad X(0) = I. \qquad (41)$$

To compute the S_k's and A_k's, we proceed as follows. First we choose S_0 as any matrix such that

$$\left[\frac{F(\mathbf{y}_0)}{\|F(\mathbf{y}_0)\|} \quad S_0 \right]$$

is orthogonal. We generate S_k for $k = 1, \ldots, N+1$ recursively by performing the QR factorizations

$$\left[\frac{F(\mathbf{y}_{k+1})}{\|F(\mathbf{y}_{k+1})\|} \quad D\phi^{h_k}(\mathbf{y}_k)S_k\right] = \left[\frac{F(\mathbf{y}_{k+1})}{\|F(\mathbf{y}_{k+1})\|} \quad S_{k+1}\right]\begin{bmatrix} \cdots & \cdots \\ 0 & A_k \end{bmatrix}$$

for $k = 0, \ldots, N$, where

$$A_k = S_{k+1}^* D\phi^{h_k}(\mathbf{y}_k)S_k.$$

Then we take S_{N+1} as the new S_0 and repeat the procedure. For chaotic systems such as the Lorenz system, we usually find that, up to sign, the columns of S_{N+1} are independent of the initial S_0. So the new S_0 will have, up to sign, the same columns as the newly generated S_{N+1}. This means that if we take A_N as $S_0^* D\phi^{h_N}(\mathbf{y}_N)S_N$ it will still be upper triangular. Hence the A_k's, as defined in Eq.(29), are upper triangular so that the invertibility of L and an upper bound for $\|L^{-1}\|$ can be determined as in the map case (see Section 11.2). The rest of the procedure is illustrated in the following example.

Example: We apply Theorem 11.11 to the Lorenz equations (28) with the classic parameter values $\sigma = 10$, $\rho = 28$ and $\beta = 8/3$. The pseudo periodic orbit $\{\mathbf{y}_k\}_{k=0}^N$ we calculate lies inside the forward invariant set

$$U = \{(x, y, z) : \rho x^2 + \sigma y^2 + \sigma(z - 2\rho)^2 \le \sigma\rho^2\beta^2/(\beta - 1)\}.$$

We find that

$$M_0 \le 5547, \quad M_1 \le 88, \quad M_2 = \sqrt{2}.$$

We take $\varepsilon_0 = .00001$ and check at each step that $\text{dist}(\mathbf{y}_k, \partial U) > \varepsilon_0$ so that if $\|\mathbf{x} - \mathbf{y}_k\| \le \varepsilon_0$ then the solution $\phi^t(\mathbf{x})$ lies in U for $t \ge 0$.

The δ pseudo periodic orbit $\{\mathbf{y}_k\}_{k=0}^N$ with associated times $\{h_k\}_{k=0}^N$ is generated using the Newton method described in Coomes, Koçak and Palmer [1997]. Here $N = 113044$ and

$$.00148 \le h_k \le .00991 \quad \text{and} \quad 1100.78 \le \sum_{k=0}^{N} h_k \le 1100.79.$$

The matrices $\{D\phi^{h_k}(\mathbf{y}_k)\}_{k=0}^N$ are generated by applying a Taylor series method of order 30 to the initial value problem (41), where $F(\mathbf{x})$ is as in Eq.(28). We find that

$$\delta \le 2.48 \times 10^{-15}.$$

Also we find that

$$\Delta \ge 10.5, \quad \overline{M}_0 \le 397, \quad \overline{M}_1 \le 39.4, \quad \theta \le 1.16, \quad \overline{M} \le 15840.$$

Next we calculate the matrices S_k and A_k as described above and find that

$$\|L^{-1}\| \le 1058, \quad C \le 1058.$$

Then we check that

$$CM_1\delta \le 1058 \times 88 \times 2.48 \times 10^{-15} \le 2.31 \times 10^{-10} < 1,$$

$$\varepsilon \le 2\times 1058\times(1-2.31\times 10^{-10})^{-1}\times 2.48\times 10^{-15} \le 5.25\times 10^{-12} < \varepsilon_0 = .00001$$

and

$$\begin{aligned} &2\overline{M}C^2(1-CM_1\delta)^{-2}\delta \\ &\qquad \le 2\times 15840\times 1058^2\times(1-2.31\times 10^{-10})^{-2}\times 2.48\times 10^{-15} \\ &\qquad < 1. \end{aligned}$$

So Theorem 11.11 enables us to conclude that the pseudo periodic orbit $\{\mathbf{y}_k\}_{k=0}^N$ is shadowed by points $\{\mathbf{x}_k\}_{k=0}^N$ on a true orbit such that

$$\mathbf{x}_{k+1} = \phi^{t_k}(\mathbf{x}_k) \quad \text{for} \quad k = 0,\ldots,N-1 \quad \text{and} \quad \mathbf{x}_0 = \phi^{t_N}(\mathbf{x}_N),$$

and such that for $k = 0,\ldots,N$

$$\|\mathbf{x}_k - \mathbf{y}_k\| \le 5.25\times 10^{-12}, \quad |t_k - h_k| \le 5.25\times 10^{-12}.$$

Also the period T of this periodic orbit satisfies

$$\sum_{k=0}^{N} h_k - (N+1)\varepsilon \le T \le \sum_{k=0}^{N} h_k + (N+1)\varepsilon$$

and so lies between 1100.77 and 1100.8.

As shown in Coomes, Koçak and Palmer [1997], error, as for example in the calculation of $D\phi^{h_k}(\mathbf{y}_k)$ and round off in some of the other calculations, does not have a significant effect on these results. In the same article, the same technique as that employed above was used to verify the existence of periodic orbits of the Lorenz system (28) with prescribed geometries, thus supplying proofs for the existence of periodic orbits found by Curry [1979].

11.5 RIGOROUS VERIFICATION OF CHAOTIC BEHAVIOUR

In this section, following Stoffer and Palmer [1998], we use shadowing techniques to rigorously verify the existence of chaos in discrete dynamical systems, where by chaos we mean the existence of a subsystem isomorphic to the shift as in Chapter 4. The basic idea, which comes from Stoffer [1988a], is that the shift can be embedded in the presence of two nearby periodic orbits, provided a certain shadowing property holds, that is, there are orbits of the dynamical system which shadow the periodic orbits with any switch from one to the other allowed. Actually the periodic orbits need only be pseudo periodic. Stoffer's idea is embodied in the following theorem.

Theorem 11.12. *Let $f : \mathbb{R}^n \to \mathbb{R}^n$ be a C^1 diffeomorphism. Let $\{\mathbf{y}_k\}_{k=0}^N$ and $\{\mathbf{z}_k\}_{k=0}^N$ be two distinct sequences of points in $\mathbb{R}^n$ and let ε be a positive number such that*

(i) $2\varepsilon < \max_{0\le k\le N} \|\mathbf{y}_k - \mathbf{z}_k\|$;

(ii) *given any sequence $\{a_i\}_{i=-\infty}^{\infty}$ of zeros and ones, there is exactly one true orbit $\{\mathbf{x}_k\}_{k=-\infty}^{\infty}$ of f such that for $i \in \mathbf{Z}$ and $0 \le k \le N$*

$$\begin{aligned} \|\mathbf{x}_{i(N+1)+k} - \mathbf{y}_k\| &\le \varepsilon \quad \text{if} \quad a_i = 0, \\ \|\mathbf{x}_{i(N+1)+k} - \mathbf{z}_k\| &\le \varepsilon \quad \text{if} \quad a_i = 1. \end{aligned}$$

Then if S is the set of all sequences $\{a_i\}_{i=-\infty}^{\infty}$ of zeros and ones endowed with the product discrete topology, there exists a homeomorphism $\phi : S \to \phi(S) \subset \mathbb{R}^n$ such that

$$\phi \circ \sigma = f^{N+1} \circ \phi,$$

where $\sigma : S \to S$ is the shift defined by

$$\sigma(\{a_i\}_{i=-\infty}^{\infty}) = \{a_{i+1}\}_{i=-\infty}^{\infty}.$$

Proof. Define

$$\phi(\{a_i\}_{i=-\infty}^{\infty}) = \mathbf{x}_0,$$

where $\{\mathbf{x}_k\}_{k=-\infty}^{\infty}$ is as in (ii).

First we show that ϕ is one to one. Let $\{a_i\}_{i=-\infty}^{\infty}$ and $\{b_i\}_{i=-\infty}^{\infty}$ be two distinct points of S. Then there exists $j \in \mathbf{Z}$ such that $a_j \ne b_j$. Now if $\{\mathbf{x}_k\}_{k=-\infty}^{\infty}$ corresponds to both $\{a_i\}_{i=-\infty}^{\infty}$ and $\{b_i\}_{i=-\infty}^{\infty}$ as in (ii), then supposing without loss of generality that $a_j = 0$ and $b_j = 1$, we have

$$\|\mathbf{y}_k - \mathbf{z}_k\| \le \|\mathbf{x}_{j(N+1)+k} - \mathbf{y}_k\| + \|\mathbf{x}_{j(N+1)+k} - \mathbf{z}_k\| \le 2\varepsilon$$

for $k = 0, \dots, N$. However this contradicts (i). Hence ϕ is, indeed, one to one.

Next note that $\{f^{N+1}(\mathbf{x}_k)\}_{k=-\infty}^{\infty}$ is an orbit of f with

$$\begin{aligned} \|f^{N+1}(\mathbf{x}_{i(N+1)+k}) - \mathbf{y}_k\| &= \|\mathbf{x}_{(i+1)(N+1)+k} - \mathbf{y}_k\| \le \varepsilon \quad \text{if} \quad a_{i+1} = 0, \\ \|f^{N+1}(\mathbf{x}_{i(N+1)+k}) - \mathbf{z}_k\| &= \|\mathbf{x}_{(i+1)(N+1)+k} - \mathbf{z}_k\| \le \varepsilon \quad \text{if} \quad a_{i+1} = 1. \end{aligned}$$

Hence, by uniqueness,

$$\phi(\{a_{i+1}\}_{i=-\infty}^{\infty})) = f^{N+1}(\mathbf{x}_0)$$

and so

$$\phi(\sigma(\{a_i\}_{i=-\infty}^{\infty})) = f^{N+1}(\phi(\{a_i\}_{i=-\infty}^{\infty})).$$

That is,

$$\phi \circ \sigma = f^{N+1} \circ \phi.$$

To show that ϕ is continuous, suppose on the contrary that it is not continuous at $\{a_i\}_{i=-\infty}^{\infty}$. Then there exists a positive number ε_1 and a sequence $\{a_i^{(m)}\}_{i=-\infty}^{\infty}$ with $\phi(\{a_i^{(m)}\}_{i=-\infty}^{\infty}) = \mathbf{x}^{(m)}$ such that $\{a_i^{(m)}\}_{i=-\infty}^{\infty} \to \{a_i\}_{i=-\infty}^{\infty}$ as $m \to \infty$ but

$$||\mathbf{x}^{(m)} - \mathbf{x}|| \geq \varepsilon_1, \tag{42}$$

for all m, where $\mathbf{x} = \phi(\{a_i\}_{i=-\infty}^{\infty})$. Now for all m either $||\mathbf{x}^{(m)} - \mathbf{y}_0|| \leq \varepsilon$ or $||\mathbf{x}^{(m)} - \mathbf{z}_0|| \leq \varepsilon$. So, by compactness, we may assume that $\mathbf{x}^{(m)} \to \bar{\mathbf{x}}$. Next we know that for $i \in \mathbf{Z}$ and $0 \leq k \leq N$

$$\begin{aligned} ||f^{i(N+1)+k}(\mathbf{x}^{(m)}) - \mathbf{y}_k|| &\leq \varepsilon \quad \text{if} \quad a_i^{(m)} = 0, \\ ||f^{i(N+1)+k}(\mathbf{x}^{(m)}) - \mathbf{z}_k|| &\leq \varepsilon \quad \text{if} \quad a_i^{(m)} = 1. \end{aligned}$$

But, for fixed i, $a_i^{(m)} = a_i$ if m is sufficiently large. So if for fixed i and k we let m tend to ∞, we obtain the fact that

$$\begin{aligned} ||f^{i(N+1)+k}(\bar{\mathbf{x}}) - \mathbf{y}_k|| &\leq \varepsilon \quad \text{if} \quad a_i = 0, \\ ||f^{i(N+1)+k}(\bar{\mathbf{x}}) - \mathbf{z}_k|| &\leq \varepsilon \quad \text{if} \quad a_i = 1. \end{aligned}$$

By uniqueness, this implies that $\bar{\mathbf{x}} = \mathbf{x}$. However, if we let m tend to ∞ in Eq.(42), we obtain $||\bar{\mathbf{x}} - \mathbf{x}|| \geq \varepsilon_1$. This is a contradiction and so we may conclude that ϕ is continuous, as asserted.

Since S is a compact Hausdorff space, we conclude that $\phi(S)$ is a closed subset of $\mathbb{R}^n$ and that $\phi : S \to \phi(S) \subset \mathbb{R}^n$ is a homeomorphism. Thus the proof of the theorem is complete.

Note that in the theorem the sequences $\{\mathbf{y}_k\}_{k=0}^{N}$ and $\{\mathbf{z}_k\}_{k=0}^{N}$ have to be δ pseudo periodic orbits with

$$||\mathbf{y}_0 - \mathbf{z}_0|| \leq 2\delta,$$

where $\delta = (1 + M_1)\varepsilon$, M_1 being a bound on $||Df(\mathbf{x})||$ over a suitable set. So to apply the theorem we first have to find two pseudo periodic orbits with closeby initial points.

However, first we deal with the problem of verifying the shadowing property. We present a shadowing theorem to be used in this verification. In the classical shadowing theorem it is shown that pseudo orbits in hyperbolic sets are shadowed by true orbits. In the situation considered below, it is not known that the pseudo orbits lie in a hyperbolic set. We give a theorem which, at least in principle, could be applied to almost any pseudo orbit of a map. A criterion for deciding whether or not such a pseudo orbit is shadowed by a true orbit is given in terms of the norm of a certain linear operator associated with the pseudo orbit.

Denote by $\ell^\infty(\mathbb{Z}, \mathbb{R}^n)$ the space of $\mathbb{R}^n$-valued sequences $\mathbf{x} = \{\mathbf{x}_k\}_{k=-\infty}^{\infty}$ with norm

$$\|\mathbf{x}\| = \sup_{k\in\mathbb{Z}} \|\mathbf{x}_k\|.$$

With a given bounded δ pseudo orbit $\{\mathbf{y}_k\}_{k=-\infty}^{\infty}$ we associate a linear operator $L : \ell^\infty(\mathbb{Z}, \mathbb{R}^n) \to \ell^\infty(\mathbb{Z}, \mathbb{R}^n)$ given for $\mathbf{z} = \{\mathbf{z}_k\}_{k=-\infty}^{\infty} \in \ell^\infty(\mathbb{Z}, \mathbb{R}^n)$ by

$$(L\mathbf{z})_k = \mathbf{z}_{k+1} - Df(\mathbf{y}_k)\mathbf{z}_k \quad \text{for} \quad k \in \mathbb{Z}. \tag{43}$$

Now we are ready to state our theorem.

Theorem 11.13. *Let $f : \mathbb{R}^n \to \mathbb{R}^n$ be a C^2 map with*

$$M = \sup_{\mathbf{x}\in\mathbb{R}^n} \|D^2 f(\mathbf{x})\|.$$

Let $\{\mathbf{y}_k\}_{k=-\infty}^{\infty}$ be a bounded δ pseudo orbit of f. Then if the operator L defined in Eq.(43) is invertible with $\|L^{-1}\| \le C$, where

$$2MC^2\delta < 1,$$

there is a unique true orbit $\{\mathbf{x}_k\}_{k=-\infty}^{\infty}$ of f, that is, $\mathbf{x}_{k+1} = f(\mathbf{x}_k)$ for all k, such that

$$\sup_{k\in\mathbb{Z}} \|\mathbf{x}_k - \mathbf{y}_k\| < 1/MC.$$

Moreover,

$$\sup_{k\in\mathbb{Z}} \|\mathbf{x}_k - \mathbf{y}_k\| \le 2C\delta.$$

Proof. We have to solve the difference equation

$$\mathbf{x}_{k+1} = f(\mathbf{x}_k), \quad k \in \mathbb{Z}$$

for $\mathbf{x}_k$ near $\mathbf{y}_k$. Set

$$\mathbf{x}_k = \mathbf{y}_k + \mathbf{z}_k.$$

Then we obtain the equation

$$\mathbf{z}_{k+1} = Df(\mathbf{y}_k)\mathbf{z}_k + g_k(\mathbf{z}_k), \quad k \in \mathbb{Z}, \tag{44}$$

where

$$g_k(\mathbf{z}) = f(\mathbf{y}_k + \mathbf{z}) - f(\mathbf{y}_k) - Df(\mathbf{y}_k)\mathbf{z} + f(\mathbf{y}_k) - \mathbf{y}_{k+1}.$$

We define the nonlinear operator $g : \ell^\infty(\mathbb{Z}, \mathbb{R}^n) \to \ell^\infty(\mathbb{Z}, \mathbb{R}^n)$ for $\mathbf{z} = \{\mathbf{z}_k\}_{k=-\infty}^{\infty}$ by

$$[g(\mathbf{z})]_k = g_k(\mathbf{z}_k) \quad \text{for} \quad k \in \mathbb{Z}.$$

Then we can write Eq.(44) as

$$Lz = g(z) \tag{45}$$

and we seek a solution $\mathbf{z} = \{\mathbf{z}_k\}_{k=-\infty}^{\infty}$ of this equation such that

$$\|\mathbf{z}\| \le \varepsilon,$$

where

$$2C\delta \le \varepsilon < 1/MC.$$

Now Eq.(45) is equivalent to the fixed point equation

$$\mathbf{z} = S\mathbf{z} := L^{-1}g(\mathbf{z}).$$

All we have to do is show that S is a contraction on the set of $\mathbf{z} \in \ell^\infty(\mathbf{Z}, \mathbb{R}^n)$ with $\|\mathbf{z}\| \le \varepsilon$. Well, if $\|\mathbf{z}\| \le \varepsilon$,

$$\begin{aligned} \|S\mathbf{z}\| &\le C\left[\tfrac{1}{2}M\|\mathbf{z}\|^2 + \delta\right] \\ &\le \tfrac{1}{2}MC\varepsilon^2 + C\delta \\ &\le \frac{\varepsilon}{2} + \frac{\varepsilon}{2} \\ &= \varepsilon. \end{aligned}$$

Furthermore, if $\mathbf{z}$ and $\mathbf{w}$ are in $\ell^\infty(\mathbf{Z}, \mathbb{R}^n)$ with $\|\mathbf{z}\| \le \varepsilon$ and $\|\mathbf{w}\| \le \varepsilon$, then

$$\|S\mathbf{z} - S\mathbf{w}\| = \|L^{-1}[g(\mathbf{z}) - g(\mathbf{w})]\| \le CM\varepsilon\|\mathbf{z} - \mathbf{w}\|.$$

Hence S is a contraction and the theorem follows.

In applying the theorem the main problem is verifying that the operator L is invertible and estimating the norm of its inverse. We deal with this problem now.

The Operator L: We restrict ourselves to the case $n = 2$. Then our operator L has the form

$$(L\mathbf{z})_k = \mathbf{z}_{k+1} - A_k\mathbf{z}_k, \quad k \in \mathbf{Z},$$

where A_k is a 2×2 matrix. It will transpire that we can assume A_k to have the upper triangular form

$$A_k = \begin{bmatrix} a_k & b_k \\ 0 & c_k \end{bmatrix}.$$

If $\mathbf{g} = \{\mathbf{g}_k\}_{k=-\infty}^{\infty} \in \ell^\infty(\mathbf{Z}, \mathbb{R}^n)$, to find $\mathbf{u} = L^{-1}\mathbf{g}$, we have to solve the equation $L\mathbf{u} = \mathbf{g}$, that is, we have to solve the difference equation

$$\mathbf{u}_{k+1} = A_k\mathbf{u}_k + \mathbf{g}_k, \quad k \in \mathbf{Z}$$

for a bounded solution $\mathbf{u}_k$. In components, this equation has the form

$$\begin{aligned}\mathbf{u}_{k+1}^{(1)} &= a_k \mathbf{u}_k^{(1)} + b_k \mathbf{u}_k^{(2)} + \mathbf{g}_k^{(1)} \\ \mathbf{u}_{k+1}^{(2)} &= c_k \mathbf{u}_k^{(2)} + \mathbf{g}_k^{(2)}.\end{aligned} \tag{46}$$

We solve the second set of equations first, using the following lemma.

Lemma 11.14. *Let $\{c_k\}_{k=-\infty}^{\infty}$ be a sequence of real numbers. For $j \geq 1$, set*

$$\alpha_j = \sup_{k \in \mathbf{Z}} |c_{k+j-1} \cdots c_k|.$$

Suppose

$$\alpha_1 < \infty$$

and for some J

$$\alpha_J < 1.$$

Then if $\{h_k\}_{k=-\infty}^{\infty}$ is a bounded sequence, the scalar difference equation

$$z_{k+1} = c_k z_k + h_k \tag{47}$$

has a unique bounded solution z_k. Moreover,

$$|z_k| \leq (1-\alpha_J)^{-1}(1 + \alpha_1 + \ldots + \alpha_{J-1}) \sup_{j \in \mathbf{Z}} |h_j| \tag{48}$$

for all k.

Proof. For a positive integer p,

$$\begin{aligned}\sum_{m=k-pJ+1}^{k} |c_{k-1} \cdots c_m| &= \sum_{i=0}^{p-1} \sum_{m=k-(i+1)J+1}^{k-iJ} |c_{k-1} \cdots c_m| \\ &= \sum_{i=0}^{p-1} |c_{k-1} \cdots c_{k-iJ}| \sum_{m=k-(i+1)J+1}^{k-iJ} |c_{k-iJ-1} \cdots c_m| \\ &\leq \sum_{i=0}^{\infty} \alpha_J^i (1 + \alpha_1 + \ldots + \alpha_{J-1}) \\ &= (1-\alpha_J)^{-1}(1 + \alpha_1 + \ldots + \alpha_{J-1}) \\ &< \infty,\end{aligned}$$

since

$$1 + \alpha_1 + \ldots + \alpha_{J-1} \leq 1 + \alpha_1 + \ldots + (\alpha_1)^{J-1}.$$

Hence, letting $p \to \infty$,

$$\sum_{m=-\infty}^{k} |c_{k-1} \cdots c_m| \leq (1-\alpha_J)^{-1}(1+\alpha_1+\ldots+\alpha_{J-1}).$$

Then we see that

$$z_k = \sum_{m=-\infty}^{k} c_{k-1} \cdots c_m h_{m-1}$$

is a bounded solution of Eq.(47) satisfying inequality (48). The difference between two such bounded solutions would be a bounded solution of

$$z_{k+1} = c_k z_k.$$

Then for all k

$$|z_k| = |c_{k-1} \cdots c_{k-J}|\,|z_{k-J}| \leq \alpha_J |z_{k-J}|.$$

Taking supremums of both sides, we conclude that $z_k = 0$ for all k.

Returning to Eq.(46), we define $\alpha_j^{(2)}$ as α_j is defined in the lemma. Then

$$\alpha_1^{(2)} < 1$$

and provided for suitable choice of J_2

$$\alpha_{J_2}^{(2)} < 1,$$

the second equation in Eq.(46) has a unique bounded solution $\mathbf{u}_k^{(2)}$ satisfying

$$|\mathbf{u}_k^{(2)}| \leq \left(1-\alpha_{J_2}^{(2)}\right)^{-1} \left(1+\alpha_1^{(2)}+\ldots+\alpha_{J_2-1}^{(2)}\right) \sup_{j \in \boldsymbol{Z}} |\mathbf{g}_j^{(2)}|.$$

We substitute this solution in the first equation in Eq.(46) and solve the resulting equation using the next lemma.

Lemma 11.15. *Let $\{a_k\}_{k=-\infty}^{\infty}$ be a sequence of nonzero real numbers. For $j \geq 1$, set*

$$\alpha_j = \sup_{k \in \boldsymbol{Z}} |a_k^{-1} \cdots a_{k+j-1}^{-1}|.$$

Suppose

$$\alpha_1 < \infty$$

and for some J

$$\alpha_J < 1.$$

Then if $\{h_k\}_{k=-\infty}^{\infty}$ *is a bounded sequence, the scalar difference equation*

$$z_{k+1} = a_k z_k + h_k \tag{49}$$

has a unique bounded solution z_k. *Moreover,*

$$|z_k| \le (1-\alpha_J)^{-1}(\alpha_1 + \ldots + \alpha_J) \sup_{j \in \mathbf{Z}} |h_j| \tag{50}$$

for all k.

Proof. For a positive integer p,

$$\sum_{m=k}^{k+pJ-1} |a_k^{-1} \cdots a_m^{-1}| = \sum_{i=0}^{p-1} \sum_{m=k+iJ}^{k+(i+1)J-1} |a_k^{-1} \cdots a_m^{-1}|$$

and so

$$\begin{aligned} \sum_{m=k}^{k+pJ-1} |a_k^{-1} \cdots a_m^{-1}| &= \sum_{i=0}^{p-1} |a_k^{-1} \cdots a_{k+iJ-1}^{-1}| \sum_{m=k+iJ}^{k+(i+1)J-1} |a_{k+iJ}^{-1} \cdots a_m^{-1}| \\ &\le \sum_{i=0}^{\infty} \alpha_J^i (\alpha_1 + \ldots + \alpha_J) \\ &= (1-\alpha_J)^{-1}(\alpha_1 + \ldots + \alpha_J) \\ &< \infty, \end{aligned}$$

since

$$\alpha_1 + \ldots + \alpha_J \le \alpha_1 + \ldots + (\alpha_1)^J.$$

Hence, letting $p \to \infty$,

$$\sum_{m=k}^{\infty} |a_k^{-1} \cdots a_m^{-1}| \le (1-\alpha_J)^{-1}(\alpha_1 + \ldots + \alpha_J).$$

Then we see that

$$z_k = -\sum_{m=k}^{\infty} a_k^{-1} \cdots a_m^{-1} h_m$$

is a bounded solution of Eq.(49) satisfying inequality (50). The difference between two such bounded solutions would be a bounded solution of

$$z_{k+1} = a_k z_k.$$

Then for all k

$$|z_k| = |a_k^{-1} \cdots a_{k+J-1}^{-1}| \, |z_{k+J}| \le \alpha_J |z_{k+J}|.$$

Taking supremums of both sides, we conclude that $z_k = 0$ for all k and so complete the proof of the lemma.

Returning to Eq.(46) again, we define $\alpha_j^{(1)}$ as α_j is defined in the lemma. Then

$$\alpha_1^{(1)} < \infty$$

and provided for suitable choice of J_1

$$\alpha_{J_1}^{(1)} < 1,$$

we can conclude that the first equation in Eq.(46) has a unique bounded solution $\mathbf{u}_k^{(1)}$ satisfying

$$|\mathbf{u}_k^{(1)}| \leq \left(1-\alpha_{J_1}^{(1)}\right)^{-1}\left(\alpha_1^{(1)}+\ldots+\alpha_{J_1}^{(1)}\right)\sup_{j\in\boldsymbol{Z}}|b_j\mathbf{u}_j^{(2)}+\mathbf{g}_j^{(1)}|.$$

Now we define

$$b=\sup_{k\in\boldsymbol{Z}}|b_k|$$

and for suitable J_1 and J_2 as chosen above,

$$L^{(1)}=\left(1-\alpha_{J_1}^{(1)}\right)^{-1}\left(\alpha_1^{(1)}+\ldots+\alpha_{J_1}^{(1)}\right) \tag{51}$$

and

$$L^{(2)}=\left(1-\alpha_{J_2}^{(2)}\right)^{-1}\left(1+\alpha_1^{(2)}+\ldots+\alpha_{J_2-1}^{(2)}\right). \tag{52}$$

Then we conclude that, if such suitable J_1 and J_2 exist, L is invertible and that (recall that we use the Euclidean norm in $\mathbb{R}^2$)

$$||L^{-1}|| \leq \sqrt{(L^{(1)})^2(bL^{(2)}+1)^2+(L^{(2)})^2}.$$

Chaotic behaviour of diffeomorphisms via periodic orbits: Let $f:\mathbb{R}^n\to\mathbb{R}^n$ be a C^2 diffeomorphism. Suppose we have two δ pseudo periodic orbits $\{\mathbf{y}_k\}_{k=0}^N$ and $\{\mathbf{z}_k\}_{k=0}^N$ such that

$$||\mathbf{y}_0-\mathbf{z}_0|| \leq \Delta.$$

Let a bi-infinite sequence $\{a_i\}_{i=-\infty}^{\infty}$ of 0's and 1's be given. We form a $(\delta+\Delta)$ pseudo orbit $\{\mathbf{w}_k\}_{k=-\infty}^{\infty}$ with 0 corresponding to $\mathbf{y}_0,\mathbf{y}_1,\ldots,\mathbf{y}_N$ and 1 corresponding to $\mathbf{z}_0,\mathbf{z}_1,\ldots,\mathbf{z}_N$. More precisely, for $i\in\boldsymbol{Z}$ and $i(N+1)\leq k<(i+1)(N+1)$, we define

$$\mathbf{w}_k=\begin{cases}\mathbf{y}_{k-i(N+1)} & \text{if } a_i=0,\\ \mathbf{z}_{k-i(N+1)} & \text{if } a_i=1.\end{cases}$$

We want to apply Theorem 11.13 to this pseudo orbit. First we examine each pseudo periodic orbit separately.

Triangularisation: Consider the difference equation

$$\mathbf{u}_{k+1} = Df(\mathbf{y}_k)\mathbf{u}_k, \quad k = 0, \ldots, N.$$

We start with an arbitrarily chosen orthogonal $n \times n$ matrix S_0 and using the Gram-Schmidt process generate a sequence $\{S_k\}_{k=0}^{N+1}$ of orthogonal matrices and a sequence $\{A_k\}_{k=0}^{N}$ of upper triangular matrices according to

$$Df(\mathbf{y}_k)S_k = S_{k+1}A_k, \quad k = 0, \ldots, N.$$

For N sufficiently large, we usually find (see the explanation at the end of Section 11.2) that, regardless of the choice of S_0, we always get the same S_{N+1} except perhaps for the sign of the columns. Then we take S_{N+1} as the new S_0 and repeat the process. So, as well as $A_k = S_{k+1}^* Df(\mathbf{y}_k)S_k$ for $k = 0, \ldots, N-1$, the matrix $S_0^* Df(\mathbf{y}_N)S_N$, which we now take to be A_N, is also upper triangular. Next let $\{T_k\}_{k=0}^{N}$ and $\{B_k\}_{k=0}^{N}$ be the similar sequences of orthogonal and upper triangular matrices found for $\{\mathbf{z}_k\}_{k=0}^{N}$. T_0 and S_0 should be close.

Estimation of $\|L^{-1}\|$*:* Let us return to $\{\mathbf{w}_k\}_{k=-\infty}^{\infty}$ as defined above. For the application of Theorem 11.13, we need to find the inverse of the linear operator $L : \ell^\infty(\mathbf{Z}, \mathbb{R}^n) \to \ell^\infty(\mathbf{Z}, \mathbb{R}^n)$ defined for $\mathbf{u} = \{\mathbf{u}_k\}_{k=-\infty}^{\infty}$ by

$$(L\mathbf{u})_k = \mathbf{u}_{k+1} - Df(\mathbf{w}_k)\mathbf{u}_k \quad \text{for} \quad k \in \mathbf{Z}.$$

To find $\mathbf{u} = L^{-1}\mathbf{g}$, we have to solve the equation

$$L\mathbf{u} = \mathbf{g}, \tag{53}$$

that is, for a bounded sequence $\mathbf{g} = \{\mathbf{g}_k\}_{k=-\infty}^{\infty}$, we have to solve the difference equation

$$\mathbf{u}_{k+1} = Df(\mathbf{w}_k)\mathbf{u}_k + \mathbf{g}_k, \quad k \in \mathbf{Z} \tag{54}$$

for a bounded solution $\mathbf{u}_k$. We make the transformation

$$\mathbf{u}_k = H_k\mathbf{v}_k, \quad k \in \mathbf{Z},$$

where

$$H_k = \begin{cases} S_j & \text{if} \quad \mathbf{w}_k = \mathbf{y}_j, \\ T_j & \text{if} \quad \mathbf{w}_k = \mathbf{z}_j, \end{cases}$$

obtaining the equation

$$\mathbf{v}_{k+1} = C_k\mathbf{v}_k + H_{k+1}^*\mathbf{g}_k, \quad k \in \mathbf{Z}. \tag{55}$$

C_k will be upper triangular except in the cases

$$C_k = \begin{cases} T_0^* Df(\mathbf{y}_N)S_N & \text{when} \quad \mathbf{w}_k = \mathbf{y}_N,\ \mathbf{w}_{k+1} = \mathbf{z}_0, \\ S_0^* Df(\mathbf{z}_N)T_N & \text{when} \quad \mathbf{w}_k = \mathbf{z}_N,\ \mathbf{w}_{k+1} = \mathbf{y}_0. \end{cases}$$

We approximate Eq.(55) by the equation

$$\mathbf{v}_{k+1} = \overline{C}_k \mathbf{v}_k + H_{k+1}^* \mathbf{g}_k, \quad k \in \boldsymbol{Z}, \tag{56}$$

where $\overline{C}_k$ is upper triangular and is given by

$$\overline{C}_k = \begin{cases} C_k & \text{if} \quad \mathbf{w}_k \neq \mathbf{y}_N \text{ or } \mathbf{z}_N, \\ S_0^* Df(\mathbf{y}_N) S_N & \text{if} \quad \mathbf{w}_k = \mathbf{y}_N, \\ T_0^* Df(\mathbf{z}_N) T_N & \text{if} \quad \mathbf{w}_k = \mathbf{z}_N. \end{cases}$$

We have

$$\|C_k - \overline{C}_k\| \leq \delta_1 \quad \text{for} \quad k \in \boldsymbol{Z}, \tag{57}$$

where δ_1 is given by

$$\begin{aligned} &\max\{\|S_0^* Df(\mathbf{y}_N) S_N - T_0^* Df(\mathbf{y}_N) S_N\|, \|T_0^* Df(\mathbf{z}_N) T_N - S_0^* Df(\mathbf{z}_N) T_N\|\} \\ &\quad = \max\{\|A_N - T_0^* S_0 A_N\|, \|B_N - S_0^* T_0 B_N\|\}. \end{aligned} \tag{58}$$

Note that for $i \in \boldsymbol{Z}$ and $i(N+1) \leq k < (i+1)(N+1)$, we have

$$\overline{C}_k = \begin{cases} A_{k-i(N+1)} & \text{if } a_i = 0, \\ B_{k-i(N+1)} & \text{if } a_i = 1. \end{cases}$$

Now if we define the operator $\overline{L} : \ell^\infty(\boldsymbol{Z}, I\!R^n) \to \ell^\infty(\boldsymbol{Z}, I\!R^n)$ for $\mathbf{v} = \{\mathbf{v}_k\}_{k=-\infty}^{\infty}$ by

$$(\overline{L}\mathbf{v})_k = \mathbf{v}_{k+1} - \overline{C}_k \mathbf{v}_k,$$

we see that Eq.(56) has the form

$$\overline{L}\mathbf{v} = \overline{\mathbf{g}},$$

where for $k \in \boldsymbol{Z}$,

$$\overline{\mathbf{g}}_k = H_{k+1}^* \mathbf{g}_k.$$

Since $\overline{C}_k$ is upper triangular, we can use the method of the subsection entitled *The Operator L* to solve the equation. In fact, in the two-dimensional case to which we now restrict ourselves, if

$$A_k = \begin{bmatrix} a_k & b_k \\ 0 & c_k \end{bmatrix} \quad \text{and} \quad B_k = \begin{bmatrix} \bar{a}_k & \bar{b}_k \\ 0 & \bar{c}_k \end{bmatrix},$$

for $j \leq N+1$, we calculate $\alpha_j^{(2)}$ as the maximum of the quantities

$$|c_{k+j-1} \cdots c_k| \quad \text{and} \quad |\bar{c}_{k+j-1} \cdots \bar{c}_k|$$

for $0 \leq k \leq N - j + 1$, and of

$$|c_{k-N+j-2} \cdots c_0 c_N \cdots c_k|, \quad |\bar{c}_{k-N+j-2} \cdots \bar{c}_0 c_N \cdots c_k|,$$

$$|\bar{c}_{k-N+j-2}\cdots\bar{c}_0\bar{c}_N\cdots\bar{c}_k| \quad\text{and}\quad |c_{k-N+j-2}\cdots c_0\bar{c}_N\cdots\bar{c}_k|$$

for $N-j+2 \le k \le N$; we calculate $\alpha_j^{(1)}$ as the maximum of the same quantities with c_k replaced by a_k^{-1}. We also define

$$b = \max\left\{\sup_{0\le k\le N}|b_k|, \sup_{0\le k\le N}|\bar{b}_k|\right\}. \tag{59}$$

Then it follows from the subsection entitled *The Operator L* that provided for some positive integers J_1 and J_2

$$\alpha_{J_1}^{(1)} < 1 \quad\text{and}\quad \alpha_{J_2}^{(2)} < 1, \tag{60}$$

the operator $\overline{L}$ is invertible and

$$\|\overline{L}^{-1}\| \le \sqrt{(L^{(1)})^2(bL^{(2)}+1)^2+(L^{(2)})^2},$$

where the $L^{(i)}$ are given in Eqs.(51) and (52).

Next define the operator $\tilde{L} : \ell^\infty(\mathbf{Z}, \mathbb{R}^n) \to \ell^\infty(\mathbf{Z}, \mathbb{R}^n)$ for $\mathbf{u} = \{\mathbf{u}_k\}_{k=-\infty}^{\infty}$ by

$$(\tilde{L}\mathbf{u})_k = \mathbf{u}_{k+1} - C_k\mathbf{u}_k.$$

It follows from inequality (57) that

$$\|\tilde{L} - \overline{L}\| \le \delta_1.$$

So if

$$\delta_1\|\overline{L}^{-1}\| < 1, \tag{61}$$

$\tilde{L}$ is also invertible and

$$\|\tilde{L}^{-1}\| \le (1-\delta_1\|\overline{L}^{-1}\|)^{-1}\|\overline{L}^{-1}\|.$$

Finally we see that Eq.(54) has the unique solution

$$\mathbf{u}_k = H_k(\tilde{L}^{-1}\bar{\mathbf{g}})_k.$$

Hence, if the conditions in Eqs.(60) and (61) are satisfied, L is invertible and

$$\|L^{-1}\| \le (1-\delta_1\|\overline{L}^{-1}\|)^{-1}\|\overline{L}^{-1}\|.$$

Application of Theorems 11.13 and 11.12: We summarise the procedure. First we triangularise the linearised systems along the δ pseudo periodic orbits $\{\mathbf{y}_k\}_{k=0}^N$ and $\{\mathbf{z}_k\}_{k=0}^N$, storing the matrices A_k, B_k $(k = 0, \dots, N)$, S_0 and T_0. Then we calculate δ_1 from Eq.(58), $\alpha_j^{(i)}$ for $i = 1, 2$ and $j = 1, \cdots, N$ in the way just described, and b from Eq.(59). Next we find J_1 and J_2 such that

$$\alpha_{J_1}^{(i)} < 1, \; \alpha_{J_2}^{(2)} < 1,$$

calculate the $L^{(i)}$ from Eqs.(51) and (52), and set

$$\tilde{C} = \sqrt{(L^{(1)})^2(bL^{(2)}+1)^2+(L^{(2)})^2}.$$

We check that

$$\delta_1\tilde{C} < 1$$

and determine

$$C = (1-\delta_1\tilde{C})^{-1}\tilde{C}.$$

Then we know the operator L associated with the $(\delta+\Delta)$ pseudo orbit $\{\mathbf{w}_k\}_{k=-\infty}^{\infty}$ corresponding to any given sequence $\{a_i\}_{i=-\infty}^{\infty}$ is invertible with

$$\|L^{-1}\| \le C.$$

Next we calculate

$$M = \sup_{\mathbf{x}\in\mathbb{R}^2} \|D^2 f(\mathbf{x})\|$$

and check that

$$2MC^2(\delta+\Delta) < 1,$$

where we recall that

$$\Delta = \|\mathbf{y}_0 - \mathbf{z}_0\|.$$

It follows from Theorem 11.13 that there is a unique orbit $\{\mathbf{x}_k\}_{k=-\infty}^{\infty}$ of f such that

$$\|\mathbf{x}_k - \mathbf{w}_k\| \le \varepsilon = 2C(\delta+\Delta)$$

for all k. In view of the definition of $\mathbf{w}_k$, this means that there is a unique orbit $\{\mathbf{x}_k\}_{k=-\infty}^{\infty}$ of f such that for $i \in \mathbb{Z}$ and $i(N+1) \le k < (i+1)(N+1)$

$$\begin{aligned} \|\mathbf{x}_k - \mathbf{y}_{k-i(N+1)}\| &\le \varepsilon \quad \text{if} \quad a_i = 0, \\ \|\mathbf{x}_k - \mathbf{z}_{k-i(N+1)}\| &\le \varepsilon \quad \text{if} \quad a_i = 1. \end{aligned}$$

Finally we check that

$$2\varepsilon < \max_{0\le k\le N} \|\mathbf{y}_k - \mathbf{z}_k\|.$$

Then we can apply Theorem 11.12 to deduce the existence of a set on which f is chaotic.

Example: As an example, we take the Hénon map

$$(x, y) \to (1 - ax^2 + y, \ -bx)$$

with $a = 1.4$ and $b = -0.3$. Using the Newton method described in Coomes, Koçak and Palmer [1998], we find two δ pseudo periodic orbits $\{\mathbf{y}_k\}_{k=0}^{42}$ and $\{\mathbf{z}_k\}_{k=0}^{42}$ with

$$\mathbf{y}_0 = (1.078449, .09588973), \ \mathbf{z}_0 = (1.078448, .09588743),$$

$$\delta = .00000019, \ \Delta = \|\mathbf{y}_0 - \mathbf{z}_0\| \leq .000003$$

and

$$\max_{0 \leq k \leq 42} \|\mathbf{y}_k - \mathbf{z}_k\| > 1.$$

We triangularise along each pseudo periodic orbit finding, for example, that

$$A_{42} = \begin{bmatrix} 1.047503 & -.8436159 \\ 0 & -.2863954 \end{bmatrix}, \ S_0 = \begin{bmatrix} .9586795 & .2844884 \\ -.2844884 & .9586795 \end{bmatrix} = S_{43}$$

and

$$B_{42} = \begin{bmatrix} 1.047478 & -.8436214 \\ 0 & -.2864021 \end{bmatrix}, \ T_0 = \begin{bmatrix} .9586775 & .2844952 \\ -.2844952 & .9586775 \end{bmatrix} = T_{43}.$$

Note that $S_0 = S_{43} = S_{N+1}$, $T_0 = T_{43} = T_{N+1}$ and S_0 and T_0 are close as expected. Then we calculate

$$\delta_1 \leq .0000098, \ b \leq 3.2.$$

We choose $J_1 = J_2 = 10$, check that $\alpha_{10}^{(1)} < 1$, $\alpha_{10}^{(2)} < 1$ and calculate

$$L^{(1)} = \left(1 - \alpha_{10}^{(1)}\right)^{-1} \left(\alpha_1^{(1)} + \ldots + \alpha_{10}^{(1)}\right) \leq 8.1,$$

$$L^{(2)} = \left(1 - \alpha_{10}^{(2)}\right)^{-1} \left(1 + \alpha_1^{(2)} + \ldots + \alpha_9^{(2)}\right) \leq 2.2$$

and

$$\tilde{C} = \sqrt{(L^{(1)})^2 (bL^{(2)} + 1)^2 + (L^{(2)})^2} \leq 65.17.$$

Then we check that $\delta_1 \tilde{C}_1 < 1$ and calculate

$$C = (1 - \delta_1 \tilde{C})^{-1} \tilde{C} \leq 65.22.$$

Then we calculate

$$M = \sup_{\mathbf{x} \in R^2} \|D^2 f(\mathbf{x})\| = 2.8$$

and check that

$$2MC^2(\delta + \Delta) \leq 5.6 \times (65.22)^2 \times .00000319 < 1.$$

Finally we set

$$\varepsilon = 2C(\delta + \Delta) \leq .000417$$

and check that

$$2\varepsilon = .000834 < 1 < \max_{0 \leq k \leq N} \|\mathbf{y}_k - \mathbf{z}_k\|.$$

Then we were able to apply Theorem 11.13 and Theorem 11.12 to deduce the existence of a set on which the Hénon map with the specified parameter values is chaotic.

Note that round-off error enters into some of these calculations but it is clear from Coomes, Koçak and Palmer [1996] that it will have no significant effect on the results here.

Finally we describe some related work. Neumaier and Rage [1993] use interval arithmetic to give a computer proof of the existence of a transversal homoclinic point (and hence chaos) in the reduced standard map. Mischaikow and Mrozek [1995b] develop a computable method using the discrete Conley index to prove the existence of shift dynamics for a homeomorphism in a locally compact metric space; they apply the method to the Hénon map with parameter values $a = 1.8$ and $-0.025 \leq b \leq 0$. Zgliczyński [1997] uses a fixed point index to show the existence of chaotic dynamics in horseshoe type mappings and uses the computer to apply his results to the Hénon map with the classical parameter values (and also the Rössler equations). Hassard, Hastings, Troy and Zhang [1994] give a rigorous numerical proof that the Lorenz equations support a form of chaos. For the same equations, Mischaikow and Mrozek [1995a] employ a computer assisted method using the Conley index to show the existence of a semiconjugacy between some invariant set and the full shift on two symbols. For further related results, see Neumaier and Schlier [1994], Spreuer and Adams [1995] and Zgliczyński and Galias [1998].

REFERENCES

Afraimovich, V.S. and Pilyugin, S.Yu. (1996). Special pseudo-trajectories of lattice dynamical systems. *Rand.and Comp.Dyn.* **4**, 29-47.

Akin, E. (1993). *The General Topology of Dynamical Systems*, Amer.Math. Soc., Providence R.I..

Albizatti, A. (1983). Sélection de phase par un term d'excitation pour les solutions périodiques de certaines équations différentielles. *C.R.Acad.Sci.Paris Sér.I* t.**296**, 259-262.

Al-Nayef, A., Diamond, P., Kloeden, P.E., Kozyakin, V. and Pokrovskii, A. (1996). Bishadowing and delay equations. *J.Dyn.Diff.Eqns.* **11**, 121-134.

Al-Nayef, A., Kloeden, P.E. and Pokrovskii, A. (1997a). Expansivity of nonsmooth functional differential equations. *J.Math.Anal.Appl.* **208**, 453-461.

Al-Nayef, A., Kloeden, P.E. and Pokrovskii, A. (1997b). Semi-hyperbolic mappings, condensing operators and neutral delay equations. *J.Diff.Eqns.* **137**, 320-339.

Angenent, S.B. (1987). The shadowing lemma for elliptic PDE, in *Dynamics of Infinite Dimensional Systems*, S.N.Chow and J.K.Hale ed., NATO ASI Ser.F 37, 7-22.

Anosov, D.V. (1967). Geodesic flows on closed Riemannian manifolds with negative curvature. *Proc. Steklov Inst. Math.* **90**, transl. by S.Feder, Amer. Math.Soc., Providence R.I., 1969.

Anosov, D.V. (1970). A certain class of invariant sets of smooth dynamical systems, in *Proc. Fifth Internat. Conference on Nonlinear Oscillations, Vol.2*, 39-45. (in Russian)

Anosov, D.V. and Solodov, V.V. (1995). Hyperbolic sets, in *Dynamical Systems IX*, Springer-Verlag, Berlin, 10-92.

Aoki, N. (1982). Homeomorphisms without the pseudo-orbit tracing property. *Nagoya Math. J.* **88**, 155-160.

Aoki, N. (1983). On homeomorphisms with pseudo-orbit tracing property. *Tokyo J. Math.* **6**, 329-334.

Aoki, N. (1989). Topological dynamics, in *Topics in General Topology*, K.Morita and J.Nagata ed., North Holland, Amsterdam, 625-740.

Aoki, N. and Hiraide, N. (1994). *Topological Theory of Dynamical Systems. Recent Advances*, North-Holland, Amsterdam.

Arrowsmith, D.V. and Place, C.M. (1990). *An Introduction to Dynamical Systems*, Cambridge University Press, Cambridge.

Aulbach, B. (1984). Continuous and discrete dynamics near manifolds of equilibria. *Lecture Notes in Mathematics* **1058**, Springer-Verlag, Berlin.

Auerbach, D., Cvitanovich, P., Eckmann, J.-P., Gunaratne, G.H. and Procaccia, I. (1987). Exploring chaotic motion through periodic orbits. *Phys. Rev.Let.* **58**, 2387-2389.

Banks, J., Brooks, J., Cairns, G., Davis, G. and Stacey, P. (1992). On Devaney's definition of chaos. *Amer.Math.Mon.* **99**, 332-334.

Barge, M. and Swanson, R. (1988). Rotation shadowing properties of circle and annulus maps. *Ergod.Th.& Dynam.Sys.* **8**, 509-521.

Barnsley, M.F. (1988). *Fractals Everywhere*, Academic Press, San Diego.

Benaïm, M. and Hirsch, M.W. (1996). Asymptotic pseudotrajectories and chain recurrent flows, with applications. *J.Dyn.Diff.Eqns.* **8**, 141-176.

Blank, M.L. (1988). Metric properties of ε-trajectories of dynamical systems with stochastic behaviour. *Ergod.Th.& Dynam.Sys.* **8**, 365-378.

Blank, M.L. (1991). Shadowing of ε-trajectories of general multidimensional mappings. *Wiss.Z.Tech.Univ.Dresden* **40**, 157-159.

Blazquez, C.M. (1986). Transverse homoclinic orbits in periodically perturbed parabolic equations. *Nonlinear Analysis* **10**, 1277-1291.

Bogenschütz, T. and Gundlach, V.M. (1992-93). Symbolic dynamics for expanding random dynamical systems. *Rand. and Comp.Dyn.* **1**, 219-227.

Botelho, F. and Chen L. (1993a). Chain transitivity and rotation shadowing for annulus endomorphisms. *Proc.Amer.Math.Soc.* **118**, 1173-1177.

Botelho, F. and Chen L. (1993b). On the rotation shadowing property for annulus maps, in *Continuum Theory and Dynamical Systems*, T. West ed., Marcel Dekker, New York, 35-42.

Botelho, F. and Garzon, M. (1994). Generalized shadowing properties. *Rand. and Comp.Dyn.* **2**, 145-164.

Bowen, R. (1970). Topological entropy and Axiom A. *Proc.Symp.Pure Math.* **14**, Amer.Math.Soc., Providence, R.I..

Bowen, R. (1972). Periodic orbits for hyperbolic flows. *Amer.J.Math.* **94**, 1-30.

Bowen, R. (1975a). ω-limit sets for Axiom A diffeomorphisms. *J.Diff.Eqns.* **18**, 333-339.

Bowen, R. (1975b). Equilibrium states and the ergodic theory of Anosov diffeomorphisms. *Lecture Notes in Mathematics* **470**, Springer-Verlag.

Bowen, R. (1978). On Axiom A diffeomorphisms. *CBMS Regional Conference Series in Math.* **35**.

Bowen, R. and Walters, P. (1972). Expansive one-parameter flows. *J.Diff. Eqns.* **12**, 180-193.

Boyarsky, A. and Góra, P. (1990). The pseudo-orbit shadowing property for Markov operators in the space of probability density functions. *Can.J.Math.* **42**, 1000-1017.

Casti, J.L. (1992). *Reality Rules I*, Wiley, New York.

Chen, L. (1991). Linking and the shadowing property for piecewise monotone maps. *Proc.Amer.Math.Soc.* **113**, 251-263.

Chen, L. and Li, S. (1992). Shadowing property for inverse limit spaces. *Proc.Amer.Math.Soc.* **115**, 573-580.

Chow, S.N., Hale, J.K. and Mallet-Paret, J. (1980). An example of bifurcation to homoclinic orbits. *J.Diff.Eqns.* **37**, 351-373.

Chow, S.N., Lin, X.B. and Palmer, K.J. (1989a). A shadowing lemma for maps in infinite dimension, in *Differential Equations*, Proceedings of the EQUADIFF conference, C.M.Dafermos, G.Ladas and G.Papanicolaou ed., Marcel Dekker, New York and Basel, 127-136.

Chow, S.N., Lin, X.B. and Palmer, K.J. (1989b). A shadowing lemma with applications to semilinear parabolic equations. *SIAM J.Math.Appl.* **20**, 547-557.

Chow, S.N. and Palmer, K.J. (1991). On the numerical computation of orbits of dynamical systems: the one dimensional case. *J.Dyn.Diff.Eqns.* **3**, 361-380.

Chow, S.N. and Palmer, K.J. (1992). On the numerical computation of orbits of dynamical systems: the higher-dimensional case. *J. Complexity* **8**, 398-423.

Chow, S.N. and Van Vleck, E.S. (1991). Algorithms for true and numerical orbits of random maps, in *Proceedings of the international symposium on functional differential equations*, Kyoto, Japan, 30 August-2 September 1990, T.Yoshizawa et al. ed., World Scientific, Singapore, 43-54.

Chow, S.N. and Van Vleck, E.S. (1992). A shadowing lemma for random diffeomorphisms. *Rand. and Comp. Dyn.* **1**, 197-218.

Chow, S.N. and Van Vleck, E.S. (1994a). Shadowing of lattice maps, in *Chaotic Numerics*, P.E.Kloeden and K.J.Palmer ed., Contemporary Mathematics **172**, Amer.Math.Soc., Providence, R.I., 97-113.

Chow, S.N. and Van Vleck, E.S. (1994b). A shadowing lemma approach to global error analysis for initial value ODEs. *SIAM J. Sci.Comput.* **15**, 959-976.

Conley, C.C. (1975). Hyperbolic invariant sets and shift automorphisms. *Dynamical Systems Theory and Applications*, J.Moser ed., *Lecture Notes in Physics* **38**, 539-549.

Conley, C.C. (1980). Isolated Invariant Sets and the Morse Index. *CBMS Regional Conference Series in Math.* **38**.

Coomes, B.A. (1997). Shadowing orbits of ordinary differential equations in invariant submanifolds. *Trans.Amer.Math.Soc.* **349**, 203-216.

Coomes, B.A., Koçak, H. and Palmer, K.J. (1994a). Shadowing orbits of ordinary differential equations. *J.Comp.Appl.Math.* **52**, 35-43.

Coomes, B.A., Koçak, H. and Palmer, K.J. (1994b). Periodic shadowing, in *Chaotic Numerics*, P.E.Kloeden and K.J.Palmer ed., Contemporary Mathematics **172**, Amer.Math.Soc., Providence, R.I., 115-130.

Coomes, B.A., Koçak, H. and Palmer, K.J. (1995a). Rigorous computational shadowing of orbits of ordinary differential equations. *Num.Math.* **69**, 401-421.

Coomes, B.A., Koçak, H. and Palmer, K.J. (1995b) A shadowing theorem for ordinary differential equations. *Z.Angew.Math.Phys.* **46**, 85-106.

Coomes, B.A., Koçak, H. and Palmer, K.J. (1996). Shadowing in discrete dynamical systems, in *Six Lectures on Dynamical Systems*, B.Aulbach and F.Colonius ed., World Scientific, Singapore, 163-211.

Coomes, B.A., Koçak, H. and Palmer, K.J. (1997). Long periodic shadowing. *Numer.Algorithms* **14**, 55-78.

Coomes, B.A., Koçak, H. and Palmer, K.J. (1998). Computation of long periodic orbits in chaotic dynamical systems. *Gazette Austral.Math.Soc.* **24**, 183-190.

Coppel, W.A. (1963). Perturbation of periodic solutions: the degenerate case, in *Qualitative Methods in the Theory of Nonlinear Vibrations*, Izdat.Akad. Nauk Ukrain.SSR, Kiev, 190-194.

Coppel, W.A. (1965). *Stability and Asymptotic Behavior of Differential Equations.* D.C.Heath, Boston.

Coppel, W.A. (1978). Dichotomies in stability theory. *Lecture Notes in Mathematics* **629**, Springer-Verlag, Berlin.

Corless, R.M. (1992). Defect controlled numerical methods and shadowing for chaotic differential equations. *Physica D* **60**, 323-334.

Corless, R.M. (1994a). What good are numerical simulations of dynamical systems? *Comput.Math.Appl.* **28**, 107-121.

Corless, R.M. (1994b). Error backward, in *Chaotic Numerics*, P.E.Kloeden and K.J.Palmer ed., *Contemporary Math.* **172**. Amer.Math.Soc., Providence R.I., 31-62.

Corless, R.M. and Pilyugin, S.Yu. (1995). Approximate and real trajectories for generic dynamical systems. *J.Math.Anal.Appl.* **189**, 409-423.

Coven, E.M., Kan, I. and Yorke, J.A. (1988). Pseudo-orbit shadowing in the family of tent maps. *Trans.Amer.Math.Soc.* **308**, 227-241.

Crandall, M.G. and Rabinowitz, P.H. (1971). Bifurcation from simple eigenvalues. *J.Funct.Anal.* **8**, 321-340.

Curry, J. (1979). An algorithm for finding closed orbits, in *Global Theory of Dynamical Systems*, Z.Nitecki and C.Robinson ed., *Lecture Notes in Mathematics* **819**, 111-120.

Daleckii, Ju.L. and Krein, M.G. (1974). *Stability of Solutions of Differential Equations in Banach Space*, Amer.Math.Soc. Translations, Providence R.I..

Dateyama, M. (1983). Homeomorphisms with the pseudo orbit tracing property of the Cantor set. *Tokyo J.Math.* **6**, 287-290.

Dateyama, M. (1989). Homeomorphisms with Markov partitions. *Osaka J.Math.* **26**, 411-428.

Dellnitz, M. and Melbourne, I. (1995). A note on the shadowing lemma and symmetric periodic points. *Nonlinearity* **8**, 1067-1075.

Deng, B. and Sakamoto, K. (1995). Sil'nikov-Hopf bifurcation. *J.Diff.Eqns.* **119**, 1-23.

Devaney, R.L. and Nitecki, Z. (1979). Shift automorphisms in the Hénon mapping. *Comm.Math.Phys.* **67**, 137-146.

Devaney, R.L. and Nitecki, Z. (1984). Homoclinic bifurcations and the area conserving Hénon mapping. *J.Diff.Eqns.* **51**, 254-266.

Devaney, R.L. (1989). *An Introduction to Chaotic Dynamical Systems.* 2nd Edition. Addison-Wesley, Redwood City.

Diamond, P., Kloeden, P.E. and Pokrovskii, A. (1995a). Cycles of spatial discretization of shadowing dynamical systems. *Math. Nachr.* **171**, 95-110.

Diamond, P., Kloeden, P.E., Kozyakin, V.S. and Pokrovskii, A. (1995b). Computer robustness of semi-hyperbolic mappings. *Rand. and Comp.Dyn.* **3**, 57-70.

Diamond, P., Kloeden, P.E., Kozyakin, V.S. and Pokrovskii, A. (1995c). Semi-hyperbolic mappings. *J.Nonlin.Sc.* **5**, 419-431.

Diamond, P., Kloeden, P.E., Kozyakin, V.S. and Pokrovskii, A. (1995d). Expansivity of semi-hyperbolic Lipschitz mappings. *Bull.Austral.Mat.Soc.* **51**, 301-308.

Diamond, P., Kloeden, P.E., Kozyakin, V.S., Krasnoselskii, M.A. and Pokrovskii, A. (1996). Robustness of dynamical systems to a class of nonsmooth perturbations. *Nonlin.Anal.TMA* **26**, 351-361.

Eirola, T., Nevanlinna, O. and Pilyugin, S.Yu. (1997). Limit shadowing property. *Numer.Funct.Anal. and Optim.* **18**, 75-92.

Eirola, T. and Pilyugin, S.Yu. (1996). Pseudotrajectories generated by a discretization of a parabolic equation. *J.Dyn.Diff.Eqns.* **8**, 281-297.

Ekeland, I. (1983). Some lemmas about dynamical systems. *Math.Scand.* **52**, 262-268.

Farmer, J.D. and Sidorowich, J.J. (1991). Optimal shadowing and noise reduction. *Physica D* **47**, 373-392.

Feckan, M. (1991). A remark on the shadowing lemma. *Funk.Ekvac.* **34**, 391-402.

Fenichel, N. (1996). Hyperbolicity and exponential dichotomy. *Dynamics Reported* **5**, 1-25.

Fontich, E. (1990). Transversal homoclinic points of a class of conservative diffeomorphisms. *J.Diff.Eqns.* **87**, 1-27.

Franke, J.E. and Selgrade, J.F. (1976). Abstract ω-limit sets, chain recurrent sets, and basic sets for flows. *Proc.Amer. Math.Soc.* **60**, 309-316.

Franke, J.E. and Selgrade, J.F. (1977). Hyperbolicity and chain recurrence. *J.Diff.Eqns.* **26**, 27-36.

Franke, J.E. and Selgrade, J.F. (1978). Hyperbolicity and cycles. *Trans. Amer.Math.Soc.* **245**, 251-262.

Franke, J.E. and Selgrade, J.F. (1979). A computer method for verification of asymptotically stable periodic orbits. *SIAM J.Math.Anal.* **10**, 614-628.

Gedeon, T. and Kuchta, M. (1992). Shadowing property of continuous maps. *Proc.Amer.Math.Soc.* **115**, 271-281.

Glasser, M.L., Papageorgiou, V.G. and Bountis, T.C. (1989). Melnikov's function for two-dimensional mappings. *SIAM J.Appl.Math* **49**, 692-703.

Grebogi, C., Hammel, S., Yorke, J.A. and Sauer, T. (1990). Shadowing of physical trajectories in chaotic dynamics: containment and refinement. *Phys.Rev.Let.* **65**, 1527-1530.

Grebogi, C., Ott, E. and Yorke, J.A.(1990). Unstable periodic orbits and the dimension of chaotic attractors. *Phys.Rev.A* **36**, 3522-3524.

Guckenheimer, J., Moser, J. and Newhouse, S.E. (1980). *Dynamical Systems*, Birkhäuser, Boston.

Gundlach, V.M. (1995). Random homoclinic orbits. *Rand. and Comp.Dyn.* **3**, 1-33.

Gundlach, V.M. and Rand, D.A. (1993). Spatio-temporal chaos: 1. Hyperbolicity, structural stability, spatio-temporal shadowing and symbolic dynamics. *Nonlinearity* **6**, 165-200.

Hadeler, K.P. (1996). Shadowing orbits and Kantorovich's theorem. *Numer.Math.* **73**, 65-73.

Hale, J.K. (1984). Introduction to dynamic bifurcation, in *Bifurcation Theory and Applications*, L.Salvadori ed., *Lecture Notes in Mathematics* **1057**, 106-151.

Hale, J.K. and Lin, X.B. (1986). Symbolic dynamics and nonlinear semiflows. *Ann.Mat.Pura Appl.* **144**, 229-259.

Hale, J.K. and Táboas, P. (1978). Interaction of damping and forcing in a second order equation. *Nonlin.Anal.TMA* **2**, 77-84.

Hammel, S.M.,Yorke, J.A. and Grebogi, C. (1987). Do numerical orbits of chaotic processes represent true orbits? *J.Complexity* **3**, 136-145.

Hammel, S.M.,Yorke, J.A. and Grebogi, C. (1988). Numerical orbits of chaotic processes represent true orbits. *Bull.Amer.Math.Soc.* **19**, 465-470.

Handel, M. (1985). Global shadowing of pseudo Anosov homeomorphisms. *Ergod.Th.& Dynam.Sys.* **5**, 373-377.

Handel, M. (1988). Entropy and semi-conjugacy in dimension 2. *Ergod.Th.& Dynam.Sys.* **8**, 585-596.

Harada, M. and Oka, M. (1989). The pseudo-orbit tracing property of c-expansive covering maps of closed surfaces, in *Stability Theory and Related Topics in Dynamical Systems*, K.Shiraiwa and G.Ikegami ed., World Scientific, Singapore, 17-22.

Hassard, B., Hastings, S.P., Troy, W.C. and Zhang, J. (1994). A computer proof that the Lorenz equations have "chaotic" solutions. *Appl.Math.Lett.* **7**, 79-83.

Hénon, M. (1976). A two-dimensional mapping with a strange attractor. *Comm.Math.Phys.* **50**, 69-77.

Henry, D. (1981). Geometric theory of semilinear parabolic equations. *Lecture Notes in Mathematics* **840**, Springer-Verlag, New York.

Henry, D. (1994). Exponential dichotomies, the shadowing lemma, and homoclinic orbits in Banach spaces. *Resenhas IME-USP* **1**, 381-401.

Hiraide, K. (1984). Manifolds which do not admit expansive homeomorphisms with pseudo orbit tracing property, in *The Theory of Dynamical Systems and its Application to Nonlinear Problems*, H.Kawakami ed., 32-34.

Hiraide, K. (1985). On homeomorphisms with Markov partitions. *Tokyo J. Math.* **8**, 219-229.

Hiraide, K. (1987). Expansive homeomorphisms with the pseudo-orbit tracing property of n-tori, in *Dynamical Systems and Applications*, N.Aoki ed., World Scientific, Singapore.

Hiraide, K. (1988). Expansive homeomorphisms with the pseudo-orbit tracing property on compact surfaces. *J.Math.Soc.Japan* **40**, 123-137.

Hiraide, K. (1989). Expansive homeomorphisms with the pseudo-orbit tracing property of n-tori. *J.Math.Soc.Japan* **41**, 357-389.

Hiraide, K. (1992). Dynamical systems of expansive maps on compact manifolds. *Sugaku Expositions*, **5**, 133-154.

Hirsch, M.W. (1994). Asymptotic phase, shadowing and reaction-diffusion systems, in *Differential Equations, Dynamical Systems and Control Science*, K.D.Elworthy, W.N.Everitt and E.B.Lee ed., Marcel-Dekker, New York, 87-99.

Hirsch, M., Palis, J., Pugh, C. and Shub, M. (1970). Neighborhoods of hyperbolic sets. *Invent.Math.* **9**, 121-134.

Hirsch, M. and Pugh, C. (1970). Stable manifolds and hyperbolic sets. *Proc.Symp.in Pure Math.* **14**, Amer.Math.Soc., Providence, R.I..

Hirsch, M., Pugh, C. and Shub, M. (1977). Invariant manifolds. *Lecture Notes in Mathematics* **583**, Springer-Verlag, New York.

Hirsch, M.W. and Smale, S. (1974). *Differential Equations, Dynamical Systems and Linear Algebra*, Academic Press, New York.

Hurley, M. (1984). Consequences of topological stability. *J.Diff.Eqns.* **54**, 60-72.

Johnson, R.A. (1987). Remarks on linear differential systems with measurable coefficients. *Proc.Amer.Math.Soc.* **100**, 491-504.

Kakubari, S. (1987). A note on linear homeomorphism on $\mathbb{R}^n$ with the pseudo-orbit tracing property. *Sci.Rep.Niigata Univ.Ser.A* **23**, 35-37.

Kato, K. (1984). Pseudo-orbits and stabilities of flows. *Mem.Fac.Sci.Kochi Univ. Ser.A* **5**, 45-62.

Kato, K. (1985). Pseudo-orbits and stabilities of flows II. *Mem.Fac.Sci.Kochi Univ.Ser.A* **6**, 33-43.

Kato, K. (1991). Hyperbolicity and pseudo-orbits for flows. *Mem.Fac.Sci. Kochi Univ.Ser.A* **12**, 43-55.

Kato, K. and Morimoto, A. (1979). Topological Ω-stability of Axiom A flows with no Ω-explosions. *J.Diff.Eqns.* **34**, 464-481.

Katok, A. (1971) Local properties of hyperbolic sets. Appendix to Russian translation of Z.Nitecki, *Differentiable Dynamics*, MIT Press. Translated as Dynamical systems with hyperbolic structure. *AMS Transl. (2)* **116** (1981), 43-94.

Katok, A. and Hasselblatt, B. (1995). *Introduction to the Modern Theory of Dynamical Systems.* Cambridge University Press, Cambridge.

Kirchgraber, U. (1982). Erratische Lösungen der periodisch gestörten Pendelgleichung. Preprint, Univ. of Würzburg.

Kirchgraber, U. and Stoffer, D. (1989). On the definition of chaos. *Z.Angew. Math. Mech.* **69**, 175-185.

Kirchgraber, U. and Stoffer, D. (1990). Chaotic behaviour in simple dynamical systems. *SIAM Review* **32**, 424-452.

Kloeden, P.E. and Ombach, J. (1997). Hyperbolic homeomorphisms and bishadowing. *Ann.Polon.Math.*, **65**, 171-177.

Komuro, M. (1984). One-parameter flows with the pseudo orbit tracing property. *Mh.Math.* **98**, 219-253.

Komuro, M. (1984). The pseudo orbit tracing property on the space of probability measures. *Tokyo J.Math.* **7**, 461-468.

Komuro, M. (1985). Lorenz attractors do not have the pseudo-orbit tracing property. *J.Math.Soc.Japan* **37**, 489-514.

Krüger, T. and Troubetzkoy, S. (1992). Markov partitions and shadowing for nonuniformly hyperbolic systems with singularities. *Ergod.Th.& Dynam.Sys.* **12**, 487-508.

Kurata, M. (1979). Markov partitions of hyperbolic sets. *J.Math.Soc.Japan* **31**, 39-51.

Lanford, O.E.III (1983). Introduction to the mathematical theory of dynamical systems, in *Chaotic Behaviour of Deterministic Systems, Les Houches, 1981*, G.Iooss, R.H.G.Helleman and R.Stora ed., North-Holland, Amsterdam, 3-51.

Lanford, O.E.III (1985). Introduction to hyperbolic sets, in *Regular and Chaotic Motion in Dynamic Systems*, NATO ASI Series, Series B, Physics, Vol. **118**, Plenum Press, New York, 73-102.

Lani-Wayda, B. (1995a). Hyperbolic sets, shadowing and persistence for noninvertible mappings in Banach spaces. *Pitman Research Notes in Mathematics* **334**, Longman, London.

Lani-Wayda, B. (1995b). Persistence of Poincaré mappings in functional differential equations (with application to structural stability of complicated behavior). *J.Dyn.Diff.Eqs.* **7**, 1-71.

Larsson, S. and Pilyugin, S.Yu. (1998). Numerical shadowing near the global attractor for a semi-linear parabolic equation. Preprint 1998-21, Dept. of Math., Chalmers Univ. of Technology, Goteborg Univ..

Larsson, S. and Sanz-Serna, J. (1999). A shadowing result with applications to finite element approximation of reaction-diffusion equations. *Math. of Comput.* **68**, 55-72.

Lerman, L.M. and Shil'nikov, L.P. (1992). Homoclinical structures in nonautonomous systems: Nonautonomous chaos. *Chaos* **2**, 447-454.

Li, S. (1992). Shadowing property for inverse limit spaces. *Proc. Amer.Math. Soc.* **115**, 573-580.

Lin, X.B. (1989). Shadowing lemma and singularly perturbed boundary value problems. *SIAM J.Appl.Math.* **49**, 26-54.

Liu, P. (1991). R-stability of orbit spaces for C^1 endomorphisms. *Chin.Ann. Math.Ser.A* **12**, 415-421.

Liu, P., Qian, M. and Jang, F. (1996). Pseudo-orbit tracing property for random diffeomorphisms. *Proc.Roy.Soc.Edin.* **126A**, 1027-1033.

Loud, W. (1957). Periodic solutions of $\ddot{x} + c\dot{x} + g(x) = \varepsilon f(t)$. *Memoirs Amer.Math.Soc.* **31**.

Marotto, F.R. (1979a). Perturbation of stable and chaotic difference equations. *J.Math.Anal.Appl.* **72**, 715-729.

Marotto, F.R. (1979b). Chaotic behaviour in the Hénon attractor. *Comm. Math.Phys.* **68**, 187-194.

Massera, J. and Schäffer, J.J. (1966). *Linear Differential Equations and Function Spaces.* Academic Press, New York.

McGehee, R. (1982). A hyperbolic invariant set for a forced pendulum. IMA preprint.

McGehee, R. and Meyer, K.R. (1974). Homoclinic points of area preserving diffeomorphisms. *Amer.J.Math.* **96**, 409-421.

Melnikov, V.K. (1963). On the stability of the center for time periodic solutions. *Trans.Moscow Math.Soc.* **12**, 3-52.

Meyer, K.R. and Hall, G.R. (1992). *Introduction to Hamiltonian Dynamical Systems and the N-body Problem*, Springer-Verlag, New York.

Meyer, K.R. and Sell, G.R. (1989). Melnikov transforms, Bernouilli bundles, and almost periodic perturbations. *Trans.Amer.Math.Soc.* **314**, 63-105.

Meyer, K.R. and Zhang, X. (1996). Stability of skew dynamical systems. *J.Diff.Eqns.* **132**, 66-86.

Millionscikov, V.M. (1969a). A proof of the attainability of the central exponents of linear systems. *Siberian Math.J.* **10**, 69-73.

Millionscikov, V.M. (1969b). Coarse properties of linear systems of differential equations. *Differential Equations* **5**, 1314-1321.

Mischaikow, K. and Mrozek, M. (1995a). Chaos in the Lorenz equations: A computer-assisted proof. *Bull.Amer.Math.Soc.* **32**, 66-72.

Mischaikow, K. and Mrozek, M. (1995b). Isolating neighborhoods and chaos. *Japan J.Indust.Appl.Math.* **12**, 205-236.

Mischaikow, K. and Mrozek, M. (1998). Chaos in the Lorenz equations: A computer assisted proof. Part II: Details. *Math. of Comput.* **67**, 1023-1046.

Misiurewicz, M. and Szewc, B. (1980). Existence of a homoclinic orbit for the Hénon map. *Comm.Math.Phys.* **75**, 285-291.

Morimoto, A. (1977). Stochastically stable diffeomorphisms and Takens' conjecture. *Kokyuroku Res.Inst.Kyoto Univ.* **303**, 8-24.

Morimoto, A. (1979). The method of pseudo-orbit tracing and stability of dynamical systems. *Seminar Note* **39**, Dept.Math.Tokyo Univ. (in Japanese).

Morimoto, A. (1981). Some stabilities of group automorphisms, in *Manifolds and Lie Groups*, J.Hano et al. ed., *Progr.Math.* **14**, Birkhäuser, 283-299.

Morinaka, M. (1983). On the existence of transversal homoclinic points of some real analytic plane transformations. *J.Math.Kyoto Univ.* **23-4**, 707-714.

Moriyasu, K. (1991). The topological stability of diffeomorphisms. *Nagoya Math.J.* **123**, 91-102.

Moriyasu, K. and Oka, M. (1995). Pseudo orbit tracing property and strong transversality of C^1 maps, in *Dynamical Systems and Chaos Vol.1*, N. Aoki, K. Shiraiwa and Y. Takahashi ed., World Scientific, Singapore, 187- 191.

Nadzieja, T. (1991). Shadowing lemma for family of ε-trajectories. *Arch. Math.* **27A**, 65-77.

Neumaier, A. and Rage, T. (1993). Rigorous chaos verification in discrete dynamical systems. *Physica D* **67**, 327-346.

Neumaier, A. and Schlier, C. (1994). Rigorous verification of chaos in a molecular model. *Phys.Rev.E* **50**, 2682-2688.

Nitecki, Z. (1971). *Differentiable Dynamics. An Introduction to the Orbit Sructure of Diffeomorphisms*, MIT Press, Cambridge.

Nusse, H. and Yorke, J.A. (1988). Is every approximate trajectory of some process near an exact trajectory of a nearby process? *Comm.Math.Phys.* **114**, 363-379.

Nusse, H. and Yorke, J.A. (1994). *Dynamics: Numerical Explorations*, Springer-Verlag, New York.

Ocken, S. (1995). Recognizing convergent orbits of discrete dynamical systems. *SIAM J.Appl.Math.* **55**, 1134-1160.

Odani, K. (1990). Generic homeomorphisms have the pseudo-orbit tracing property. *Proc.Amer.Math.Soc.* **1101**, 281-284.

Ombach, J. (1986). Equivalent conditions for hyperbolic coordinates. *Topology and its Applics.* **23**, 87-90.

Ombach, J. (1987). Consequences of the pseudo orbits tracing property and expansiveness. *J.Austral.Math.Soc.Ser.A* **43**, 301-313.

Ombach, J. (1991a). Sinks, sources and saddles for expansive flows with pseudo orbits tracing property. *Ann.Polon.Math.* **53**, 237-252.

Ombach, J. (1991b). Saddles for expansive flows with the pseudo orbits tracing property. *Ann.Polon.Math.* **56**, 37-48.

Ombach, J. (1992). The pseudo-orbits tracing property for linear systems of differential equations. *Glasnik Matem.* **17(47)**, 49-56.

Ombach, J. (1993a). The simplest shadowing. *Ann.Polon.Math.* **58**, 243-258.

Ombach, J. (1993b). Shadowing for linear systems of differential equations. *Publ.Mat., Barc.* **37**, 245-253.

Ombach, J. (1994). The shadowing lemma in the linear case. *Univ.Iagell. Acta Math.* **31**, 69-94.

Ombach, J. (1996). Shadowing, expansiveness and hyperbolic homeomorphisms. *J.Austral.Math.Soc.Ser.A* **61**, 57-72.

Osipenko, G. and Komarchev, I. (1995). Applied symbolic dynamics: construction of periodic trajectories, in *Dynamical Systems and Applications*, World Sci.Ser.Appl.Anal. **4**, World Scientific, River Edge N.J., 573-587.

Palmer, K.J. (1984). Exponential dichotomies and transversal homoclinic points. *J.Diff.Eqns.* **55**, 225-256.

Palmer, K.J. (1987). A perturbation theorem for exponential dichotomies. *Proc.Roy.Soc.Edin.* **106A**, 25-37.

Palmer, K.J. (1988). Exponential dichotomies, the shadowing lemma and transversal homoclinic points. *Dynamics Reported* **1**, 265-306.

Palmer, K.J. (1996). Shadowing and Silnikov chaos. *Nonlin.Anal.TMA* **27**, 1075-1093.

Palmer, K.J. and Stoffer, D. (1989). Chaos in almost periodic systems. *Z.Angew.Math.Phys.* **40**, 592-602.

Papaschinopoulos G. (1988). Some roughness results concerning reducibility for linear difference equations. *Int.J.Math.Math.Sci.* **11**, 793-804.

Park, J.S., Lee, K.H. and Koo, K.S. (1995). Hyperbolic homeomorphisms. *Bull.Korean Math.Soc.* **32**, 93-102.

Peitgen, H.-O. (1992). *Fractals for the Classroom II*, Springer-Verlag, New York.

Pennings, T. and Van-Eeuwen, J. (1990/91). Pseudo-orbit shadowing on the unit interval. *Real Anal.Exchange* **16**, 238-244.

Peterson, Ivars. (1990). *Islands of Truth*, W.H.Freeman, New York.

Pilyugin, S.Yu. (1992) *Introduction to Structurally Stable Systems of Differential Equations*, Birkhäuser, Basel.

Pilyugin, S.Yu. (1994a) The space of dynamical systems with the C^0 topology. *Lecture Notes in Mathematics* **1571**, Springer-Verlag, New York.

Pilyugin, S.Yu. (1994b). Complete families of pseudo trajectories and shape of attractors. *Rand.and Comp.Dyn.* **2**, 205-226.

Pilyugin, S.Yu. (1997). Shadowing in structurally stable flows. *J.Diff.Eqns.* **140**, 238-265.

Pilyugin, S.Yu. and Plamenevskaya, O.B. (1999). Shadowing is generic. *Topology and its Applics.*, to appear.

Pliss, V.A. and Sell, G.R. (1991). Perturbations of attractors of differential equations. *J.Diff.Eqns.* **92**, 100-124.

Poon, L., Dawson, S.P., Grebogi, C., Sauer, T. and Yorke, J.A. (1995). Shadowing in chaotic systems, in *Dynamical Systems and Chaos Vol.2*, Y. Aizawa, S. Saito and K. Shiraiwa ed., World Scientific, Singapore, 13-21.

Robinson, C. (1977). Stability theorems and hyperbolicity in dynamical systems. *Rocky Mt.J.Math.* **7**, 425-437.

Robinson, C. (1995). *Dynamical Systems. Stability, Symbolic Dynamics, and Chaos*, CRC Press, Boca Raton.

Rogers, T.D. and Marotto, F.R. (1983). Perturbations of mappings with periodic repellers. *Nonlin.Anal.TMA* **7**, 97-100.

Sacker, R.A. and Sell, G.R. (1974). Existence of dichotomies and invariant splittings for linear differential systems I. *J.Diff.Eqns.* **15**, 429-458.

Sacker, R.A. and Sell, G.R. (1978). A spectral theory for linear differential systems. *J.Diff.Eqns.* **27**, 320-358.

Sakai, K. (1988). Quasi-Anosov diffeomorphisms and pseudoorbit tracing property. *Nagoya J.Math.* **111**, 111-114.

Sakai, K. (1992). Diffeomorphisms with pseudo orbit tracing property. *Nagoya J.Math.* **126**, 125-140

Sakai, K. (1994). Pseudo orbit tracing property and strong transversality of diffeomorphisms on closed manifolds. *Osaka J.Math.* **31**, 373-386.

Sakai, K. (1995a). Hyperbolic metrics of expansive homeomorphisms. *Topology and its Applics.* **63**, 263-266.

Sakai, K. (1995b). Shadowing property and transversality condition, in *Dynamical Systems and Chaos Vol.1*, N. Aoki, K. Shiraiwa and Y. Takahashi ed., World Scientific, Singapore, 233-238.

Sakai, K. (1996). Diffeomorphisms with the shadowing property. *J.Austral. Math.Soc.Ser.A* **61**, 396-399.

Sakamoto, K. (1990). Invariant manifolds in singular perturbation problems for ordinary differential equations. *Proc.Roy.Soc.Edin.* **116A**, 45-78.

Sakamoto, K. (1994). Estimates on the strength of exponential dichotomies and application to integral manifolds. *J.Diff.Eqns.* **107**, 259-279.

Sanz-Serna, J.M. and Larsson, S. (1993). Shadows, chaos and saddles. *Appl.Numer.Math.* **13**, 181-190.

Sasaki, K. (1978). Some examples of stochastically stable homeomorphisms. *Nagoya Math.J.* **71**, 97-105.

Sauer, T. and Yorke, J.A. (1990). Shadowing trajectories of dynamical systems, in *Computer-Aided Proofs in Analysis*, K.R.Meyer and D.S.Schmidt ed., Springer-Verlag, Berlin, 229-234.

Sauer, T. and Yorke, J.A. (1991). Rigorous verification of trajectories for computer simulations of dynamical systems. *Nonlinearity* **4**, 961-979.

Sawada, K. (1980). Extended f-orbits are approximated by orbits. *Nagoya Math.J.* **79**, 33-45.

Scheurle, J. (1986). Chaotic solutions of systems with almost periodic forcing. *Z.Angew.Math.Phys.* **37**, 12-26.

Schwartz, I.B. (1983). Estimating regions of unstable periodic orbits using computer-based techniques. *SIAM J.Numer.Anal.* **20**, 106-120.

Shimomura, T. (1984). Topological entropy and the pseudo-orbit tracing property, in *The Theory of Dynamical Systems and Its Application to Nonlinear problems*, H.Kawakami ed., World Scientific, Singapore, 35-41.

Shimomura, T. (1987). Chain recurrence and P.O.T.P., in *Dynamical Systems and Singular Phenomena*, G.Ikegami ed., World Scientific, Singapore, 224-241.

Shimomura, T. (1989). On the structure of discrete dynamical systems from the view point of chain components and some applications. *Jap.J. Math.* **15**, 99-126.

Shub, M. (1987). *Global Stability of Dynamical Systems*, Springer-Verlag, New York.

Sigmund, K. (1974). On dynamical systems with the specification property. *Trans.Amer.Math.Soc.* **190**, 285-299.

Silnikov, L.P. (1967). On a Poincaré-Birkhoff problem. *Math.USSR-Sb.* **3**, 353-371.

Silverman, S. (1992). On maps with dense orbits and the definition of chaos. *Rocky Mt.J.Math.* **22**, 353-375.

Sinai, Ja.G. (1972). Gibbs measures in ergodic theory. *Russ.Math.Surveys* **27**, 21-69.

Sinai, Ja.G. and Vul, E.B. (1980). Discovery of closed orbits of dynamical systems with the use of computers. *J.Stat.Phys.* **23**, 27-47.

Slackov, S.V. (1985). A theorem on ε-trajectories for Lorenz mappings. *Funkts.Anal.Prilozh.* **19**, 84-85. (in Russian)

Slackov, S.V. (1992). Pseudo-orbit tracing property and structural stability of expanding maps of the interval. *Ergod.Th.& Dynam.Sys.* **12**, 573-587.

Slyusarchuk, V.E. (1983). Exponential dichotomy for solutions of discrete systems. *Ukrain.Math.J.* **35**, 98-103.

Smale, S. (1965). Diffeomorphisms with many periodic points, in *Differential and Combinatorial Topology*, S.S.Cairns ed., Princeton Univ. Press, 63-80.

Smale, S. (1967). Differentiable dynamical systems. *Bull.Amer.Math.Soc.* **73**, 747-817.

Spreuer, H. and Adams, F. (1995). On the strange attractor and transverse homoclinic orbits for the Lorenz equations. *J.Math.Anal.Appl.* **190**, 329-360.

Steinlein, H. and Walther, H.-O. (1989). Hyperbolic sets and shadowing for noninvertible maps, in *Advanced Topics in the Theory of Dynamical Systems*, G.Fusco, M.Iannelli and L.Salvadori ed., Academic Press, Boston, 219-234.

Steinlein, H. and Walther, H.-O. (1990). Hyperbolic sets, transversal homoclinic trajectories and symbolic dynamics for C^1-maps in Banach spaces. *J.Dyn.Diff.Eqns.* **2**, 325-365.

Stoffer, D. (1988a). Some geometric and numerical methods for perturbed integrable systems. *Diss. ETH* **8456**.

Stoffer, D. (1988b). Transversal homoclinic points and hyperbolic sets for non-autonomous maps I. *Z.Angew.Math.Phys.* **39**, 518-549.

Stoffer, D. (1988c). Transversal homoclinic points and hyperbolic sets for non-autonomous maps II. *Z.Angew.Math.Phys.* **39**, 783-812.

Stoffer, D. and Palmer, K.J. (1999). Rigorous verification of chaotic behaviour of maps using validated shadowing. *Nonlinearity*, to appear.

Szmolyan, P. (1991). Transversal heteroclinic and homoclinic orbits in singular perturbation problems. *J.Diff.Eqns.* **92**, 252-281.

Thomas, R.F. (1982). Stability properties of one-parameter flows. *Proc.Lond. Math.Soc.* **45**, 479-505.

Thomas, R.F. (1985). Topological stability: some fundamental properties. *J.Diff.Eqns.* **59**, 103-122.

Thomas, R.F. (1987). Entropy of expansive flows. *Ergod.Th.& Dynam.Sys.* **7**, 621-625.

Thomas, R.F. (1991). Canonical coordinates and the pseudo orbit tracing property. *J.Diff.Eqns.* **90**, 316-343.

Van Vleck, E.S. (1995). Numerical shadowing near hyperbolic trajectories. *SIAM J.Sci.Comput.* **16**, 1177-1189.

Walters, P. (1978). On the pseudo orbit tracing property and its relationship to stability. *Lecture Notes in Mathematics* **668**, 231-244.

Wilkinson, J.H. (1963). *Rounding Errors in Algebraic Processes*, Prentice-Hall, Englewood Cliffs N.J..

Wilkinson, J.H. (1965). *The Algebraic Eigenvalue Problem*, Clarendon Press, Oxford.

Yano, K. (1980). Topologically stable homeomorphisms of the circle. *Nagoya Math.J.* **79**, 145-149.

Yano, K. (1987). Generic homeomorphisms of S^1 have the pseudo orbit tracing property. *J.Fac.Sci.Univ.Tokyo Sect.1A Math.* **34**, 51-55.

Yuri, M. (1983). A construction of an invariant stable foliation by the shadowing lemma. *Tokyo J.Math.* **6**, 291-296.

Zgliczyński, P. (1997a). Computer assisted proof of the horseshoe dynamics in the Hénon map. *Rand.and Comp.Dyn.* **5**, 1-17.

Zgliczyński, P. (1997b). Computer assisted proof of chaos in the Roessler equations and the Hénon map. *Nonlinearity* **10**, 243-252.

Zgliczyński, P. and Galias, Z. (1998). Computer assisted proof of chaos in the Lorenz system. *Physica D* **115**, 165-188.